California Mammals

Bighorn Sheep (*Ovis canadensis*)

California Natural History Guides: 52

California Mammals

E. W. Jameson, jr., and
Hans J. Peeters

Illustrations by Hans J. Peeters

UNIVERSITY OF CALIFORNIA PRESS
Berkeley Los Angeles London

University of California Press
Berkeley, California

University of California Press, Ltd.
London, England

10 9 8 7 6 5 4 3 2 1

Library of Congress Cataloging-in-Publication Data
Jameson, E. W. (Everett Williams), 1921–
California mammals.

(California natural history guides; 52)
Includes index.
I. Mammals—California. I. Peeters, Hans J.
II. Title. III. Series.
QL719.C2J35 1986 599.09794 85-28933
ISBN 0-520-05252-8 (alk. paper)
ISBN 0-520-05391-5 (pbk.: alk. paper)

Dedication

The first four volumes of the California Natural History Guides were published in May 1959. In August 1959 a Technical Advisory Committee (now Advisory Editorial Committee) was established to advise the General Editor and the Press on the future scope and content of the series. Over the years the committee has performed a very valuable service for the publisher and for the more than a million users of the guides. A. Starker Leopold was appointed to this committee at the outset in 1959, continued his participation after retirement, and rarely missed a meeting during the twenty-four years he served. Other original members of the committee were Robert C. Stebbins and the late Robert L. Usinger.

Starker always took a great deal of interest in the guides under preparation; despite an extremely busy schedule, he always found time for consultation and assistance on the multitude of projects and problems the series entailed. For some time prior to his death, Starker had been working with us on final revisions of the manuscript of *California Mammals*. Indeed, only a few days before his death I reported to him on the implementation of his recommendations on the project.

I met Starker Leopold March 20, 1941, when he gave a characteristically informative and interesting talk, "Turkey Management in the Ozarks," at a Grinnell Society meeting in Berkeley. Our paths crossed and recrossed many times during the years since. Work on any project with Starker was always a rewarding experience. We miss him, but we have much to remember him by. It is with great pleasure that I join with the authors of this book, E. W. Jameson, jr., and Hans J. Peeters, in dedicating *California Mammals,* Volume 52 in the California Natural History Guide Series, to the memory of

A. STARKER LEOPOLD, 1913–1983

Arthur C. Smith
General Editor
California Natural History Guides

A. Starker Leopold

It is appropriate that *California Mammals* be dedicated to A. Starker Leopold, who devoted most of his life to studies of California wildlife and, up to the time of his death, was actively involved in efforts to conserve it. Although Starker's work took him to many parts of the world, California was his adopted home. He and his students at the University of California, Berkeley, inquired into the lives and ways of many mammal species, including a long-continuing series of studies of California deer populations.

Starker was born in Burlington, Iowa, in 1913, the son of Aldo Leopold and Estella Bergere. His education took him to the universities of Wisconsin, Yale, and Berkeley—where he received his Ph.D. in 1944 and joined the faculty as an assistant professor of zoology in 1946. Before coming to Berkeley he had done extensive research on wild turkeys in Missouri and had begun his extensive surveys of Mexican wildlife.

Starker was active in many organizations, serving a term as president of the Wildlife Society and fifteen years as president of the California Academy of Sciences. In his role as chairman of the advisory committees on predator control, national parks, and wildlife refuges for the Department of the Interior he was highly influential in changing federal policies concerning the management of wildlife and natural areas in the United States. He was the recipient of many awards and honors, including the Aldo Leopold Medal of the Wildlife Society (1965). In 1970 he was elected to the National Academy of Sciences.

Among Starker's writings, his books *Wildlife in Alaska* with Sir Frank Fraser Darling, *Wildlife of Mexico,* and *The California Quail* will be particularly rewarding to readers.

It was my privilege to be among Starker's first students at Berkeley. He was an inspiring teacher and later a good friend until his death in 1983. Although many people confused

Starker with his father, Starker carved his own niche in the field of wildlife management and conservation. It will be a difficult one for any other person to fill.

Raymond F. Dasmann
Member, Advisory Editorial Committee

Contents

Acknowledgments

It is a pleasure to indicate the many friends who have provided invaluable help. They have assisted us in numerous ways, and to them we tender our most profound thanks. These include H. L. Cogswell, R. Cole, D. F. Hoffmeister, W. J. Houck, G. S. Jeffers, H. Leach, W. Z. Lidicker, Jr., R. D. Mallette, F. Marino, B. Nelson, J. L. Patton, R. L. Rudd, B. Schulenberg, J. Schultz, V. Simpson, and P. Q. Tomich. J. A. Junge and R. S. Hoffmann have generously allowed us to reproduce their drawings of shrew skulls. In addition, we should like to express our appreciation for the editorial assistance of O. P. Pearson, A. C. Smith, and the late A. S. Leopold. All of these friends have given generously of their time and have made unique contributions to this book.

Mammal Ecology

STUDYING MAMMALS

Introduction

The diversity of the California mammal fauna reflects the wide variety of climates and habitats as well as the vast area and longitudinal range of our state. Much of the climatic diversification is the result of mountain building over the last 3 to 4 million years and concurrent drastic fluctuations in weather. The high elevation of the Cascades and the Sierra Nevada traps rain sufficient to support extensive coniferous forests on the western slopes and creates rain shadow deserts to the east. Along the coast, the cool winds from the north account for the persistent remnant redwood forest and the associated flora and fauna.

The elevational variations alone provide for a rich fauna and the close occurrence of such lowland forms as skunks separated by only a few miles from such alpine mammals as Pikas and Wolverines. The many rivers are the habitat of Beaver, Otter, and Mink; formerly the Central Valley was populated by herds of Pronghorn and Tule Elk.

The Pleistocene, or Ice Age, was a time of dramatically alternating climatic changes which increased the diversity of our mammalian fauna. During times of glaciation, many mountain slopes were deeply buried in ice, and the Great Basin was a humid region of many freshwater lakes. In these cold, wet periods some northern mammals moved south into our area; in the interglacial periods, when the climate ameliorated and ice melted, faunas moved northward and also into the mountains. During these glacial/interglacial fluctuations, many species became alternately separated and rejoined as they moved along north–south "pathways" parallel to the major mountain ranges. These environmental changes with their concurrent isolation and remixture of populations undoubtedly produced rapid genetic changes and account for the rich and varied mammalian fauna we have today.

Among California mammals are some large groups which are bewildering in their complexity. Among chipmunks, pocket mice, kangaroo rats, pocket gophers, and others, many species present major difficulties in identification. We hope that this volume will assist in the recognition of all California mammals. The keys, illustrations, and descriptions, when used together, will serve to identify all but a few of the most similar species.

The brief discussions of mammalian biology establish a base for interpreting mammalian behavior as it is seen among wild species. We apologize for the occasional but necessary use of unfamiliar or technical terms. Most are defined at their first occurrence in the text, and all are included in the Glossary.

Geographic distributions in the species accounts are given in general terms; topographic features (Map 2) will aid in associating ranges with habitat. The ranges for each species account of land mammal represent an approximation based on known occurrences. Although we intend the range maps to be accurate, the reader may fail to find some species where they are reported to occur and encounter others where they have not previously been known.

Some Hints

The world of mammals is essentially one of darkness and smells. Because most of the abundant species are entirely or mostly nocturnal, they are unfamiliar to all but the few people who seek them out. Many species move from daytime resting places only after dark and return before dawn. Much of their communication is through odors, most of which go undetected by the human mammal, and through sounds, which are frequently high pitched and of low volume. We are, therefore, usually aware only of those few species that regularly move abroad in daylight hours: chipmunks, ground squirrels, tree

MAP 1. California Counties.

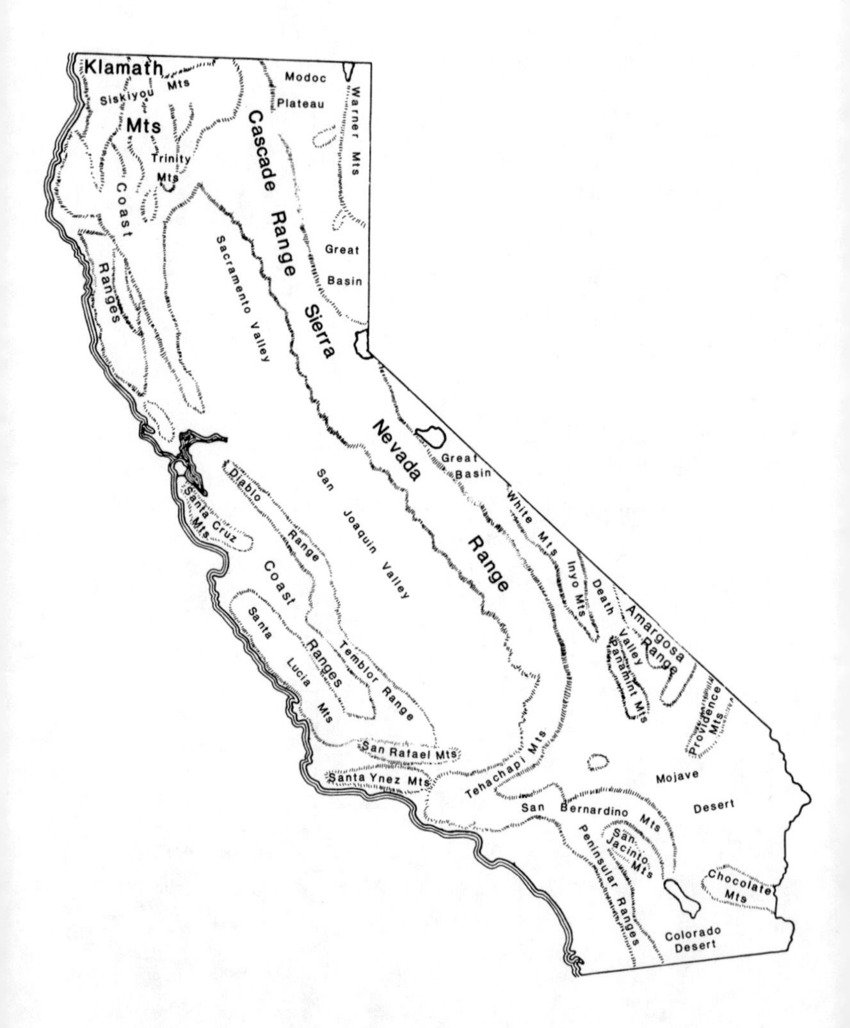

MAP 2. California Topography.

squirrels, and a few others. Occasionally we have a glimpse of a fox or an Opossum as it is briefly illuminated by automobile headlights, but few people have really observed kangaroo rats, flying squirrels, or water shrews.

Nevertheless, to learn about wild mammals it is necessary to learn to recognize indications of their presence and, when possible, to see them move about naturally. Armed with a pair of binoculars and a wealth of patience, you can begin to understand the behavior of ground squirrels and other diurnal mammals. Generally, you cannot watch wild mammals as readily as you can identify and observe wild birds.

If you are interested in observing nocturnal species, there are infrared light sources and binoculars developed to convert infrared light to visible wavelengths. Because the mammalian eye does not detect infrared light, wild mammals are unaware that they have become illuminated. With such equipment you can observe a fox, Raccoon, or flying squirrel under natural conditions. This equipment is available from scientific supply houses. A spotlight covered with red plastic is sometimes almost as satisfactory. Many species are relatively undisturbed by illumination from a hand-held flashlight, provided that the beam is not directed at the subject. Some observers have found that such creatures as kangaroo rats and pocket mice pursue their activities without inhibition in dim artificial light.

Small mammals that remain in the open can be observed at night with the aid of luminescent capsules. Translucent capsules containing Cyalume (which produces a cold light) can be temporarily attached to fur with the aid of rubber cement and produce light for up to six hours. These capsules fall off after a day or less. Cyalume is available from American Cyanamid Company, Bound Brook, NJ 08805.

Sophisticated hunters have long known that battery-powered hearing aids make audible many low-volume sounds made by both birds and mammals. A hearing aid is of greatest benefit when there are no competing "cultural" sounds from freeways and urban areas. Hearing aids are very useful at night, especially in open areas where there is little vegetation to disrupt sound transmission.

In contrast to bird songs, which have been studied in detail, the voices of mammals have not been extensively explored. Most mammals utter sounds that are brief and usually constitute "call notes," which may bear a superficial similarity among several species. Some careful studies have, however, demonstrated a variety of utterances produced under different circumstances (see Vocal Communication). Mammalian vocalization is an area of research in which talented amateurs may make original contributions to the understanding of mammalian behavior. Portable tape recorders are now of excellent quality and operate over a broad range of wavelengths. Vocalizations thus recorded may be converted (or transduced) to visual signals on an oscilloscope and their details compared. Specific sounds can then be correlated with specific sorts of behavior of both free-ranging and captive mammals.

Small mammals can be captured in a variety of homemade or commercially available traps. You can make a simple live trap by wiring an ordinary mouse trap to a small tin can and tying a piece of hardware cloth to the loop of the trap so that the can is closed off with the wire mesh when the trap is sprung (fig. 1). More refined live traps are sold in many sizes and are adequate for the capture of mammals from the size of a shrew to a fox. Good-quality traps for small mammals can be obtained from M. J. Spenser & Son, P.O. Box 131, Gainesville, FL 32601. These traps are suitable for species from the size of a small mouse to a chipmunk. Larger live traps are often advertised in sporting magazines.

Many students of small mammals use ordinary snap traps. They are cheap and readily available, but animals caught in snap traps are killed instantly. Nevertheless, snap traps may be preferable to live traps. Mice, wood rats, and squirrels frequently enter empty cabins and can cause extensive damage to blankets, pillows, and food. Although white-footed mice (*Peromyscus* spp.) and House Mice (*Mus musculus*) are the most common intruders, wood rats (*Neotoma* spp.) and the Northern Flying Squirrel (*Glaucomys sabrinus*) frequently take up residence in attics. When owners are absent snap traps are more humane; under these circumstances a live trap would hold the occupant until it starved.

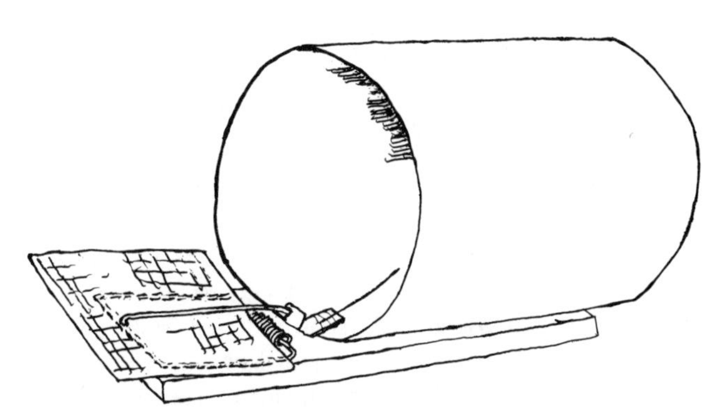

FIG. 1. Homemade trap.

You can create a simple, cheap, and usually effective trap by burying a large tin can (half gallon or larger) in the ground, leaving the top exactly flush with the surface of the ground. Small mice and shrews will often stumble into such a hole, and the smooth sides prevent escape. The can may be baited with rolled oats, walnut meats, sunflower seeds, or some other attractant, but the trap may be effective even without bait. Escape is more difficult if a lip is placed around the inside of the top, but a lip reduces the size of the hole through which the small mammal is to fall. Mice and shrews captured in these pitfalls are unharmed and can be observed in a glass jar and later released. Cans sunk in the deep sand of a desert canyon may capture a great variety of small mammals. Be sure to remove trap cans before leaving the area.

Regulations may limit or prohibit the capture of some or all wild mammals. Prior to trapping mice and shrews, you should become familiar with local restrictions. Some mammals are protected because they are scarce (see Table 1), and others are potential reservoirs of disease. Trapping is usually prohibited in county, state, and national parks.

Why capture mammals? The study of small species begins with their identification, which usually depends on a critical examination of structural details. With a knowledge of some major external features and locality, you may not need details of dentition and cranial aspects. Thus, a mouse or a chipmunk may be captured in a live trap, examined while restrained in a glass gallon jar, and then released at the capture site. The release should be prompt, and handling of the mouse or chipmunk should be minimal. (See Mammals as Carriers of Pathogens.) Not only can some kinds inflict painful bites, but their bones are fragile and easily broken.

The trap should be placed to maximize the likelihood of capture and to minimize the risk of danger to whatever might enter the trap. Choose a shady spot protected from rain, and never leave a trap unattended for more than four or five hours. Live traps set in late afternoon should be examined several hours after dark and again at daybreak. If the same traps are left set for diurnal species, they must not be exposed to direct sunlight.

You can observe some species at close range without trapping. Meadow voles, which make discrete trails through grass, and deer mice often make their nests under boards lying on the ground, and a quick hand may capture a startled individual when the nest roof is suddenly removed. Be cautious when grabbing dazed creatures, however, for they can bite. You can place boards on the ground, and eventually mice will nest beneath them. A small "nest box" sunk in the ground may also eventually be occupied by wild mice. Provided with a loose lid, the box allows you to observe the presence of young in the spring and food stores in the autumn. Minor disturbances may be tolerated by wild mice, especially if your movements are slow and quiet.

Individual mammals can be marked for later recognition. A deer mouse, for example, can be restrained in a thick sock while a color (such as a food coloring) is applied to the head or back. This mark will remain until the next molt and will enable you to recognize that individual if seen again. By such means you can have a local population of mice or chipmunks with green heads or red backs in a variety of combinations so that many individuals can be recognized. The survival of these unnaturally marked rodents indicates that they are not especially vulnerable to predators.

Although wild mice and other small mammals are naturally

secretive, it is possible to create situations in which you can observe their behavior. At a campsite, where there is likely to be a resident population of scavenging rodents, a spotlight covered with red paint or red plastic will not disturb their movements. Mice may be already conditioned to search for scraps of leftover food, and a little encouragement will bring them to a dimly lit site. (This practice is not to be encouraged where bears occur.) Then, by marking individuals, you may learn individual behavior traits.

Small mammal tracks, in addition, can be observed at such feeding stations. By sweeping the ground clean of twigs and small stones, you can prepare a clean slate on which the foot imprints of mice and the like can be found (fig. 2). A thin layer of wheat flour, lightly dusted over the smooth dirt, will leave the tracks distinctly outlined. If the tracks are sufficiently clear, you can make plaster casts of them by pouring fluid plaster of paris on the imprints. With a collection of track casts, the consistent and the variable features of tracks can be learned.

A similar sort of feeding station can be used for larger species, such as Raccoons and foxes. Remember, however, that the same fare that draws a Raccoon or fox will also attract a skunk. Both Spotted Skunks and Striped Skunks are fearless, and their extremely offensive scent, which can be squirted 5 ft (about 2 m) or more, is very difficult to remove from fabric. Rats (*Rattus* spp.) will also be drawn to feeding stations, especially in suburban areas, and are generally undesirable for many reasons.

You can learn to recognize some of the more distinctive mammal tracks by observing animals as they move about. The origin of the tracks is then certain, and, moreover, you can associate the spacing of the tracks with the gait of the individual that made them. Also, with some knowledge of tracks, you may frequently get a clue as to the species which left feces, or scats. It is far more difficult to identify scats than mammal tracks, for the scats vary in size and shape according to the diet and age of the mammal that made them (fig. 3).

Bats cannot be enticed to a feeding station, but they occasionally do concentrate and can then be easily observed. In

FIG. 2. Mammal tracks: *a,* Raccoon; *b,* Opossum; *c,* Muskrat (note tail mark); *d,* Black Bear; *e,* Striped Skunk; *f,* Badger; *g,* Elk; *h,* Mule Deer; *i,* Mountain Lion; *j,* Bobcat; *k,* California Ground Squirrel; *l,* Long-tailed Weasel; *m,* Gray Fox; *n,* Coyote (note large outer toes of hind foot). Each scale bar equals 1 inch.

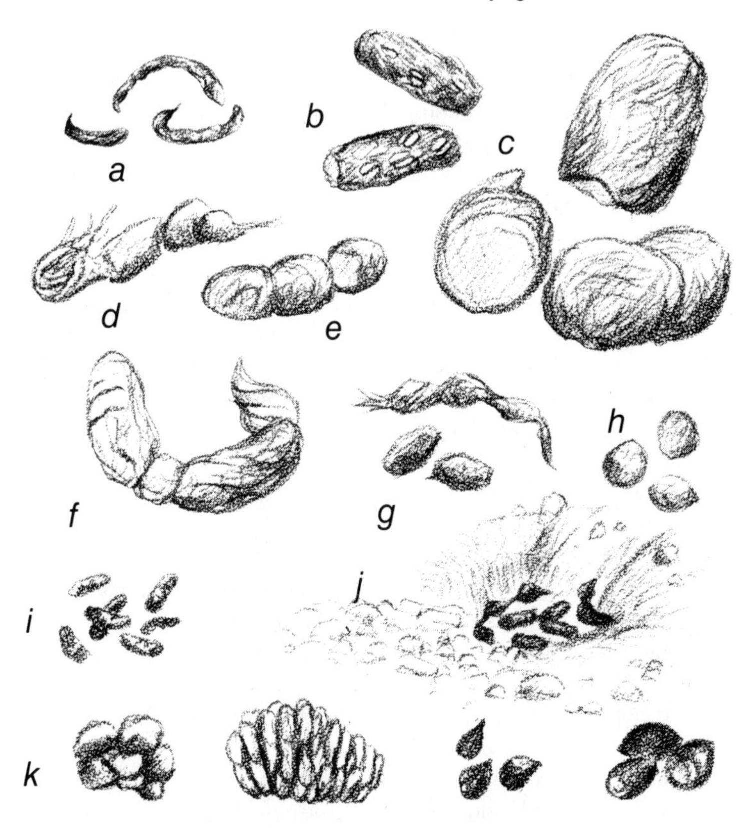

FIG. 3. Mammal scats: Species of cats frequently deposit their rather uniformly segmented feces in so-called scrapes, often near rocks; the irregularly segmented scats of foxes and the Coyote are often found along paths and roads. *a,* Long-tailed Weasel; *b,* Raccoon; *c,* Mountain Lion; *d,* Gray Fox; *e,* Bobcat; *f,* Coyote; *g,* Yellow-bellied Marmot; *h,* Black-tailed Jackrabbit; *i,* Desert Wood Rat; *j,* California Ground Squirrel scats in "latrine"; *k,* four forms of Mule Deer droppings. Not to scale.

arid regions, bats gather at water troughs, where they drop to the surface of the water to drink. Moderate light usually will not frighten them away and will assist in identifying them. Bats may also be found roosting in small groups or large aggregations in caves or mine shafts. You should take pains, however, to avoid disturbing resting bats, for movement of one may cause others to take flight. Sometimes you may find the night-time roost of one or several bats, perhaps under the eaves of a porch roof; by examining the fragments of insects under such roosts, you can learn the diet of these bats. In relatively open habitats, bats can be followed at night by the aid of luminescent Cyalume capsules glued to their fur.

The more sophisticated student of mammals may wish to follow individuals by radiotelemetry. Tiny radios that emit a distinctive signal can be detected at least several hundred yards away (about 100 m) and sometimes at more than a mile. These radios provide an accurate means of following an animal's movements. Using such a device, you can eventually determine a mammal's home range and homesite as well as where and when it forages. Small radios and receivers can be purchased from Custom Electronics, 2009 Silver Court West, Urbana, IL 61801.

Direct observation in the daylight is feasible for only a few species, such as chipmunks, ground squirrels, tree squirrels, and some other mammals under certain circumstances. Muskrats, for example, rarely move far from water, and you can watch them from a car parked next to a slough or irrigation canal. They are active in early morning and late afternoon and can be attracted to a small raft, about 1 m square, anchored in the open. Muskrats will voluntarily take their food to such floating objects and sometimes cover small rafts with shells of freshwater clams on which they have fed. They can also be baited with carrots and other succulents.

Some of the shiest mammals can be observed while rather young, when they spend much time in play and before fear has developed. When young foxes first venture from the burrow, they readily play in the open. You can enjoy watching a litter of kits as they chase one another, catch grasshoppers, wait for their parents to bring food, or just sleep. You should watch from a respectful distance with the aid of binoculars or a spotting telescope to avoid alarming the parents.

Many mammals, but especially carnivores, can be brought into view with a device known as a predator call. These "whistles" produce the sound of a stricken rabbit and have a magnetic effect on foxes, Coyotes, Bobcats, and raptorial birds. Even bears may respond to this sound. The use of predator calls is most effective when the caller is concealed in a blind.

These are a few of the ways in which you can learn about the habits and behavior of wild mammals. The clever field naturalist will think of many more. The basic ingredients are patience and the ability to remain quiet for long periods. You should always remember to avoid disturbing wild mammals. An alarmed animal may become careless and expose itself to danger—and if the animal is large, the observer may also be placed in danger.

Certain additional methods may be used to study dead animals. The species descriptions in the second part of this book include comments on the structure and shape of the skull and the nature of the teeth. In many kinds of mammals, the teeth hold some of the most distinctive characters. The teeth, moreover, are extremely hard and durable and therefore endure as fossils when many other bones have been crushed or broken. It is logical, then, to give special attention to dentition—the number and form of the teeth. Teeth are designated in order of position—for example, 2/1, 1/0, 1/0, 3/3 refers to the incisors, canines, premolars, and molars, upper and lower. If you find unfamiliar dead specimens, you should save the skull for identification. When it is dried, the flesh can be removed by small beetles of the family Dermestidae. These beetles come naturally to a skull that is hung outdoors from a tree or clothesline, where it is out of reach of dogs, cats, and small children. Dermestid beetles do a fine job of cleaning the dry meat, and then the critical features of the skull and teeth become apparent (fig. 4).

The species descriptions also include measurements, in millimeters, of the main parts of the body, usually total length (TL), tail (T), hind foot (HF), sometimes the ear (E), and, in bats, the forearm (F). The ear is commonly measured from the lower and outer notch to the tip; the hind foot is measured on the bottom surface; the tail is measured from its origin at the back to the tip of the last tail vertebra (fig. 5). The weight is given in grams or kilograms. In the event that you find a freshly killed animal, these measurements greatly facilitate identification. Be careful to avoid fresh blood and ectoparasites.

You may wish to preserve freshly killed specimens in good condition. Many texts give instructions for preparing specimens for subsequent identification and study. One of the best is *Methods of Collecting and Preserving Vertebrate Animals* by R. M. Anderson, published by the National Museum of Canada (Bulletin No. 69, Biological Series No. 18); this is avail-

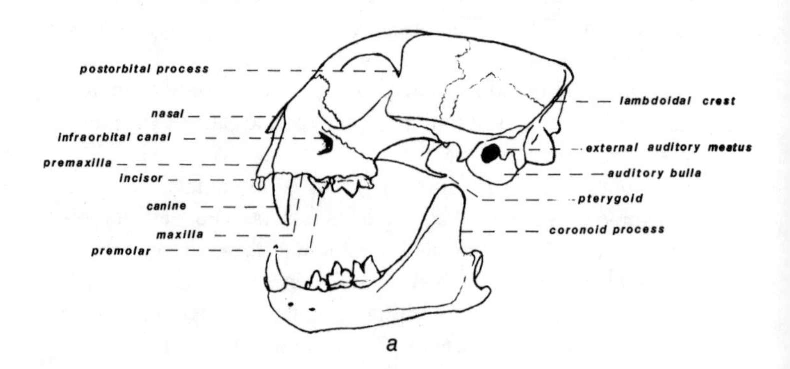

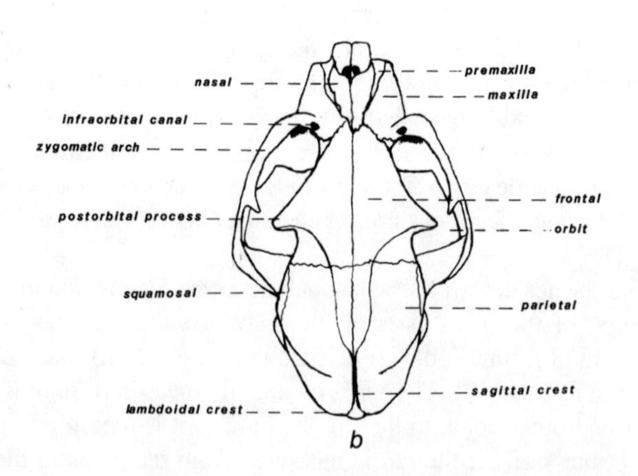

FIG. 4. Skull of Bobcat (*Lynx rufus*) showing important diagnostic features; *a*, lateral view; *b*, dorsal view.

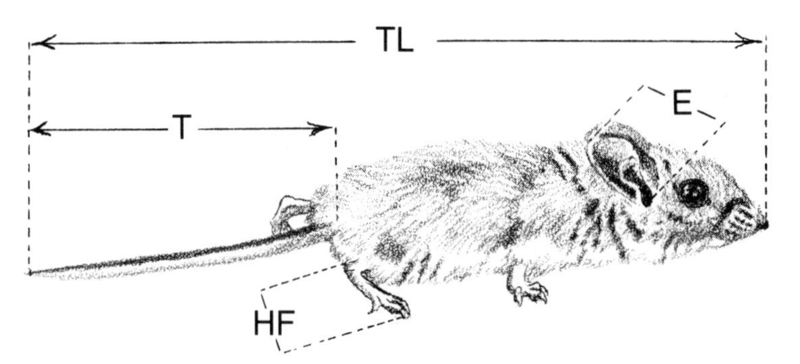

FIG. 5. Measurements of mammals: *TL,* total length; *T,* tail (to tip of last vertebra); *HF,* hind foot (heel to end of longest claw); *E,* ear (notch to tip).

able from the Queen's Printer, Ottawa, Canada. Another is *Directions for Preparing Specimens of Mammals* by Gerrit S. Miller, Jr., U.S. National Museum (Part N of Bulletin 39); this is available from the U.S. Government Printing Office, Washington, D.C.

References

Anderson, S., and Jones, J. K., Jr. 1984. *Orders and families of recent mammals of the world.* New York: John Wiley & Sons.

Chapman, J. A., and Feldhamer, G. A. (Eds.) 1982. *Wild mammals of North America: Biology, management and economics.* Baltimore: Johns Hopkins University Press.

Cheeseman, C. L., and Mitson, R. B. (Eds.) 1982. *Telemetric studies of vertebrates.* Zoological Society of London, series 49. New York and London: Academic Press.

Cockrum, E. L. 1962. *Introduction to mammalogy.* New York: Ronald Press.

Gunderson, H. L. 1976. *Mammalogy.* New York: McGraw-Hill Book Company.

Hall, E. R. 1981. *The mammals of North America.* 2d ed. 2 vols. New York: John Wiley & Sons.

Hamilton, W. J., Jr. 1939. *American mammals: Their lives, habits and economic relations.* New York: McGraw-Hill Book Company.

Ingles, L. G. 1965. *Mammals of the Pacific states: California, Oregon and Washington.* Stanford: Stanford University Press.

Matthews, L. H. 1969. *The life of mammals.* 2 vols. London: Weidenfeld & Nicolson.

Murie, O. J. 1954. *A field guide to animal tracks*. Boston: Houghton Mifflin Company.

Savage, A., and Savage, C. 1981. *Wild mammals of northwest America*. Baltimore: Johns Hopkins University Press.

Stoddard, D. M. (Ed.) 1979. *Ecology of small mammals*. New York: Chapman & Hall and John Wiley & Sons.

Van Gelder, R. G. 1982. *Mammals of the national parks*. Baltimore: Johns Hopkins University Press.

Vaughn, T. A. 1978. *Mammalogy*. 2d ed. Philadelphia: W. B. Saunders Co.

RARE AND ENDANGERED SPECIES

The expansion and increased intensity of human activities bring greater pressure on populations of wild vertebrates. Even when there is no direct destruction of wild animals, habitat alterations may reduce the *carrying capacity* of the land—its ability to provide adequate food and cover. When the carrying capacity for a given species is reduced, there is a decline in its abundance. In some situations, habitat alteration may *increase* the carrying capacity for other species. For example, the removal of a large stand of firs and pines, allowing the development of a rich cover of grasses, forbs, and shrubs, increases the carrying capacity for deer and chipmunks while decreasing that for the Northern Flying Squirrel and the Douglas' Squirrel.

The vast majority of mammals can adjust to small environmental changes: there may be fluctuations in abundance and local distributions, but extinction is not usually an immediate danger. Some species, however, either because of specific environmental requirements or sometimes because of slow reproductive rates, are especially vulnerable to disturbances. In such species, habitat changes can result in either a reduction in numbers or a great restriction in geographic range. To preserve the populations of wild species at levels compatible with concepts of multiple use of land, regulatory agencies have given scarce species particular recognition and protection.

The state legislature in 1970 passed the Endangered Species Act, which provides for recognition of rare and endangered species by the Fish and Game Commission. These species have been reviewed by the Department of Fish and Game in a series of documents entitled *At the Crossroads,* published biennially beginning in 1974. Similar legislation was enacted by Congress in 1966; in 1973 the federal Endangered Species Act included invertebrate as well as vertebrate animals and provided for federal recognition and protection of rare and endangered species of plants.

Currently defined, an *endangered species* is a form faced with an immediate danger of extinction in at least a significant part of its geographic range. A *threatened species* is one which is rare and approaching the status of an endangered species. Thus, both state and federal government designate animals whose numbers are reduced, and there are two degrees indicating the relative dangers of extinction.

Among California mammals, some are identified by the state as either rare or endangered, including some not placed in any special status by the federal regulatory agencies. California mammals are currently (1985) designated as either *rare* (that is, threatened) or *endangered.* Table 1 indicates the category designated by the state and federal authorities.

Mammals listed in Table 1 are totally protected. In many instances there are suitable habitats of public land set aside for one or more of these species. For some of them, populations are very low and survival is uncertain. As numbers dwindle, vulnerability increases. Some populations of Bighorn Sheep, for example, are dangerously low and perhaps unable to survive in the face of increasing numbers of Burros.

Some rare or endangered mammals occur on private lands, and provision for their total protection becomes sociologically complex. Large mammals, such as Bighorn Sheep and the Tule Elk, move freely from public to private land. Small rodents may occur on private land and are not always an unmixed blessing.

When totally protected mammals occur on private land, there is a possible conflict of interest, because their presence may involve expense to the landowner. Small rodents do compete with grazing domestic stock, although the problem may not be a costly one. Ground squirrels not only take some for-

TABLE 1. Rare or Endangered California Mammals: 1980

Mammal	State	Federal
Island Fox *(Urocyon littoralis)*	Rare	
San Joaquin Kit Fox *(Vulpes velox mutica)*	Rare	Endangered
Sierra Nevada Red Fox *(Vulpes vulpes necator)*	Rare	
Wolverine *(Gulo gulo)*	Rare	
Guadalupe Fur Seal *(Arctocephalus townsendi)*	Rare	Endangered
California Bighorn Sheep *(Ovis canadensis californiana)*	Rare	
Peninsular Bighorn Sheep *(Ovis canadensis cremnobates)*	Rare	
Mojave Ground Squirrel *(Spermophilus mohavensis)*	Rare	
San Joaquin Antelope Squirrel *(Ammospermophilus nelsoni)*	Rare	
Fresno Kangaroo Rat *(Dipodomys nitratoides exilis)*	Endangered	
Giant Kangaroo Rat *(Dipodomys ingens)*	Endangered	
Morro Bay Kangaroo Rat *(Dipodomys heermanni morroensis)*	Endangered	Endangered
Stephens' Kangaroo Rat *(Dipodomys stephensi)*	Rare	
Salt Marsh Harvest Mouse *(Reithrodontomys raviventris)*	Endangered	Endangered
Amargosa Vole *(Microtus californicus scirpensis)*	Endangered	

age and their seeds, but also are potential reservoirs of disease, especially plague; in any event, they are usually unpopular with landowners. Although the Wolverine quite properly receives total protection, it has a long history of breaking into mountain cabins, where it does extensive damage.

Although landowners are legally bound to honor the designations of rare and endangered forms, they are not compensated for the trouble or expense of complying with the law. Conservationists and the general public usually agree on the need for special protection for scarce animals, but the cost of this protection is sometimes not carried by conservationists or taxpayers. Although these socioeconomic problems are usually minor, they are real and constitute a problem which is commonly neglected.

Mammals Recently Extinct in California

Within historic times, four species of mammals have disappeared from within our state boundaries. Each remains elsewhere, however, so the elimination of these mammals from California is properly called *extirpation* and not *extinction*. The Wolf, Grizzly Bear, Jaguar, and Bison no longer exist as wild species in California, and their departure is a matter of historic record. One can express regret that these magnificent beasts no longer exist in our state, but were they to roam freely over our lands today, they would cause some serious problems. They all demand large pieces of land, and all have habits that conflict with certain aspects of human activity.

The Wolf (*Canis lupus*) occurred along the eastern edge of the state and in the Central Valley. Early travelers mentioned Wolves in many parts of California, but some of these animals may have been Coyotes. In his diary of his gold rush experiences in 1850, J. Goldsborough Bruff frequently wrote about Wolves. He mentioned both Wolves and Coyotes and distinguished them on the basis of both size and color. Bruff noted that Wolves occurred both near Sacramento and in Calaveras County. In the early part of this century the Bureau of Biological Survey employed trappers and hunters whose task was to eliminate the Wolf from cattle ranges. The Wolf disappeared from Nevada in 1923. In California, Wolves were seen southwest of Tule Lake in 1922, and the last one was captured in that area in 1924. The last Wolf in Oregon was killed in 1974. More recent reports of Wolves in the Sierra Nevada are based on an animal which was of Asiatic origin, presumably an escaped captive or pet.

The Grizzly Bear is the New World representative of the Brown Bear (*Ursus arctos*) of the Old World. The Grizzly once occurred widely throughout California and was a constant threat to humans and domestic stock. Its total lack of fear combined with its destructive habits rendered it a dangerous member of our fauna. Up to the 1850s or 1870s the Grizzly roamed the Central Valley, Coast Ranges, Cascades, and Sierra Nevada. Generally, it reached its greatest concentrations in the Central Valley and in chaparral at low elevations. Vivid accounts of this huge bear are to be found in *California Grizzly* by Tracy Storer and Lloyd Tevis, Jr. This bear was genuinely abundant, and there were numerous places where travelers risked meeting groups of from ten to twenty or more Grizzlies together. Because its centers of density coincided with ranching activities, it was persecuted whenever encountered. The last Grizzly in our state was killed in the early 1920s.

The Jaguar (*Felis onca*) is the largest cat in the New World and, like the Grizzly, an awesome creature to have as a neighbor. A large Jaguar may exceed 110–120 kg and can bring down prey up to the size of a grown horse or a large bull. The occasional individuals that have recently occurred within the United States have been conspicuously destructive to domestic stock. Unlike the Grizzly, however, the Jaguar is rarely, if ever, a threat to human life. It is, in fact, a rather shy creature and seldom to be seen without the aid of hunting dogs. Today the Jaguar is sometimes found in the states bordering Mexico, but it is not clear if these are occasional strays or residents. The Jaguar roamed the South Coast Ranges between San Francisco and Monterey up to at least 1826. The last known individual was killed in Palm Springs about 1860. Outside of California, the Jaguar once ranged north to Colorado, but it must now be regarded as extremely rare in the United States.

The Bison (*Bison bison*) once lived in the northeastern corner of California and the adjacent regions to the north and northeast; it disappeared from that area early in the nineteenth century. It died out in our state about the time Indians acquired horses, but before they had firearms. Europeans apparently had only a secondary role in the demise of the Bison in California. It was most common about Malheur Lake, in Oregon, and in the area between Eagle and Honey lakes in California.

References

Allen, G. M. 1942. *Extinct and vanishing mammals of the Western Hemisphere, with the marine species of all the oceans.* Special Publication No. 11. Washington, D.C.: American Committee for International Wild Life Protection.

Bruff, J. G. 1949. *The journals, drawings and other papers of J. Goldsborough Bruff, April 2, 1849–July 20, 1851.* Edited by Georgia Willis Read and Ruth Gaines. New York: Columbia University Press.

California Department of Fish and Game. 1980. *At the crossroads.* Sacramento.

Craighead, F. C., Jr. 1979. *Track of the grizzly.* San Francisco: Sierra Club Books.

Klinghammer, E. (Ed.) 1979. *The behavior and ecology of wolves.* New York and London: Garland STPM Press.

Roe, F. G. 1970. *The North American buffalo.* 2d ed. Toronto: University of Toronto Press.

Rorabacher, J. A. 1970. *The American buffalo in transition: A historical and economic survey of the bison in America.* Saint Cloud, Minn.: North Star Press.

Storer, T. I., and Tevis, L. P., Jr. 1955. *California grizzly.* Berkeley and Los Angeles: University of California Press.

Young, S. P., and Goldman, E. A. 1944. *The wolves of North America.* Washington, D.C.: American Wildlife Institute.

REPRODUCTION

Reproduction embraces a variety of events, all relating to the production and establishment of a new generation. In many small short-lived mammals, such as mice, embryonic development, or *gestation,* is brief; mating and birth may be separated by only three weeks or less. In most larger mammals, such as deer and bears, reproduction occupies months, and in discussing the breeding season of these species we must consider the period from mating to birth.

Seasonality

Breeding cycles usually have an annual rhythm. This reproductive seasonality is a combined response of the body to age (sexual maturity) and environmental factors or cues. Until the body has a mature sexual system, it cannot respond to sexually stimulating signals from the environment, and virtually all adult wild mammals in temperate areas remain sexually quiescent until seasonal stimuli are encountered. Thus, reproductive activity represents the combined effects of sexual maturity (resulting from internal stimuli) and the fluctuating environment (or external stimuli which trigger internal stimuli).

Most biologists believe that the breeding season is adjusted so that the young are born at the season most favorable for survival and rapid growth. Young of most mammals are born in the spring, the time of melting snow, vegetative growth, and increasing warmth. For small species, such as mice, mating too occurs in the spring. In contrast, some small mammals, notably bats, may mate in the autumn and produce young the following spring; and some carnivores mate in the summer or autumn and give birth to young the following winter. Such diverse schedules of mating and birth (or *parturition*) usually constitute adaptations to various other (usually nonreproductive) aspects of the species' life cycle.

Reproductive Factors or Cues

Mammals are adapted to respond to several different sorts of environmental changes, and these changes are most commonly seasonal in occurrence. The daily change in day length (or *photoperiod*) is an important environmental cue for annual biological events in many, and probably most, plants and animals. The annual *photocycle* (the change in length of day and night) is dependable and repeats itself precisely every year. Thus, from late December until late June there is a daily increase in the duration and intensity of sunlight. Many mammals, having experienced six months of decreasing daylight, are very sensitive to even slight increases in day length; and as winter blends into spring, long days are often increasingly stimulating to the reproductive system. Mammals that breed in late winter or spring are frequently species whose *gonads* (testes and ovaries) respond to long days. Laboratory studies have shown that some mice (*Peromyscus maniculatus*) and squirrels (*Glaucomys volans*) are long-day breeders, whereas other mammals, such as domestic sheep and Black-tailed Deer (*Odocoileus hemionus*) are short-day breeders. Contrary to what one might guess, nocturnal mammals may be just as responsive to day length as are certain diurnal species.

The sexual activity of some mammals is apparently only partly cued by day length. Laboratory observations on several kinds of voles (Arvicolidae) revealed that after exposure to short days, the gonads shrink or regress, but not to the degree seen in wild voles. This situation suggests that although the natural photocycle may affect reproductive seasonality in voles, other environmental cues may also promote their sexual activity. The reproductive season of some voles, at least, is not always closely associated with the regular seasonal changes in day length but is often related to plant growth. In the Great Basin, for example, the Montane Vole (*Microtus montanus*) breeds when autumn rains produce a lush growth of annual herbs.

Some other desert rodents, especially kangaroo rats (*Dipodomys* spp.) and pocket mice (*Perognathus* spp.), are apparently stimulated by the fresh growth of annual plants. In years when fall rains produce a luxuriant plant growth, kangaroo rats and pocket mice are sexually active by late winter, but in dry years these desert rodents may forgo reproduction entirely. Because kangaroo rats and pocket mice feed on the fresh leaves of desert annuals, plant material is suspected of stimulating their gonadal growth. A possible effect of fresh leaves on mammalian reproduction is seen in pocket gophers, voles, and jackrabbits on agricultural land in the Central Valley. In irrigated pastures these mammals tend to breed for longer periods than on adjacent nonirrigated land.

This stimulatory effect of plants on wild mammals, sometimes apparent in nature, has been duplicated in the laboratory. A number of plants, especially certain legumes, contain substances similar or even identical to the sexual hormones produced by vertebrate gonads. Other materials stimulatory to the reproductive system and found in sprouting grain are chemically different from vertebrate hormones. These "phytoestrogens," as they are sometimes called, may well provide the additional reproductive stimulation to account for the surges of breeding in some desert mammals during years of lush plant growth, and their absence may result in the occasional failure of entire populations to breed in dry years.

Hormonal Controls of Reproduction

An internal system of chemical controls regulates the reproductive process. Certain tissues, called *endocrine glands,* produce chemicals that circulate in the blood plasma and are carried throughout the body. These substances, called *hormones,* have specific effects on certain tissues, called *target organs,* while not altering the activity of most other tissues. Some hormones stimulate activity in other endocrine glands to produce still different hormones; but hormones also influence nonendocrine tissues.

One of the most important endocrine glands is the *hypothalamus,* situated in the midventral part of the brain. It receives, either directly or indirectly, environmental stimuli and influences many bodily activities. It produces a hormone that is carried by a short blood vessel directly to the *anterior lobe of the pituitary gland,* or *hypophysis,* which lies directly below the hypothalamus. This hormone, *gonadotropin-release hormone* (Gn-RH) stimulates the anterior pituitary to release two *gonadotropins—follicle-stimulating hormone* (FSH) and *luteinizing hormone* (LH)—which are carried thoughout the body.

The two gonadotropins increase the circulation of blood to the gonads and also stimulate their growth and metabolic activity. The female gonads (the ovaries) produce two hormones: *estrogen* and *progesterone*. Both estrogen and progesterone enter the circulation and are carried throughout the body in blood serum. They not only stimulate the growth and metabolic activity of the uterus and breasts but have behavioral effects, increasing the receptivity of the female to sexual advances of a male. As these hormones circulate throughout the body, they reach both the hypothalamus and the anterior pituitary, where they have important effects. At low levels these two ovarian hormones tend to depress the release of Gn-RH from the hypothalamus and FSH and LH from the anterior pituitary; this is sometimes referred to as a negative feedback effect. At high levels, however, these same hormones stimulate the hypothalamus and the anterior pituitary so that the gonadotropin LH is released in greater amounts. High levels of LH cause the ovary to release one or several eggs, or *ova;* this process is called *ovulation.*

The Estrous Cycle

Many female mammals experience cycles in which preparation of the uterus and release of one or more ova are synchronized with sexual acceptance of a male. This is called *estrus,* or "heat." If mating occurs, sperm meet the ova in the upper *oviduct* (between the ovary and the uterus), a sperm enters an ovum, an act known as *fertilization,* and the fertilized ova slowly move down the oviduct to the uterus. If mating does not occur, the unfertilized ova pass into the uterus and eventually die. Then the entire process, called the *estrous cycle,* may repeat itself.

In some mammals, such as voles, ovulation results only after a mating; when the act of mating stimulates the release of one or more ova, this is called *induced ovulation.* In species in which there is induced ovulation, estrus is prolonged. Induced ovulation has the advantage of timing the release of ova with the arrival of sperm, thus eliminating potential loss of ova. Because estrus is extended in species with induced ovulation, there is a greatly increased period during which the female will accept a male.

Delays in Birth

After mating there are four possible sequences of events. Most commonly an egg is fertilized in the oviduct shortly after mating, and the fertilized egg then attaches to the wall of the uterus. This stage is called *implantation*. The fertilized egg starts to divide immediately, and the growing egg together with the uterine wall form a *placenta*. Thereafter embryonic growth proceeds at a steady rate until birth. This direct development is the most common pattern in California mammals. Such direct development occurs also in the Opossum; but, as in most marsupials, the placenta is not well developed.

A second pattern is seen in some members of the weasel family. The Marten, Fisher, and Spotted Skunk, for example, have a gestation period extended by *delayed implantation* or *embryonic diapause*. In these mammals, fertilization and cell division occur shortly after mating but then cease, and implantation may be postponed for weeks or even months. The adaptive significance of delayed implantation in these mammals is not always clear. Delayed implantation is seen in the Black Bear; mating takes place in the summer, and the small infant is born in midwinter. Early birth provides an extended period for growth before the young bear's first hibernation. Adults must mate prior to hibernation, because after that time they are solitary.

A third pattern is seen in some bats: mating may occur in the fall, but without ovulation or fertilization. In such mammals the testes become active in late summer and sperm mature in autumn. Unlike most mammals, gonads of both sexes in these bats do not mature in the same season; but, after mating, sperm are stored in the oviduct of the female, and there they remain viable during the winter while both sexes hibernate. Upon emergence from the hibernation in the spring, females are sexually active. Probably because females of some kinds of bats hibernate together, they are subjected to the same cues, to which they respond simultaneously. Although it is not known what these cues are, it is known that ovulation and fertilization are synchronized in the spring among a group of female bats that have overwintered together; the ova are then fertilized by sperm stored from mating in the previous autumn. This is called *delayed fertilization.* Females of some species move to old barns, empty attics, or a hollow tree and establish "nursery colonies"; subsequently, birth of the young is synchronized.

In a few bats there is *delayed development,* a fourth pattern, which differs physiologically from both delayed implantation and delayed fertilization with sperm storage. In the California Leaf-nosed Bat (*Macrotus californicus*), which neither migrates nor hibernates, mating, fertilization, and implantation occur in rapid sequence in the fall. Following the autumnal mating there is a very slow and gradual development with the result that gestation in this bat occupies some nine months.

In addition to hibernation, many California bats also migrate, and in migratory species the sexes may be separate for much of the year. Delayed fertilization or sperm storage might have arisen as an adaptation to the separation of male and female bats in the spring. This is a firmly fixed phenomenon in many bats of temperate regions and undoubtedly originated in ancient geological time. The adaptive significance of delayed fertilization for some species may therefore lie deep in the past.

Condition of Young at Birth

Many infant mammals are *altricial,* born nude and helpless, usually within a protective nest. Young mice (*Peromyscus* spp.), voles (*Microtus* spp.), shrews (*Sorex* spp.), and squirrels (*Sciuridae*) are blind and naked, with the earflap folded over the ear opening. In contrast, other species have a longer gestation period after which the *precocial* young are born well formed with eyes open and ears erect. Young Porcupines, for example, have a long gestation and are born well developed, even with quills. Jackrabbits, like all hares (*Lepus* spp.), are fully furred and alert at birth and can feed on tender grass even before their fur is dry. Likewise, deer and antelope are precocial. The distinction between the altricial and precocial conditions is not always clear-cut, for there are intermediates: cats are born with some fur but with their eyes and ears closed, and the Opossum is born in a virtually embryonic state after a gestation of only twelve and a half days.

The speed of postnatal development varies among different species and is probably an important physiological adaptation. Among California ground squirrels (*Spermophilus* spp.), hibernating species have a brief activity period and from birth grow more rapidly than do the nonhibernating Antelope Ground Squirrels.

Litter Size

Each species of mammal has its characteristic litter size, but always with some variation. Even among humans, who usually produce one infant from a single ovum, there are occasional twins or triplets. Similarly, most bats produce one young annually, but there are some twins born. Among mammals that regularly have several young in a single litter there is usually substantial variation. This variation in litter size has received much attention by ecologists, for presumably there is an underlying pattern of adaptive significance.

There are perhaps three basic patterns of variations in litter size: variation among species (interspecific); local variation within a single species (intraspecific); and geographic intraspecific variation. Litter size tends to compensate for mortality in the long term, but birth rate and death rate are not always in balance. On the one hand, a large litter size and overall high birth rate (resulting from two or more litters) can be viewed as compensating for a high mortality rate; or, on the other hand, a low mortality rate (or great longevity) can be regarded as an adaptation to low productivity. These concepts, however, are more easily formulated than confirmed; theories on birth rate (*natality*) and death rate (*mortality*) are based more on assumptions than on facts.

Milk and Lactation

Mammals nourish their young by a secretion from the mammary glands of the female. Nursing (or *lactation*) is a uniquely mammalian feature and should be carefully studied by anyone wishing to compare the life patterns of different species.

Mammary tissue develops in the growing embryo, apparently from modified sweat glands, and sexual differentiation is apparent in very early development. Although present in both sexes, mammary glands normally function only in females. Some mammary tissue is present in early growth, but the greatest development occurs during the period of sexual maturity, and the sexual disparity increases during pregnancy.

In contrast to the pubertal breast development in the human female, mammary tissue in almost all other mammals undergoes most of its development during pregnancy. With the approach of sexual maturity and ovarian cyclicity, hormones from the ovaries and the anterior pituitary stimulate mammary growth, or *mammogenesis*. In addition, the placenta, through its release of a hormone called *placental lactogen,* plays a major role in mammogenesis during pregnancy. Histologically, the *mammary "gland"* is composed of secretory hollow spheres called *alveoli,* which drain into a branching system of *ducts* (fig. 6), which in turn lead into the nipple. In ungulates, the ducts lead into a collecting chamber, the cistern, housed in the udder; in these mammals the milk leaves the cistern through a false nipple, or teat, at the time of nursing. About each alveolus are *myoepithelial strands,* fine muscles that contract under hormonal stimuli. This complex structure is supplied with nerves, especially from the nipple, leading to the brain and specifically to the hypothalamus.

Since the role of hormones in mammogenesis of wild mammals is poorly known, we must study the process in laboratory and domestic species. In the laboratory rat the development of the ducts progresses under the stimulation of estrogen, and alveolar growth is induced by progesterone and estrogen from the ovary, *prolactin* from the anterior pituitary, and also placental lactogen. Moreover, *insulin* (from the pancreas) promotes the stimulatory action of the ovarian hormones. The roles of these hormones not only vary from one kind of mammal to another but also change through the duration of both pregnancy and lactation.

As the embryo or embryos develop, progesterone and placental lactogen levels rise and are presumably increasingly important in mammogenesis. At the same time, the level of estrogen, and perhaps its role in mammary development, gradually decline.

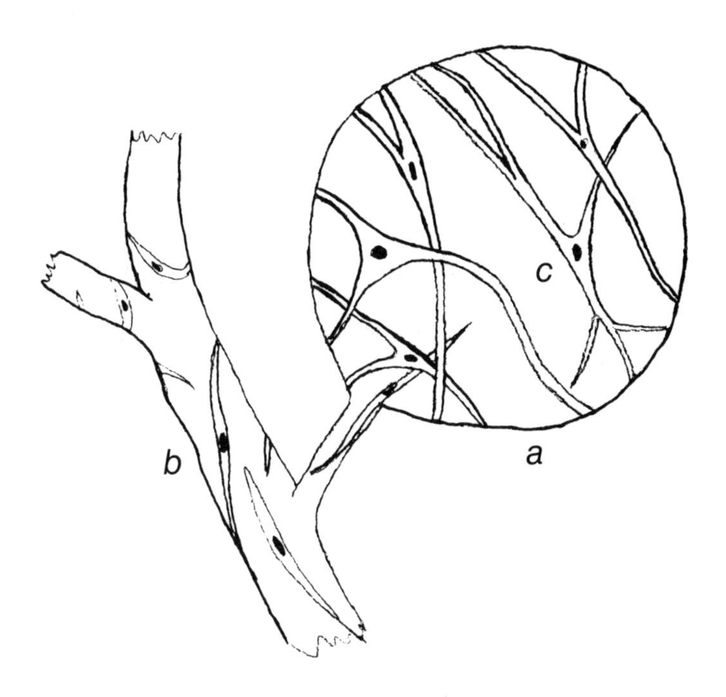

FIG. 6. *a,* Mammary alveolus; *b,* adjacent duct of the mammary gland, showing *c,* the covering of contractile cells, which force milk from the alveolus to the ducts and eventually to the nipple.

The hormones that induce the growth of mammary glands also prevent the synthesis and flow of milk until the infant is born. At the time of birth of the young there are major changes in circulating hormones; these include a drastic decline in progesterone, elimination of placental lactogen because of the discharge of the placenta, and a rise in estrogen. In addition, the stimulation of the nipple by the infant is essential to full function of mammary tissue. Thus, milk production and flow are responses to the altered hormonal balance following birth and also the stimulation brought about by suckling.

Suckling is instinctive for all mammals, and the infant usually finds a vacant nipple shortly after birth. Since the number of nipples is greater than the mean litter size for any given species, finding a nipple is seldom a problem. The degree of mammary development, moreover, varies with the size of the litter, presumably because of a greater amount of placental lactogen with a larger number of embryos. Thus, a female with a large litter is prepared to nurse a large litter.

Sucking on the nipple has two very distinct and separate effects. Because the nipple has a rich supply of sensory nerve endings, it is very responsive to touch, and stimulation initiates nervous signals to the hypothalamus at the base of the brain. This stimulation of the hypothalamus causes it to produce and release *oxytocin,* which flows into the posterior pituitary, where it enters the circulatory system. Oxytocin affects the myoepithelial strands about the alveoli, and the contraction of these strands forces accumulated milk into the ducts and the nipple.

This nervous stimulation of the hypothalamus and the resultant release of oxytocin with its stimulation of the myoepithelial strands about the alveoli constitute a well-known *neuroendocrine loop*. The nervous message from the nipple to the hypothalamus is the neural part, and the release of oxytocin and its response in the alveoli is the endocrine part. The contraction of the myoepithelial strands with the ejection of milk into the ducts is a pleasurable experience for a woman nursing her baby. It seems highly likely that the suckling of a brood of mice provides physical pleasure to the lactating female mouse as well.

In domestic cattle, milk flow or "letdown" is known to be stimulated by the sight and sound of a milking machine, and the same effect is produced in mice by high-pitched squeals of infant mice. Thus, it appears that several kinds of stimuli may cause the release of oxytocin. Moreover, the failure of milk to flow may result from fear, a phenomenon known in both domestic mammals and nursing women. Difficulties in rearing wild mammals from females captured when pregnant may well be related to fear-induced failure in milk flow.

Thus, the neuroendocrine effect of nursing accounts for the ejection of milk into the ducts and nipples. The nursing infant, however, accounts for most of the removal of the milk from the nipple. As the young feeds, milk is withdrawn by two actions. Firstly, by gripping the nipple between its tongue and palate, the infant removes the milk by a stripping action, much as in the manner a person strips milk from the teat of a cow by hand. Secondly, sucking reduces pressure within the mouth so that the infant literally sucks from the nipple as if it were a straw. These two very different actions together constitute suckling.

The first material to be made by the mammary gland is a thin, almost colorless fluid called *colostrum*. Although colostrum is low in lactose (milk sugar), it is rich in sodium, calcium, and fat-soluble vitamins. Perhaps most important of all are *immunoglobulins,* which provide immediate protection from disease microorganisms. In some mammals, such as rabbits and humans, these antibodies are transmitted across the placenta, but in most species colostrum provides the initial protection from disease organisms. Colostrum flows for only about one day.

The manufacture or synthesis of milk (*lactogenesis*) follows colostrum, and in most species of mammals the infant subsists on milk until weaning—the time at which it is capable of obtaining adequate nourishment from solid food alone. Normally the young begins to take solid food before weaning, but some mammals (seals, for example) stop nursing before they take their first solid food.

The composition of milk varies widely among different species and also with the stage of lactation. Presumably the nature of the milk is adapted to the species as well as to the stage of dietary independence. Variations occur in the four major constituents of milk: fat, protein, sugar, and water. There seems to be no general rule to explain the known variations in milk composition. In some domestic species, lactose levels may be high early in lactation, declining gradually while fat and protein increase.

Specific patterns of milk intake also exist. The infant Opossum (*Didelphis virginiana*) virtually fuses to a nipple but still has an irregular pattern of suckling, as indicated by the sporadic occurrence of milk visible through its translucent skin. Rabbits nurse but once a day. Most mammals probably have rather regular frequencies of nursing, but there is little real information on nursing rates, growth rates, and caloric intake of the young and that of the lactating mother. There are very likely adaptive features to lactation of native mammals, and an enterprising student could make major contributions to behavioral ecology by exploring early food and growth patterns of native mice, bats, and squirrels. How does a lactating bat adjust its need for increased food intake to the nutritional demands of its young, for example, and still follow its usual foraging pattern?

It is indeed ironic that milk, lactation, and seasonal changes in mammary tissue—the central features which unite the class Mammalia—have been well studied in only a small number of wild species.

Parental Care

Variation in mammalian parental care rests first of all in the permanence of the pair bond. In *polygynous* species, those in which one male mates with more than a single female, the male parent is not likely to take an active or direct role in caring for its offspring. In a species in which the parents form a pair bond that persists until the young are independent, the male is likely to participate in their feeding and protection. Paternal care in pair-bonded species may follow indirectly by territorial behavior of the male, but this behavior is more likely to result in exclusion of potential competitors rather than in care as such. Care in some mammals (as in certain species of foxes) is provided by half-siblings, young of an older generation, but this is clearly not parental care.

Parental care is parental behavior that enhances growth and survival of offspring; in mammals, parental care virtually always includes lactation. Maternal care includes prenatal nourishment in both placental and nonplacental (marsupial) mammals, but parental care in this context is assumed to be postnatal care.

After birth, the infant mammal receives both heat and milk from its mother. In a few species, such as hares and Porcupines, the young might survive and grow solely on solid food from birth, but milk is normally the initial food for all kinds of mammals. As the infant takes milk, it also receives heat from the mother—not only through radiation and contact but also from the milk itself, which is at very nearly the inner or core temperature of the mother and warmer than the body of the infant. In addition to providing materials for growth, milk contains calories that are converted to heat. Thus, in some mammals at least, milk not only contains the materials for development but also, along with suckling behavior, raises the body temperature of the newborn until it can control its own body temperature, or *thermoregulate*.

The significance of heat from both milk and body contact with the mother has been shown in captive rats and mice. The body temperature of newborn young declines between suckling periods, with a substantial saving of energy. As the growth rate slows, the infant develops a better thermoregulatory ability, concurrent with growth of fur and increase in body size. The thermoregulatory aspect of maternal behavior is undoubtedly very important for such altricial mammals as mice, shrews, and bats, but perhaps less so for such precocial species as Porcupines, hares, and ungulates.

Because the nipples lie on the ventral part of the body in almost all mammals, the mother's body constitutes a cover for her nursing offspring. In this way, the mother not only protects her brood from hostile aspects of the physical environment but also shields them from potential predators as well.

Clearly, lactation is the basis of parental care in mammals. The early mother–offspring contacts have important effects on the behavioral development of the young—and, to some extent, the young elicit maternal behavior in the mother. The hormonal basis for lactation is naturally bound to other reproductive events, especially the birth of the young. Thus, the development of the breasts occurs during gestation and the release of milk is stimulated by the activities of the newborn young.

Maternal Aggressive Behavior

In addition to providing comfort, heat, and nourishment to her newborn, the mother is usually aggressive toward other adults. This well-known behavior should dispel the myth that aggression is a trait solely of males. The hostility of females with suckling young occurs in most, if not all, orders of mammals. Maternal behavior is produced by the hormones oxytocin and estrogen. Indeed, custodial and aggressive maternal behavior may be a constellation of responses to the same hormonal stimuli. Maternal aggressiveness is strongest just after young are born and tends to decline as they grow. Aggression disappears, moreover, with the experimentally early removal of suckling young.

Oxytocin and estrogen administered to virgin laboratory rats result in prolonged maternal behavior, such as nest building and picking up newborn young. The rise of these two hormones at the time the young are born presumably accounts for the appearance of maternal instincts; and inasmuch as suckling by the young induces a prompt secretion of oxytocin in the mother, the infant is responsible for preserving the maternal behavior of its mother.

Although in most mammals a female with young is indifferent or even hostile toward the young of other females, exceptions to this generality seem to occur among certain colonial bats. In colonies of the Pallid Bat (*Antrozous pallidus*), a cluster of ten very small young (naked and with eyes still unopened) may be "guarded" by a single female, which seems to have temporary custody of the nursery of newborn. Robert L. Rudd and his associate Albert J. Beck noticed another cluster of somewhat older young (perhaps two weeks of age) which were guarded by two adult females; a still older group (about four weeks of age) was found to be segregated, each individual young with its mother. These observations suggest that in colonial bats, female aggression and intraspecific hostility may somehow be suppressed when the offspring are very small. This pattern is facilitated by the simultaneous ovulation and birth within a given colony.

The duration of parental care in mammals is thus loosely tied to the duration of lactation, but there is a diverse pattern of associations of suckling and parental–offspring contact. A foal feeds frequently, for example, and is seldom far from the mare, whereas young rabbits take milk but once a day and the mother may leave then unattended for hours. Parental care is also linked to the age of *dispersal,* which normally occurs prior to sexual maturity. Thus, parental care is brief in a vole, which may be sexually mature at four weeks of age. Young Beavers, however, remain in their parents' lodge for nearly two years, though overt parental care probably ceases long before dispersal. These events—lactation, dispersal, and sexual maturity—although sequential, differ from one species to another and probably constitute adaptations to many sorts of environments.

References

Asdell, S. A. 1964. *Patterns of mammalian reproduction.* 2d ed. Ithaca: Cornell University Press.

Dewsbury, D. A. (Ed.) 1981. *Mammalian sexual behavior: Foundations for contemporary research.* Stroudsburg, Penn.: Hutchinson Ross Publishing Co.

Sadleir, R. M. F. S. 1969. *The ecology of reproduction in wild and domestic mammals.* London: Methuen & Co., Ltd.

Stoddart, D. 1980. *The ecology of vertebrate olfaction.* London and New York: Chapman & Hall.

van Teinhoven, A. 1983. *Reproductive physiology of vertebrates.* 2d ed. Ithaca: Cornell University Press.

Vandenbergh, J. G. (Ed.) 1983. *Pheromones and reproduction in mammals.* New York: Academic Press.

SEASONAL DORMANCY

Many mammals experience seasonal changes in activity. These shifts are profound and reflect great alterations in many bodily activities such as breathing, oxygen consumption, the respiration of cells, and kidney activity. These seasonal changes in the biology of mammals are correlated with environmental fluctuations in temperature and rainfall. Because the seasonal decline in bodily activity or metabolism occurs during hostile seasons, usually winter, the period of inactivity is called *hibernation*. In some California mammals, inactivity begins in July and is then referred to as *estivation* (or aestivation).

Regardless of whether the decline in bodily activity begins in a warm, dry summer or at the onset of winter weather, the physiological changes are very much the same. The body of the hibernator changes as the season for winter sleep approaches. It first becomes rather sensitive to the lowering of the temperature of the environment and may become dormant for a few hours or even for a day or more. During these dormant periods the heart rate slows to a range from two to eight beats a minute, and breathing may be reduced to twice a minute. The body temperature drops to a point just slightly above that of the surrounding air; a ground squirrel in its underground nest, for example, may have a body temperature of 6° C when the air is at 4° C. As the season progresses, the duration of the dormant period increases so that by midwinter a ground squirrel may remain in a dormant or torpid condition for from ten days to two weeks.

Thus, hibernation is not a long spell of continuous dormancy but rather a series of rather brief dormant periods interrupted by arousals. These periodic arousals begin with an increase in body temperature and are shortly followed by a rise in heartbeat and breathing rate; these changes continue until the normal (active) rate of metabolism has returned. A little food may or may not be eaten at this time, but invariably the individual urinates. Possibly the accumulation of metabolic waste in the blood somehow stimulates the periodic arousals. As spring approaches, the frequency of periodic arousals increases and the periods of dormancy become briefer.

Prior to hibernation (or estivation), body fat accumulates. This body fat is sufficient to maintain the life of the hibernator until spring—for some mammals, such as the Black Bear and the Golden-mantled Ground Squirrel, do not eat between the onset of hibernation in late summer or early fall until emergence the following spring. Some species, certain chipmunks, for example, store large amounts of food and do eat during periodic arousals. Stored body fat in a female bear provides nourishment not only for her own metabolic needs but also for the growth of the embryo, which is born in midwinter, and for the production of milk, which will support the development of the cub until it emerges some months later.

In addition to the fat which provides for the nutrition of the hibernating mammal, there are special pads of fat that are richly provided with nerve endings and a blood supply. This fat is tan in color and is commonly known as *brown fat* or brown adipose tissue. Brown fat functions in periodic arousal by a sudden creation of heat, which is quickly carried to such critical regions as the brain, heart, and lungs. As a result, the anterior part of the body warms first. When the heart and lungs work at an increased speed, the other body fat (or white fat) may provide additional warmth. When the hibernating mammal emerges in the springtime, stores of both brown and white fat are depleted but not entirely exhausted.

The environmental cues for hibernation are unknown. In the Golden-mantled Ground Squirrel, males enter hibernation in August or early September while the weather is warm and food is plentiful. Females are slower to enter dormancy, perhaps because they must provide some fat for the manufacture of milk in the springtime, and lactation probably delays the accumulation of body fat in late summer. The young are last to enter seasonal dormancy. The Townsend and Belding ground squirrels enter dormancy in July while the weather is warm. They are, however, extremely fat. While fat itself does not constitute an environmental cue, adequate fat is closely correlated with hibernation in ground squirrels, pocket mice, jumping mice, and bats. Emergence from hibernation in the spring appears to be affected by temperature and snow cover. Ground squirrels emerge earlier on south-facing slopes, where the snow melts early and the sun warms the soil, than on north-facing slopes. Hibernators generally emerge earlier in years with an early spring.

The Red Bat (*Lasiurus borealis*) is one of the most cold-tolerant bats in the United States. Its dense pelage reduces heat loss, and the furred interfemoral membrane can be drawn over the ventral surface of the body as additional protection from the cold. This species survives in temperatures well below freezing and compensates for extremely low temperatures by an increase of its heartbeat. Presumably Red Bats and Hoary Bats (*Lasiurus cinerea*) hibernate in trees. Whereas cave bats arouse from dormancy spontaneously within the chilled air of a cave or mine shaft, tree bats remain dormant until air temperatures reach nearly 20° C; this behavior probably precludes their using caves as hibernacula.

Several groups of California mammals are well known to be hibernators, and this physiological pattern has clearly developed at different times in the evolution of mammals. Almost all California bats are known to hibernate, and many rodents do so. Among closely related groups, such as species of chipmunks and pocket mice, there are both hibernating and non-hibernating species. Some carnivores, such as the Black Bear, hibernate, while others, such as the various kinds of weasels (Mustelidae), do not. There is obviously no relationship of this adaptation to mammalian phylogeny.

The Black Bear (*Ursus americanus*) experiences a hibernation rather unlike that seen in ground squirrels. Bears retreat to a den for the winter and may remain there for up to six or seven months, during which time they not only do not urinate or defecate but also apparently do not arouse. Dormant bears also differ from hibernating ground squirrels in that their body temperature is maintained at about 31°–32° C. The dormant bear has a greatly slowed heartbeat, which may drop to twelve beats a minute, with a consequent drop in metabolic rate.

Although one may call this state "hibernation," it is physiologically very different from hibernation in ground squirrels and chipmunks. It is not surprising that the bear preserves a core temperature not greatly lower than its active temperature, for parturition occurs during the coldest part of the winter, and from that time the offspring take milk. It is unthinkable that bears would give birth and synthesize milk at any but a slightly reduced body temperature.

Almost all California bats hibernate, usually moving into caves in autumn. They are fat at this time and very sensitive to the lowered temperatures they encounter underground. The Guano Bat (*Tadarida brasiliensis*) has been found roosting in buildings as far north as Oregon and survives temperatures as low as 5° C.

Bats of many species seem to suffer from loss of body water during hibernation, even though metabolism of stored fat releases some water into the body. Bats urinate upon their periodic arousal and frequently ingest water droplets from their fur before becoming dormant again. Cave bats frequently roost in clusters, and this habit tends to reduce heat loss. Studies have shown that dormant bats in aggregation maintain higher body temperatures than do bats of the same species roosting individually.

References

Kayser, C. 1961. *The physiology of natural hibernation.* Oxford: Pergamon Press.

Lyman, C. P.; Willis, J. S.; Malan, A.; and Wang, L. C. H. 1982. *Hibernation and torpor in mammals and birds.* New York: Academic Press.

Musacchia, X. J., and Jansky, L. 1981. *Survival in the cold: Hibernation and other adaptations.* Amsterdam: Elsevier/North Holland.

SENSES

It is difficult to imagine the world in which wild mammals find themselves, for their sensory development involves specializations and abilities quite different from human sensory perception. Mammalian perception almost always includes an excellent olfactory sense, and vision is usually less well developed than it is in birds (most of which have a poor sense of smell). Sensory perception is generally accompanied by a corresponding signal. Insectivorous bats, for example, which can detect sound waves well in excess of 20,000 cycles per second (20 kilohertz), can also produce these high-frequency sounds.

The type of sensory specialization, moreover, is suited to the activity pattern and habitat of the mammal in question, and abilities that are not frequently employed tend to deteriorate. Nocturnal mammals usually have well-developed senses of smell and hearing but may have much poorer vision than we enjoy. The vision of moles and pocket gophers is poor, and these mammals spend much of their time in the darkness of their burrows. On the other hand, mammals such as tree squirrels possess excellent vision suited for arboreal activity and the visual detection of predators. For the most part, however, mammals live in a world of odors—a world of many sorts of self-produced and alien smells. Odors are produced by a variety of dermal glands from which a mammal can establish signposts within its territory. Distinctive scents are contained in urine, and many mammals customarily discharge small amounts of urine to renew their olfactory signposts frequently. Odors in urine reveal not only the species and sex of the mammal in question; they also indicate its sexual condition and individual identity. These odors usually are detected by both sexes and all ages in contrast to insect pheromones, which are specifically sexual attractants.

The word *pheromone* came from entomology and was first applied to odors produced by one insect to elicit a specific response in another. A pheromone is usually a single compound that causes a very precise behavior. In contrast, the odors synthesized by mammals usually consist of several compounds, frequently modified by bacteria, and cause various responses depending upon the age, sex, and experience of the recipient. For these reasons some students of animal behavior prefer not to apply "pheromone" to olfactory signals of mammals.

Olfaction

An acute olfactory sense in mammals is apparent from their large olfactory nerve tract. The nasal passage is expanded and folded in a way that increases its sensory surface; in addition, there is a branch leading to a pair of orifices in the anterior region of the roof of the mouth, the *vomeronasal organ,* a specialized olfactory receptor. The olfactory ability of mammals is paralleled by a variety of scent-producing structures; together these two systems form the basis for much mammalian communication and behavior.

Urine Marking

Urine marking is a major activity in many species of mammals. Adult males place urine throughout their territories and constantly explore scents left on rocks, pathways, or tree trunks; thus, they continuously search for intruders while renewing their own boundary markers. A male can also recognize the scent from a female and detect whether she is in estrus. This odor can be carried many miles in the air, a phenomenon well known to anyone who has kept a female dog in heat. Usually the urine not only contains metabolic wastes released from the kidney but also includes sexual hormones and species-specific scents from glands that discharge into the urine. Although most research on scent marking concerns dogs and laboratory animals, the importance of scents is well known to trappers of furbearing mammals.

Among the various species of dogs (Canidae), olfaction is a

major means of social organization. Foxes, wolves, Coyotes, and domestic dogs all secrete substances into the urine and vaginal discharge, and these odors communicate aspects of territoriality, reproductive state, and social dominance. Scent marking is not solely a property of dogs. Wood rats have ventral glands that produce a secretion which rubs off on rocks and branches when they run. Deer and antelope possess tarsal glands and sometimes facial glands that produce scents with which they mark their territories. Individual markings of newborn young is an essential activity of many mothers and serves as a means of later recognition in at least some species. Finally, some mammals may mark themselves with urine, as bull Wapiti do when in rut, and in some species, such as Porcupines, the male may urinate on a female, a phenomenon known as a "urinal shower."

Dirt bathing by some rodents may constitute an effort to mark the area with their scent. Dirt bathing has been observed in the California Ground Squirrel and some kinds of kangaroo rats; in the process, the dorsal gland in these species seems to be applied to the ground.

The Role of Olfaction in Reproduction

The intraspecific effects of odors in mammalian reproduction can be divided into those that cause prompt behavioral responses (*releasers*) and those that alter a hormonal state of the individual receiving the odor (*primers*). The role of odors as releasers in most species involves the attraction of males to a female in estrus, but in some species (such as sheep) females in estrus are attracted to a male.

Among many mammals, olfactory signals announce the presence and reproductive state of the individuals occupying the area. Odors of conspecific individuals not only transmit information but may also constitute intersexual stimuli and intrasexual suppressants. Although much of the current information on these effects of odors is based on behavior of laboratory populations of the House Mouse (*Mus musculus*), there are similar responses in species of deer mice (*Peromyscus* spp.) and voles (*Microtus* spp.). These intersexual and intrasexual effects of scents are called *primer effects* because they modify sexual development and performance.

Generally there are five well-studied primer effects in laboratory mice. The "Bruce effect" is the failure of a *blastocyst,* or fertilized egg, to implant when a pregnant mouse has been exposed to the odor from urine of a male other than the one to which she mated. Apparently odors from an "alien" male modify hormonal levels in the early pregnancy of the female so that implantation fails. One can speculate on the role of the Bruce effect in nature. Research on this olfactory response is necessarily performed on laboratory populations, and replication in the field would be extremely difficult.

Odors in male urine have two stimulatory effects on the reproductive performance of females. When female mice are subjected to male urine, their estrous cycles become shorter and more frequent. This effect has been shown not only for laboratory mice but also in a vole (*Microtus ochrogaster*) of the central United States and may also exist in California voles. Female mice, moreover, reach sexual maturity at an earlier age when exposed to the odor of urine from adult males than they do when they are kept away from male odors. These effects of male urine occur only from urine from sexually active males and are not seen when the male has been castrated. They are clearly due to odors and not to some behavior on the part of the males, because there is no such response in females which have had their olfactory nerve tract destroyed.

The urine of females has two opposite effects. When two or more female mice are kept together without a male, their odors tend to retard the frequency of estrous cycling and also to retard sexual maturity in young females. Again, one may speculate on the significance of these phenomena in wild populations. These primer effects are, to date, known only in mice and, except for the Bruce effect, would seem to require a preponderance of one sex or the other. It would be premature, however, to suggest that they lack an important role in wild populations.

Most mammals have accessory reproductive glands, called *preputial glands,* that release materials into urine. Preputial glands are under control of gonadal hormones and are usually larger in males than in females. Among laboratory mice of known social structure, the preputial gland is largest in the dominant males, which also do more of the urine marking. Sexually experienced females respond most strongly to the odor from the preputial products, but response is weak in virgin females and absent in pregnant females. Thus, it appears that preputial secretions have an important role in social organization and sexual activity in mice.

The olfactory tract is large in most species of mammals, and in some there is a branch leading to the two small openings located anteriorly in the roof of the mouth. These orifices are lined with odor-sensitive tissue and provide for the detection of smells that enter the mouth. This structure, the vomeronasal organ, is a modification of the Jacobson's organ of some reptiles. The oral detection of odors can be observed when an animal such as a deer or horse raises its lips in the position known as *Flehmen*. Flehmen exposes the anterior region of the mouth to airborne odors, facilitating the use of the vomeronasal nerve tract. In some kinds of wild mice, the male nuzzles or licks the genital region of a female in estrus and seems to receive vaginal scents through the vomeronasal tract. In some species, these olfactory signals from the female may be essential to stimulate the male to mate. In hamsters, for example, which are related to deer mice, voles, and wood rats, if the male's vomeronasal tract is anesthetized, he is rendered sexually incapable.

There appears to be extensive use of olfactory signaling in Mule Deer, Black-tailed Deer, and Pronghorns. The male Pronghorn begins marking by scraping a small area with the forefeet; it then urinates on the pawed ground and defecates on the urine. Because this species moves about from one area to another in small groups, odors probably do not have a territorial significance. Deer and Wapiti (or elk) may urinate on themselves, either on their belly or on the area of their tarsal glands. Moreover, deer and elk may also apply secretions from their tarsal and facial glands directly onto the trunks or branches of trees. The secretions of facial glands are seasonal and may increase with gonadal activity. Tarsal glands in the Mule Deer secrete odors that may serve for individual identification.

Pigs are renowned for their ability to detect truffles in the ground, and sows possess a special skill for recognizing the unusual fragrance of truffles growing as much as 3 ft (1 m) beneath the surface. The basis of this special talent stems from the production of a peculiar substance in truffles. This substance is also produced in the testes of the boar and secreted in his saliva; it is the musklike odor of this material which is attractive to the sow. Human testes produce the same material, and it is secreted in the underarm perspiration of adult males.

General Roles of Olfaction

In addition to the specific functions of odors and their detection mentioned above, there are numerous nonspecific examples of the importance of olfaction. The location of water in the desert is, to some degree, a matter of recognizing water vapor in the air. Probably food is located primarily by its odor. Even to our relatively insensitive olfactory endings most plants have distinctive odors, and undoubtedly most herbivores recognize a broad spectrum of plants, some of which may be only mildly aromatic. Not only are odors used by mammalian predators to detect their prey, but the prey may be warned by the scent of a predator. Even young ground squirrels recognize and avoid the scent left on the trail of a rattlesnake.

Visual Communication

Visual signals require light. Inasmuch as most mammalian groups are largely nocturnal, visual communication is less common than it is in birds, and most communication among mammals is by sound and odor. One would assume, then, that visual acuity is reduced among most nocturnal and burrowing mammals. Although this assumption seems generally to be valid, some nocturnal species, such as bats, may employ environmental light in navigation, and the light/dark cycle is essential to the timing or entrainment of daily rhythms. Except for flying squirrels, all squirrels are diurnal, and it is logical that they communicate by visual and vocal means. Visual intraspecific signals occur in chipmunks and ground squirrels, but the significance of their various tail flicks is poorly understood.

Most visual signals in mammals are of a passive nature. Although one might not think of color patterns as signals, they are visual transmissions and do have meaning. Some of these patterns have a clear significance, whereas the meaning of others is not always apparent. The stark black-and-white warning pattern of skunks is a permanent announcement of their potential danger, and this signal is obvious even in dim light. The color patterns of many tree squirrels may not carry an obvious meaning, but the fur of ground squirrels and chipmunks is usually colored so that these animals are clearly visible when at rest in their normal habitat. The winter pelage of the Snowshoe Rabbit, White-tailed Jackrabbit, and some weasels conceals them from possible predators or, in the example of the weasels, from possible prey.

An animal's posture, too, may represent a visual signal. A submissive individual, such as a puppy, can indicate its social ranking by lowering its head or rolling on its back, whereas an aggressive animal might raise its head or bare its teeth. Although these behaviors may be expressed with vocal signals, the visual aspects are very important. One of the most startling posturing signals is the handstand of the Spotted Skunk: when alarmed, this skunk sometimes stands on its forefeet so that the black-and-white dorsum is entirely directed toward its potential enemy.

The Big Brown Bat (*Eptesicus fuscus*) uses the glow of the evening sky in determining the direction in which it begins its evening foraging; the directional movement of captives can be altered by changing the source of light. It is conceivable that they recognize polarized light at dusk, a feature employed in orientation by migrating sparrows.

Probably because most California mammals retreat from bright light, light usually does not have a major role in communication. Visual signals have been intensively studied in elephants and primates, and such communication may have hitherto undetected roles in California mammals.

Vocal Communication

Sound production is very common among most vertebrates, and virtually all mammals use sound under certain conditions. Although sound in mammals is more varied and extensive than it was thought to be a few years ago, it is still poorly understood in contrast to the detailed information that researchers have assembled on sound in birds. The importance of sound depends partly on the low energy needed to produce it. There is little cost to a bark compared to the metabolic expense of producing scents.

Many mammals produce ultrasonic sounds (frequencies above 20,000 cps, or 20 kHz), sounds above the range of human hearing. Ultrasonic frequencies are important because of the broad range of frequencies available, but sounds at extremely high frequencies dissipate quickly. For this reason, ultrasonic frequencies are used for communication over very short distances. Because ultrasonic vocalization must be determined by extremely sensitive instruments, the full extent to which mammals use such high frequencies may not be known for some time. Small young of some rodents make ultrasonic squeals of alarm that quickly draw the mother to the nest. This maternal response is strong in nursing females, which are attracted even to young not their own.

Laboratory rats and mice produce characteristic sounds during courtship and mating. In rats, these copulating calls reach 22 kHz and in the laboratory mouse they are approximately 70 kHz. The pitches vary sexually and have a premating stimulatory effect. They are, moreover, under the control of the testicular hormone *testosterone* and disappear after removal of the testes. In the Little Brown Bat (*Myotis lucifugus*), there seem to be no premating tactile or visual overtures so characteristic of most mammals. As the male mounts the female, however, he utters a "copulation call," a highly distinctive sound that induces acceptance by the female.

Various species of insectivorous bats are distinctive in the frequency and duration of their lower-pitched calls. These lower frequencies carry farther than do the ultrasonic or very high frequency calls; although they are less useful for echolocation, lower frequencies are probably more important for intraspecific recognition. Vocal communication among Pallid Bats, for example, is believed to help flying individuals in locating their roosts and also to maintain the association of females and fledged young.

Simple calls, such as the clear two or three-note whistle of the Pika, may be given by both sexes and may be alarm calls as well as territorial announcement. Moreover, male Pikas utter prolonged songs during the breeding season. These songs are individually distinctive and may provide the basis for sexual as well as individual identification. Some mice are known to produce songs. This ability has been reported for species of voles (*Microtus*), the House Mouse (*Mus musculus*), and the Grasshopper Mouse (*Onychomys leucogaster*). The last has a complex of sonic patterns of both moderate and very high frequency parts.

Distress or alarm calls are made by many kinds of mammals. These calls are always short and usually consist of a single note. Among different species, moreover, there are similarities among alarm calls; the alarm call given by a California Ground Squirrel is recognized as a danger signal by the California Valley Quail.

Apart from the precise vocal signals that have specific meanings, some mammalian sounds appear to express rather general emotions. These sounds are commonly given together with a characteristic body movement, and the two kinds of signals (vocal and visual) clearly reinforce one another. Low-pitched sounds usually indicate hostility or aggression. Typical examples are the growls and other guttural sounds that are accompanied by grimaces or a show of teeth—characteristic mammalian expressions of hostility. In contrast, the high-pitched squeals of a submissive puppy are often expressed as it assumes a crouching or supine position. Such high- and low-pitched vocalizations can be heard in many orders of mammals.

Echolocation

Some mammals locate food and detect the presence of obstacles through the echoes from highly pitched or ultrasonic sounds. Such echoes reveal the distance and size of the object from which the sound waves bounce; if the object is moving, echolocation may also indicate the speed and direction of movement. Echolocation has been most intensively studied in bats, which can maneuver and capture flying insects in the dark, but it is known to occur in shrews, porpoises, and several species of birds.

Echolocation depends on the relationship between wavelength and frequency (or pitch) and also on the constant speed of sound through air. Because the speed of sound is constant (under the same air conditions), as frequency increases, the wavelength decreases. For this reason, the very short sound waves of ultrasonic emissions can bounce back from very small objects, whereas the sound wave of a lower frequency might well be longer than, for example, a moth; small objects, such as flying insects, therefore do not reflect sound waves of low frequencies. In addition, high-frequency sounds dissipate rather quickly, and therefore sound emissions from one foraging bat are not likely to "jam" the high-pitched squeaks from another bat. Thus, ultrasonic emissions are well suited for the echolocation of small objects.

Typically the high-frequency emissions of North American bats vary over a brief period of time: the emission frequently ends at a lower pitch than the beginning frequency, a phenomenon known as *frequency modulation*. If the sound is directed at a flying moth and the brief pulse is of a descending frequency, the distance can be determined within the complexity of a bat's brain—and if the sound is heard by both ears, the bat can also determine the direction from which the echo is coming. In frequency modulation the echos are of declining pitch and therefore indicate the movement of the source of the echo. That is, if the moth is moving away from the bat, the lower-frequency echos (those emitted nearer the end of the pulse) will take longer to reach the ears of the bat. In bats of the family Vespertilionidae, the frequency-modulation pulse may begin near 100 kHz and drop to near 20 kHz. For the echo not to cross with the bat's emission, the emission must be very brief because some objects from which echos might return might be very close. To avoid confusion of signals, a hunting bat might emit high-pitched squeaks lasting only from 0.2 to 0.5 millisecond.

References

Doty, R. L. (Ed.) 1976. *Mammalian olfaction, reproductive processes and behavior.* New York: Academic Press.

Brown, R. E., and Macdonald, D. W. (Eds.) 1985. *Social odours in mammals.* 2 vols. Oxford: Clarendon Press.

Griffin, D. R. 1958. *Listening in the dark.* New Haven: Yale University Press.

Kellogg, W. N. 1961. *Porpoises and sonar.* Chicago: University of Chicago Press.

Purves, P. R., and Pilleri, G. (Eds.) 1983. *Echolocation in whales and dolphins*. New York: Academic Press.

Sales, G. 1974. *Ultrasonic communication by animals*. London: Chapman & Hall.

Stoddart, D. M. 1980. *The ecology of vertebrate olfaction*. London: Chapman & Hall.

Vandenbergh, J. G. (Ed.) 1983. *Pheromones and reproduction in mammals*. New York: Academic Press.

Walther, F. R. 1983. *Communication and expression in hoofed mammals*. Bloomington: Indiana University Press.

MIGRATION AND MOVEMENTS

There are two distinct types of movements characteristic of wild mammals. Movements which involve an eventual return to the place of origin are called *migrations,* regardless of the distance or the time required for the return. We commonly think of annual migrations of mammals, but some are seasonal and some are daily, and they all include a return. On the other hand, there appears to be a stage in the lives of all animals in which they depart from the homesite and assume residence in another area; this sort of movement is *dispersal.* The two concepts might seem similar, but the distinction becomes clear if one remembers that the origin of a migration is seldom the exact place of birth, although it is sometimes very close.

Because the daily foraging activities of animals frequently end in a return to a nest or resting site, they meet the requirements of a migration. Nevertheless, we have come to consider such short-term excursions as *movements,* and it is convenient to discuss them separately from annual or seasonal migrations. We shall therefore discuss short-term movements under Home Ranges, bearing in mind, however, that even very brief movements may constitute a migration, and at times it is hard to distinguish between daily foraging, nomadic wandering, and clear-cut annual migrations.

Migrations

Probably in the earliest human societies aborigines followed migrating ungulates, and a knowledge of the factors affecting the times of arrival and departure of these species was essential to the well-being of the hunter groups. Important today are the migration routes and schedules of the herds of wild American ungulates such as the Black-tailed Deer and Mule Deer. Such data provide part of the basis for determining the times and locations of legal hunting of these animals. The short-range and long-range movements of bears are still poorly known, but many details are revealed by radiotelemetry. This information, which has begun to show the movements of individual male and female Black Bears as well as where bears are prone to den for the winter, assists game managers in minimizing damage from these potentially dangerous carnivores and in planning the times and places for public hunting of bears.

Throughout the length of the Sierra Nevada, deer migrate every autumn from the higher elevations. In Tulare County they spend the summer between 2000 and 3000 m and move downhill to between 500 and 1200 m in the winter. In a study by the Department of Fish and Game, one doe was followed by radiotelemetry for a year. She had a summer home range with a radius of approximately 1/2 mi (1 km). In early October she returned to lower elevations for the winter. She was located weekly throughout the winter but stayed within an area of 1 km in radius. In late May she returned to the summer range, a distance of about 6 mi (10 km) in a direct line, in less than three days. In this study some individuals, marked by bells, moved annually to summer and winter ranges roughly 18 mi (30 km) apart.

The North Kings deer herd in eastern Fresno County has discrete migratory paths. Movements are slight except during the period of migration, and they are adjusted to weather and forage conditions. Some individuals make exploratory excursions to higher elevations in late winter, at which time the winter range and forage are deteriorating. Actual movements are irregular in the spring, for deer sometimes pause for several days or even weeks along their narrowly prescribed migration paths. The uphill migration takes the deer to several elevations, and individuals tend to return to the home range they occupied the previous year. They return to lower elevations with early storms in October, using the same routes they followed in the spring; but the distance traveled and the time taken in full migrations depend on the severity of the fall weather.

Bats are migratory, but there are conspicuous differences between species, and within a species there are frequently distinctive sexual patterns of migrations. These variations are characteristic of species but exhibit patterns which seem unnecessarily diverse under modern climatic conditions.

The annual movements of tree bats (species of the genera *Lasiurus* and *Lasionycteris*) are the most extensive of any North American bats. Tree bats are found far into Canada in the summer, and the Hoary Bat (*Lasiurus cinereus*) is found from the north end of Hudson Bay south to central Mexico. Tree bats are powerful fliers and move far to the south of the summer range every autumn. Not uncommonly, they are found on ships 100 mi (160 km) or more from the continent, and insular populations of *Lasiurus* occur in the Hawaiian Islands, Galápagos, the West Indies, and Bermuda. In parts of their winter range, tree bats may be active on rather cold nights, and they are much more resilient to the effects of low temperatures than are species of *Myotis*. (Species of *Lasiurus* are not known to hibernate in nature, although it is possible.) The Hoary Bat (*L. cinereus*) can be found in California in winter and is sometimes seen flying in late afternoon.

In contrast, the various species sometimes referred to as "cave bats" (species of the genera *Myotis, Eptesicus, Pipistrellus,* and others) make migrations which involve a movement from a cave (or a mine shaft) in the spring to a summer range and a return in autumn to the cave, the hibernaculum. These migrations are not necessarily north-south in direction, and some individuals may not cover much ground. These migrational patterns have been extensively studied in the United States and Europe by banding bats when they are hibernating in caves and recapturing them subsequently in their summer quarters. In these species, migration is associated with hibernation; the autumn migration of cave bats removes them from a hostile winter climate without the departure from a hostile geographic region.

Migratory patterns of the Guano Bat (*Tadarida brasilien-*

sis), one of the cave bats, have been studied by E. Lendell Cockrum of the University of Arizona. Some populations of this bat, which summers from Arizona east to Texas, migrate south into Mexico for the winter, flying more than 1000 mi (1600 km) at the end of summer and an equal distance in the spring. California populations of this species seem to make only local movements in spring and fall, and some individuals in southern Nevada and western Arizona may winter in southern California. Females move north in the spring and assemble in large nursery colonies, where the young are born. At least some males remain in the south, for one summer colony in the Mexican state of Chiapas consisted of 40,000 males and no females.

Much of our knowledge of the biology of bats, and especially of their migrations, is due to their danger as *vectors* and reservoirs of rabies. Because bats can transmit rabies to each other and also migrate vast distances, they have great potential for the dissemination of this fearful disease. Banding of bats has not only revealed some general patterns of migrations but has also provided an opportunity to explore their homing abilities. Banding and the subsequent recapture of bats require a substantial amount of money, usually over a period of years.

Migratory patterns evolved probably as an escape from an adverse environmental feature and as a pursuit of a beneficial one. Thus, on the one hand, there seems to be a powerful genetic component to migratory patterns, but they remain, nevertheless, clearly sensitive to minor environmental changes today, as is apparent in the preceding examples. One can look at migratory mammals today and speculate on the environmental pressures behind the evolution of these rhythmic movements. The Mule Deer (*Odocoileus hemionus columbianus*), as we have seen, ascends to high mountain meadows in the summer and withdraws to lower elevations in the winter. Fresh food occurs throughout the summer only at higher elevations, but deep snow in the winter precludes grazing there, so the Mule Deer's descent in autumn is easily understood. A differential ascent in the spring and early summer results in bucks reaching higher elevations, above most of the does. This separation of the sexes is of no significance in the spring and summer; but in the downhill migration gonads are growing, and the sexes remain together in the winter range, where mating occurs. The annual migration of this mammal is clearly adjusted to both the occurrence of fresh forage and the need for joining of the sexes.

Quite different patterns fit the conventional definition of migration. The movement of a cave bat such as *Myotis* to and from its hibernaculum annually involves a radiation from the cave in many different directions with no single geographic trend. This movement is profoundly unlike the general north-south movements of a *Lasiurus* or a *Tadarida;* the latter two movements themselves appear to be rather different, with *Tadarida* entering a cave and *Lasiurus* remaining at least somewhat active in the southern part of its range. It seems quite possible that migrations in these three groups evolved separately and independently of one another.

It is well to remember, however, that modern migratory patterns may have evolved under markedly different climatic conditions. We may still be in the Pleistocene with drastically fluctuating climates, and biological patterns that exist today may have evolved under quite different conditions. In the rather recent past, from about 1450 to 1850, the Little Ice Age was a time for lowered temperatures and reduced agricultural productivity in the mid-latitudes. Moreover, prior to about the year 1000 there was a period of mild temperatures, at which time successful agricultural establishments existed in Greenland. It is difficult to imagine that patterns of migration were not modified to allow survival under a rather broad climatic spectrum. At least it helps to view some natural patterns with such possibilities in mind and to remember that those species surviving today have survived the drastic climatic changes that occurred during the Pleistocene. Inasmuch as the Pleistocene is known to have seen the disappearance of many mammals, small as well as large, we can regard living species as those which were sufficiently flexible to accommodate to sudden environmental changes. This background is offered as a partial explanation of why some wild mammals may not seem to be perfectly adapted to conditions that exist at this point in geological time. We can see that migrations of mammals reveal a diversity that is not easily explained as finely tuned adaptations to current conditions.

Home Ranges

Most small mammals, such as mice, squirrels, and rabbits, forage daily over more or less the same area, each individual (or sometimes family) remaining essentially on the same ground every day. This area provides adequate food and cover for one individual and is called its *home range*. Sometimes a mouse or squirrel is hostile when its area of residence is invaded by another of its own kind, and the resident drives out the alien. The part of the home range that is defended against intruders is called a *territory*. Territories may be quite apparent to observers of such mammals as squirrels and sea lions, which are obvious in displaying their social relationships, but bats appear to forage harmoniously over the same ground. The distinctions between territories and home ranges are not always clear, and there are important differences in these concepts among various species of mammals.

Although a theoretical function of male territoriality is the exclusion of other males from his mate or mates, a defending male is frequently only partly effective in this effort. A bull sea lion or fur seal may defend his surroundings against intrusion by other bulls, but the cows may move from one territory to another during their fertile periods. A dominant male Douglas' Squirrel (*Tamiasciurus douglasii*) is often less than 100 percent successful in preventing the encroachment of subordinate males into his territory, and females may mate with the intruders.

Although most studies of home ranges of small mammals are based on live-trapping and releasing animals, some students have attached small radio transmitters to mammals from the size of a rat or even a bat up to a deer or bear. A small citizen band receiver with a directional antenna can identify radio signals from individual mammals; from two signals received close together from two positions, one can determine the source of the signal. Small transmitters emit a signal that can be picked up a mile or more away. Because radio telemetry reveals the actual location of an animal during its natural meanderings, details of its home range as well as seasonal changes can be carefully plotted by anyone willing to spend the necessary time in the field.

Some mammals move very little throughout a daily period of activity and are very closely bound to their home. The California Ground Squirrel (*Spermophilus beecheyi*), for example, forages close to its burrow entrance; it rarely moves far, nor does it remain away for long. This squirrel was studied in great detail at the Hastings Natural History Reservation in the Carmel Valley by the late Jean Linsdale and his associates. Like many rodents, the California Ground Squirrel makes most of its brief trips along well-worn paths. Young individuals may stray no more than 5 m or so from the burrow entrance, but adults frequently move out more than 30 m. The abundance of fresh grass and small forbs allows the squirrels to feed close to their burrow entrance; as they deplete this food supply, they move farther away. Professor Linsdale observed that almost all daily movements were less than 100 m and that the few long journeys of 320 m or more involved a change in residence.

Similarly, wood rats have rather small home ranges. The Desert Wood Rat (*Neotoma lepida*) seldom goes more than 7 to 10 m from its home. Because home ranges overlap only slightly, there seems to be a clear element of territoriality in their boundaries. Home ranges of the Dusky-footed Wood Rat (*Neotoma fuscipes*) have been meticulously plotted by radiotelemetry. As with small mammals generally, adult males have larger home ranges (averaging more than 2200 m^2) than those of adult females (1900 to 2000 m^2) and juveniles (about 1700 m^2). The home range is not a fixed feature, however, for when vegetation dries up after the end of spring rains, wood rats move widely, presumably in search of food. On the other hand, females nursing young remain close to their nests and have home ranges of less than 1500 m^2. Other seasonal changes in home range may relate to the reproduction, for the male/female overlap in home ranges increases during the breeding season. At other times, males and females tend to have mutually exclusive home ranges.

Movements of the Kit Fox in the San Joaquin Valley were followed by releasing individuals with small radio transmitters attached to their necks. Individuals seem to share their hunting areas and hunt in family groups. They may forage over common ground with other family groups, but not at the same time; that is, they seem not to defend territories from which they exclude other Kit Foxes, but they do consistently use a home range. They do not move extensively, and a Kit Fox may spend its entire life within a 2-sq-mi (5-km^2) area.

In California, researchers in the Department of Fish and Game have studied movements of the Black Bear (*Ursus americanus*) by radiotelemetry and by tagging. In the summer an adult bear ranges over an area of from 5 to 10 sq mi (12 to 25 km^2). Movements increase in the fall, but there is no well-marked migration. Autumnal movement may be for food or in search of a place to hibernate.

Among larger herbivores, social organization modifies the nature of home range. An animal as large as an ungulate cannot disappear in a hole in the ground but rather depends on running to escape predators. Among these mammals, the herd uses common ground, employing eyes, ears, and noses for enemy detection.

In the Roosevelt Elk (*Cervus elaphus roosevelti*), home range is an area occupied by a herd of cows and their offspring, and at all seasons their home range is concentrated near a good supply of food. The bulls remain separate except during the breeding season. In the Pronghorn (*Antilocapra americana*), the herd instinct also prevails, and the herd moves to the most favorable food supplies. Less productive land requires the Pronghorn herd to change its home frequently.

Although individual home ranges of Black-tailed Deer are the same from year to year, at higher elevations they occupy greater areas. The forage is less abundant and the seasonal plant growth later at 2200 to 3200 m, and the higher elevations may make greater metabolic demands than do habitats below 1800 m. In the highest summer ranges, a deer might range over a 4-sq-mi (10-km^2) area, whereas below 1800 m the home range is usually less than half that much.

Homing

An important aspect of both daily movements and migrations over long distances is an animal's homing ability. Although little is known about the sensory input that indicates to a mammal its geographic location, homing or displacement experiments have shown that many kinds of bats can find their way home from up to several hundred miles away in a few days. After placing numbered bands on bats so that individual recognition is possible, students have transported them various distances and directions from their home caves and then waited for their return. It has been learned that the Guano Bat (*Tadarida brasiliensis*) can return from up to nearly 400 mi (640 km) away, traveling about 20 mi (32 km) a night. The Pallid Bat (*Antrozous pallidus*) has been known to cover up to 32 mi (51 km) in one night, and the Little Brown Bat (*Myotis lucifugus*) can return 60 mi (96 km) in the same night it is released, covering about 4 mph (6 kmh).

As interesting as these experiments are, they do not reveal the means by which a bat finds its way home. The annual movement away from a cave in the spring may be hundreds of kilometers, and individual bats return to the same caves the following autumn. Thus the homing ability, as demonstrated by displacement experiments, has an important role in the annual cycle of many kinds of bats. Some of these distances are so great as to preclude the likelihood of familiarity with landmarks en route. Presumably the same mechanisms that aid a bat in returning home from unfamiliar territory also constitute the basis for its navigational ability.

To understand the ability of an animal to find its way home from an unfamiliar region, one must separate the several components or mechanisms involved. The animal first must be able to determine its new, or displaced, position. This process is called *orientation*. Anyone who has been lost in a forest is aware that most humans have a poorly developed sense of orientation. Orientation itself, however, is not sufficient for homing, for the animal must also have the ability to determine direction. This process is called *navigation*. Different kinds of animals use various sorts of navigational cues, which vary with the animal's sensory ability. Orientation and navigation together provide the bat with its ability to return home from an unfamiliar region.

Although many bats clearly have the ability to navigate, our knowledge is far less advanced than is our understanding of the navigation of birds. Some students have released blinded bats and normal bats at varying distances from their diurnal resting sites and observed the speed with which they return. Generally, bats deprived of their vision by means of some covering over the eyes seem to return less frequently and less quickly than those with normal eyesight. Experiments have also compared homing of bats when their eyes were covered with opaque goggles and clean goggles. At distances of 10 mi (16 km) or more, the lack of vision impairs the homing of some species. This pattern suggests that some bats have a familiar home range with a radius of approximately 10 mi (16 km) and can find their way home by echolocation, but that at greater distances vision is required for effective navigation.

Homing ability is known to be rather limited in some mammals that do not naturally make extensive movements. Among Deer Mice (*Peromyscus maniculatus*), for example, individuals displaced only 100 m from their home range may not return; there are always some individuals that fail to return home and assume a new residence near the point of release or in another area. The extreme differences between the homing ability of mice and bats doubtless reflects not only a difference in motivation but also a fundamental lack of ability in mice to orient and navigate.

Dispersal

Unlike migratory movements, dispersal is a departure of an individual from its birthplace without a return. Dispersal occurs at some point in the life cycle of all plants and animals and seems to be an innate drive in all species. Dispersal almost always occurs prior to sexual maturity; in some species it is known to take place after mating but prior to birth of young (or *oviposition,* in the case of egg-laying animals). This innate drive, which characterizes the youth, is assisted by the increasing hostility of the parents as their offspring approach adulthood. Beaver colonies, for example, typically consist of an adult male and adult female, the young of the year, and young of the previous year. As the older offspring reach sexual maturity, at about two years of age, they are driven away by the adult male. In the Beaver, therefore, dispersal is bound to territoriality, but dispersal is seen early in some mammals and is not always associated with parental behavior. Dispersal may take a mouse only a few meters from its birth site, but a young Coyote may travel many kilometers before settling in a new area.

In most individuals, dispersal involves movement from the natal area to another well within the geographic range of the species. In this way, dispersal tends to reduce inbreeding and, instead, enhances the genetic heterogeneity of the population. Thus, dispersal contributes to the genetic vigor of a population.

Dispersal has an additional effect at the edge of a species' geographic range, for there dispersing individuals may move into areas not already occupied by others of the same species. In this situation, dispersal may promote the expansion of the geographic range of the species. Most commonly, such dispersal into new areas is not permanent. If peripheral regions were favorable for the existence of the species in question, these areas would already have been populated from dispersing individuals in previous years. Nevertheless, around the margin of a species' geographic range dispersal constitutes a continuous or pulsating pressure for expansion. A fluctuation in climatic conditions (or man-made changes in the environment) may alter a habitat so that it becomes suitable whereas previously it was not. The species' innate tendency for dispersal may in this way take some individuals into habitats marginal for their survival, which is enhanced or depressed by small climatic changes. Because climatic changes occur on a small scale every year, small shifts in the geographic ranges of mammals do occur. When climatic changes are major and progress over extended periods, such as the alterations that occurred during the Pleistocene, there may be vast shifts in the geographic ranges of mammals. During major climatic changes, entire faunas may move. Moreover, different species may move in separate directions, for their environmental demands are seldom the same.

Shifts in geographic ranges of mammals due to natural changes in the environment occur slowly and are not frequently witnessed during the human life span. Man-made changes may be rapid and profound, however, and some of these account for alterations in the geographic ranges of mammals. In California, the Cotton Rat (*Sigmodon hispidus*) has spread from its original home along the Colorado River into the Imperial Valley with the advent of irrigation. This region was previously desert, a hostile and deadly environment for Cotton Rats. But when irrigation created a suitable habitat for the Cotton Rat, dispersal accounted for its spread into the Imperial Valley.

Dispersing mammals normally follow routes through their typical habitat; that is, when moving from its birthplace, a mammal tends to remain in the environment to which it is adapted. This tendency is well illustrated by the spread of the Muskrat (*Ondatra zibethicus*) in California. This aquatic rodent was brought into the state for its value as a furbearer and was maintained captive in California in the 1920s. Previously this rodent occurred only along margins of the Colorado River and along the eastern slope of the Sierra Nevada in California, although the species is widespread along slow streams and marshes in much of North America. Because the value of Muskrat skins does not justify the cost of rearing them to maturity, disappointed managers of these "Muskrat farms" released them in the Central Valley and a wild population soon became established. Subsequently the Muskrat dispersed along irrigation canals and other slow-moving watercourses throughout the valley; today its distribution within central California is almost totally restricted to the immediate margins of irrigation ditches and sluggish streams. Dispersal along any other route is exceptional.

Clearly, animal movements are not random and disorganized. Throughout an individual's life, there are phases of dispersal, occupancy of a home range, defense of a territory, and sometimes a migration. The life cycle can be viewed as a series of successive types of movements, each with its own special function, and comparisons among species aid in appreciating the roles of movements in nature.

References

Aidley, D. J. (Ed.) 1981. *Animal migration.* New York and Cambridge: Cambridge University Press.

Gauthreaux, S. A., Jr. 1980. *Animal migration, orientation and navigation.* New York: Academic Press.

Lidicker, W. Z., and Caldwell, R. L. 1982. *Dispersal and migration.* Stroudsburg, Penn.: Hutchinson Ross Publishing Co.

Schmidt-Koenig, K., and Keeton, W. T. 1978. *Animal migration, navigation and homing.* New York: Springer-Verlag.

MAMMALS AND CALIFORNIA SOCIETY

Since the arrival of aborigines in the New World, populations of wild mammals have had a critical role in the welfare of California societies. Prey species, such as deer and rabbits, provided food and clothing for the first Indians and today continue to draw many people to the outdoors. California furbearers were the stimulus for much early exploration and continue to occupy the efforts of a small number of dedicated trappers. Some species have negative aspects, for some damage field crops or domestic stock, and many rodents are reservoirs of diseases transmissible to humans.

Predation on Domestic Livestock

Most attacks by predatory mammals on domestic stock are on sheep. Cattle suffer rather small losses, and poultry are usually well protected. Most of the domestic sheep in California are more or less unprotected from predators; in the winter they are concentrated in fenced pastures at lower elevations, and in summer many are moved to open ranges on public lands in the mountains. The overall loss of sheep, from all causes, amounts to an annual mortality of some 5 to 10 percent. Some of this loss is due to predation, and the Coyote is the major predator of domestic sheep in the western United States.

The Coyote is the most versatile carnivore in North America, and its adaptability renders it capable of great harm as well as genuine benefit. Although Coyotes may, at times, feed largely on mice, ground squirrels, and rabbits, they kill substantial numbers of deer and destroy many sheep. It is realistic to view the predation of the Coyote on deer as an ecologically desirable relationship, but the killing of domestic sheep amounts to a large monetary loss to the rancher.

Detailed studies of the diets of Coyotes indicate some interesting differences from one individual to another. Coyotes that have managed to escape from steel traps frequently do so at the loss of a foot. Such animals can be recognized by their distinctive tracks in the dirt and are referred to as "peg-leg" Coyotes. These individuals are apparently handicapped in their pursuit of natural quarry and are more prone than others to attack domestic sheep. Other Coyotes, possibly some that are aged and losing vigor, seem to develop the habit of preying on easily captured domestic animals. When the troublesome individual is captured, predation usually ceases. Specialists in predator control therefore believe that the most efficient protection from damage by Coyotes is that which is directed at the specific individual inflicting the damage.

Several other predators attack domestic stock, but their effect is local and limited. The Bobcat sometimes destroys lambs in remote regions, but Bobcats are scarce in most of the winter lambing areas of the Central Valley. Bears and Mountain Lions also have been known to kill domestic sheep, but such predation is not common.

The Feral Dog

Dogs are unlike other predators in California. Although dogs may *seem* to be feral at times, these same dogs belong to someone and are fed and cared for at someone's home or ranch; they do not survive and reproduce in the wild, as does the domestic cat, for example, and in this sense the dog is not truly feral. Depredation by dogs is less a problem of destructive dogs than it is one of irresponsible dog owners.

Depredations most frequently occur in concentrations of wildlife or in areas where groups of dogs roam uncontrolled in sheep ranching areas, areas in which dogs are loose at night. Because the offending dogs are someone's property, the responsibility for damage lies with the dog's owner. Responsibility becomes difficult to establish, however, because many uncontrolled dogs are also unlicensed. The problem of appraising damage inflicted by dogs is further complicated by the reluctance of ranchers and conservation, animal control, and law enforcement officers to release information on dogs they have dispatched in the act of molesting domestic stock or wildlife. Thus, depredation by feral dogs is very difficult to measure with any real precision. Nevertheless, dogs account for an estimated 4 to 45 percent of the depredations on domestic sheep; the degree varies with the concentration of sheep near human habitation. In some regions, damage attributed to dogs may equal that assigned to Coyotes.

In spite of the problems in evaluating the activities of uncontrolled dogs, knowledgeable officials agree that serious problems exist in many parts of the state. The severity of depredation seems to be directly related to the local abundance of wildlife or domestic stock. If deer migrate to a community of foothill homes in the winter, for example, they are locally vulnerable. Uncontrolled dogs kill an estimated 1300 deer annually in California; this is a very small part of the total deer population, but it could be locally important. With the encroachment of human settlements into foothill areas, this sort of damage will increase. There are examples, also, of roaming dogs killing Antelope when fences and deep snow handicapped the Antelopes' movements and prevented their escape.

Although loose dogs damage domestic stock and wild game, the greatest danger is to human health. The risk of rabies is ever present in a society with a high population of free-ranging dogs, and the risk is great where the disease is endemic in skunks and Raccoons. Although most of the 1 to 2 million dog-bite victims annually have wounds that are trivial, the frequent follow-up of the rabies treatment is both expensive and extremely painful. Finally, the danger from the dog bite itself can be dangerous. At the very least, a dog bite is unpleasant and painful, and the wound can be serious. The death of a six-year-old girl in Colorado in 1977 is a tragic reminder of the potential danger of uncontrolled dogs.

Although feral dogs are not truly part of the mammal fauna of California, they do constitute a population of free-ranging mammals and are frequently far more dangerous than our native mammals.

Furbearing Mammals

From the earliest periods of the European invasion into North America, furbearers have always been a stimulus for humans to move onward. California is blessed with a variety of mammals with beautiful fur, mostly carnivores and rodents. Some species are probably less common than in former years, and others are probably at their original population levels, if not even more common. Several species are introduced mammals in our fauna—for example, the Muskrat, Red Fox, and Opossum in the Central Valley.

Early History

The principal furbearers in the early days were the Sea Otter and the California Fur Seal. Early trappers fanned out from eastern North America in search of Beaver and began operating in California early in the nineteenth century. Not only were Beaver found along both mountain and lowland watercourses, but River Otters and Mink were also taken. In that period, hunters contributed skins of deer and elk, two hooved mammals not included among furbearers today. The gold rush in the middle of the nineteenth century attracted the energies of potential trappers, but professional trapping probably never ceased.

For the latter half of the nineteenth and the first half of the twentieth centuries, prices and demand for pelts were low and trappers pursued high-quality furbearers. Early regulations allowed their capture, but at a level designed to prevent the disappearance from our fauna. Some furbearers today are totally protected and are common or even locally abundant. Populations of the River Otter, for example, are high in some areas, and controlled trapping might well be tolerable. The Marten and Fisher, two of our most prized furbearers, are totally protected, as are the Wolverine, Kit Fox, Red Fox, and Ringtail.

Current Fur Trapping

Since the early 1970s the increase in fur prices has stimulated commercial trapping, and the sale of wild furs has once again become an important source of income for a relatively small number of people. The total value of furs taken commercially has reached all-time highs. In the 1977–1978 season, fur sales in California were $1,159,126, passing the $1 million mark for the first time. This sum reflects an increase in licensed trappers and animals taken, as well as the high price of furs. Several species contributed to the bulk of the California fur trade that season: the value of skins of Raccoon, Coyote, and Gray Fox came to more than $100,000 for each species; Muskrat pelts brought more than $200,000; and the Bobcat, with a fur of mediocre quality, brought more than $400,000.

The Bobcat achieved prominence because today it is the only spotted cat legally sold in the fur trade. Despite the poor quality of the fur and the very thin skin, the demand for Bobcat fur remains high. Fortunately, the Bobcat is a prolific animal and its populations hold up well under trapping pressure.

In the 1978–1979 trapping season, the total value of furs of wild mammals in California was $2,399,565; the value of Bobcat pelts alone was more than $1,130,000. This figure represents the increase in average price of a Bobcat skin from $106 in 1977–1978 to $190 in 1978–1979. In the early 1940s, a good Bobcat pelt brought the trapper $6. Other major furbearers in the 1978–1979 season—the Coyote, Gray Fox, Muskrat, and Raccoon—each had a value of from $200,000 to almost $400,000.

The Significance of Commercial Trapping

The role of furbearers in California culture is extremely complex. The value of the raw pelts is important, especially in view of the economic position of the average trapper, who is either an adult augmenting a poverty-level income or a teenager trying to earn money. Throughout the United States and Canada, thousands of young boys earn their college expenses by trapping Muskrats and the bonanza of an occasional Mink.

There is no question that the money from trapping furbearers goes mostly to people in need of additional income.

Several furbearing mammals are sometimes extremely destructive to agricultural activities. Most prominent are the Coyote, the Black Bear, and the Muskrat. The major Coyote depredations are against domestic sheep, especially in the winter lambing region of the Sacramento–San Joaquin Delta. Muskrat and Beaver frequently create serious problems, for their burrows through small levees as well as through irrigation ditches account for occasional water losses and the general deterioration of levees. Major levee breaks are sometimes the work of Beavers. Other furbearers can be local or sporadic pests. The Badger, for example, is infrequently encountered, but one Badger digging in pursuit of a ground squirrel can ruin a levee and thus counter the good it does as a destroyer of ground squirrels. And skunks, who have no real friends at all, are sometimes major reservoirs of rabies.

Furbearers also have an important role as predators. The same Coyote that kills lambs in the winter will prey on ground squirrels, mice, and possibly a few ducks and Muskrats from spring until autumn. The same skunk that is a potential disseminator of rabies subsists mainly on soil insects and mice, with an occasional bird's egg in the spring. The Mink, which contributes its fur to elegant stoles and coats, may destroy many young Muskrats. Both Bobcats and Coyotes account for the disappearance of some fawns, a natural event which outrages deer hunters, but predation is perhaps preferable to starvation, the fate of deer populations totally protected from predators.

Finally, but politically of major importance, is the protection of furbearers for their aesthetic value. Our inability to place an economic value on the thrill of glimpsing a Coyote in a mountain meadow or encountering a fearless Bobcat creates conflicts among the many groups interested in wildlife. Nor can we entirely discount the aesthetic value felt by a hunter whose aim may be to shoot a Bobcat. This is the conflict encountered by game managers, conscientious professionals who are dedicated to the concept of multiple use of land and other natural resources. This concept defines a policy of democratic use of land and resources for all people, not for one group to the exclusion of others.

Effects of Trapping on Wild Populations

Between 1977 and 1980, from 5000 to 6000 California Bobcats entered the commercial fur trade. This is the number of export tags the Department of Fish and Game issued to trappers and does not include animals shot by hunters or those destroyed under damage control permits. Bobcats killed in the act of molesting domestic stock declined from 347 in 1976–1977 to 50 in 1979–1980. In addition to the limitation of 6000 Bobcats exported annually from California by permit, noncommercial hunters take another 5000 to 10,000. Although this is a substantial number and amounts to an estimated maximum of 16,000 a year, the Department of Fish and Game calculates that a total of 20,000 Bobcats amounts to roughly one-third of the spring population of an estimated 60,000; this is 40 percent of the probable annual increase.

Such estimates lack precision, but the continued take of Bobcats supports the appraisal of the professional game managers. The current annual capture of Bobcats is not likely to lower present populations, but it does seem to have depressed predation on domestic sheep.

Some conservationists would prefer total protection for all furbearers, including Bobcats. While total protection would very likely result in increased densities, it would also increase attacks on domestic stock. The Bobcat, fortunately, remains a common carnivore which one may encounter on mountain trails, cutover forest lands, and mountain pastures. Throughout North America the Bobcat, like the Coyote, has been persistently persecuted, yet both have been able to thrive under this pressure.

Many carnivores have been protected for years and are today as common as they probably were in the past. The Marten and the Fisher are fully protected and today occur as common predators in favored habitats. They are most common in remote areas, and their numbers could tolerate regulated trapping. The River Otter, another valuable but protected furbearer, is a common animal along many California watercourses. Although it is not commonly observed, its tracks and its slides indicate local abundance of this graceful predator. Its numbers today would certainly support limited commercial capture. Although River Otters are a joy to observe, these carnivores are not universally appreciated. Many fishermen resent the River Otter for the occasional trout in its diet.

Wild Mammals as Carriers of Pathogens

Several important pathogenic organisms, transmissible to humans, occur in some wild mammals. Several of these pathogens are common enough to justify special precautions in handling mammals, especially rodents, or sometimes in simply being near populations of wild mammals. Among the more important of the diseases common to wild mammals and humans are rabies, plague, tularemia, and relapsing fever.

Rabies

Rabies occurs in a broad variety of wild mammals. The rabies virus is generally transmitted in saliva and for this reason is most likely to be acquired as a result of bites from a carnivore, such as a skunk or a fox. Rabies has also been found in bats throughout the state and may occur in rodents, deer, and also domestic dogs and cats. The typical symptoms in a wild mammal include loss of fear; the appearance of a usually shy species in a strange place and showing no fear, suggests rabies. An apparently tame fox walking around in circles should be captured by an animal control officer. A bat abroad in daylight should not be handled. Although most bats are small, innocuous beasts and their bite can scarcely break the skin, a strangely acting bat could be rabid and carry the rabies virus in its saliva.

Plague

Plague is a bacterial disease common in many wild rodents, especially rats and ground squirrels. It is found throughout the world in both humans and rodents. When in wild rodents, such as ground squirrels, chipmunks, and wood rats, plague is referred to as sylvatic plague. The plague bacterium is passed from one mammal to another by fleas. If the squirrel or rat dies of plague, the fleas may leave its body to attack another mammal or may remain in the host's nest to infest the next individual which may enter. Thus, the disease is difficult to eradicate. Not only can a person become infected by flea bites, but one can also contract plague by eating inadequately cooked meat of an animal ill with plague.

Plague has been found in many species of small mammals in California and is especially common in the Golden-mantled Ground Squirrel as well as other ground squirrels and chipmunks in the Sierra. Deer mice (*Peromyscus* spp.) and voles (*Microtus* spp.) are sometimes reservoirs of the plague bacterium. Public health representatives poison rodents about resort areas to lessen the likelihood of vacationers encountering plague. Plague-infected rodents have been found about Lake Tahoe, for example, and plague has occurred in the Bushy-tailed Wood Rat (*Neotoma cinerea*) and the Deer Mouse (*Peromyscus maniculatus*) in Lava Beds National Monument. Between 1970 and 1980 there were thirteen reported cases of human plague in California, with five fatalities, and it is very likely that some cases go unrecognized and unreported. Exposure to sylvatic plague is certain to increase with greater use of outdoor recreational areas.

When contracted by humans, sylvatic plague is called bubonic plague, for it affects lymph glands and produces swellings known as bubos. The swollen lymph nodes are commonly under the arms or in the groin but may occur elsewhere in the body. Bubonic plague is normally transmitted by fleas. When the plague enters the respiratory system, it is called pneumonic plague; this form can be transmitted by coughing and is highly contagious.

Tularemia

Tularemia is a bacterial infection common in rabbits, hares, and many kinds of mice and voles. Tularemia exists not only in the tissues of an infected mammal; Muskrats have been known to contaminate water supplies, presumably through their urine, as well. Although tularemia is commonly thought of as an arthropod-borne disease, the organism can enter our bodies through our skin or orally—by eating incompletely cooked infected meat or by drinking infected water. The disease is more likely to be contracted from a rabbit, squirrel, or other small mammal when its blood comes into contact with our skin. It is rarely, if ever, passed from one human to another.

Tularemia is probably capable of infecting virtually all wild mammals, and it has been found throughout most of temperate North America and Eurasia. The disease is widespread in California but seems to be declining in importance. It has been most frequently contracted in Kern and Los Angeles counties.

Relapsing Fever

Relapsing fever in California is a tick-borne bacterial disease associated with chipmunks, ground squirrels, and tree squirrels. In this disease, the mammal is not only a reservoir but also an agent which maintains the ticks that are the vector. Humans get the disease from the feeding of the tick. These ticks (species of *Ornithodorus*) live in the nests of mammals. If the host, such as a chipmunk, has a nest beneath the floor of a mountain cabin, the ticks may move from the chipmunk's nest to the bedding of the human occupants. These ticks do not attach to their hosts, and they feed quickly, usually at night. Therefore, one can be unaware of these ticks and still, while sleeping, be infected by the relapsing fever spirochete through a tick bite.

Rabies, plague, tularemia, and relapsing fever are all diseases associated with California mammals. The pathogens producing these illnesses can all be transmitted from wild mammals to humans, and all can be fatal. Knowledge of these diseases and the methods of avoiding them enables us to enjoy observing wild mammals without endangering our health.

References

Davis, J. W.; Karstad, L. H.; and Trainer, O. (Eds.) 1970. *Infectious diseases of wild mammals.* Ames: Iowa State University Press.

Maloney, A. B. (Ed.) 1945. *Fur brigade to the Bonaventura: John Work's California Expedition 1832–1833 for the Hudson's Bay Company.* San Francisco: California Historical Society.

Meyer, K. F. 1955. *The zoonoses in their relation to rural health.* Berkeley and Los Angeles: University of California Press.

Rich, E. E. 1967. *The fur trade and the Northwest to 1857.* Toronto: McClelland & Stewart.

Saum, L. O. 1965. *The fur trader and the Indians.* Seattle: University of Washington Press.

Wishart, D. J. 1979. *The fur trade of the American west, 1807–40: A geographical synthesis.* Lincoln: University of Nebraska Press.

FOOD OF MAMMALS

Among the various groups of California mammals one sees many feeding patterns. Some mammals are extremely versatile in their diet, whereas others confine their feeding to single categories such as plants, insects, or other vertebrates. Still others are extremely restricted and may feed on only a single kind of food. Diversity in feeding is one way in which different species live in the same area compatibly, without interfering with each other in their search for food or, perhaps, without competing for food.

There are, then, both generalists and specialists with respect to diet, and there is a gradation from one extreme to the other. A species with a restricted diet may compete with no other species, but few specialists are that narrow in their selection of food. Nevertheless, the specialist possesses some adaptation which enables it to find nourishment in food that others might disdain; thus, the specialist has achieved a high degree of independence in the constant scramble for food. A dietary generalist, on the other hand, is a potential competitor of other generalists. One should remember, however, that dietary overlap, in itself, does not constitute competition. The element of competition does not become real until the object for which animals are competing, food in this example, becomes limited in supply or availability. When the food supply is so short that there is not enough for all the generalized feeders, then competition may become a reality. Because a generalist is versatile, however, when one food item becomes limited in supply, the animal may well be able to subsist on another.

Specialized Diets

Before there can be an advantage in restricting its energy intake to a single kind of food, an animal must develop the capacity to obtain and digest a material that is not available to generalists. By selecting foods not sought after by others, a species has taken a big step in preserving its food for itself alone. Moreover, the specialist must concentrate on a kind of food that will not be temporarily eliminated by variations in weather, for even a temporary food shortage might wipe out a population of mammals depending on it. For example, one of the best-known specialized feeders among California mammals is the Red Tree Vole (*Arborimus longicaudus*), a small species that lives only along the coasts of Oregon and northern California. This little mouse feeds on the soft tissue in the needles of Douglas Fir (*Pseudotsuga menziesii*), a substance not normally eaten by other mammals. Thus, the Red Tree Vole does not have competitors among mammals, nor does it compete with them for space. It has an abundant supply of food which other mammals do not seek. This advantage involves some risk, however, for the Red Tree Vole has become largely dependent on one species for its food; if the Douglas Fir should disappear, the Red Tree Vole would perish with it. Thus, it is an *obligate specialist*. Caterpillars of the Pine White (*Neophasia menapia*), a butterfly in the family Pieridae, have been known to defoliate pine trees and sometimes also to attack the Douglas Fir. The distinctiveness of the Red Tree Vole indicates that it has survived as a specialist for a long time, and occasional inroads of the caterpillars on its food supply have not been damaging.

There are varying degrees of specialization involving anatomical adaptations that give a mammal species an advantage in seeking certain foods without its becoming an obligate specialist. The Marten (*Martes americana*) and the Badger (*Taxidea taxus*) are two moderately specialized carnivores. The Marten is light and agile and travels easily through the trees, where it captures tree squirrels and birds. The Badger is heavy and stout, with powerful forelimbs for digging, and pursues a variety of small rodents, especially such soil-dwelling forms as pocket gophers (*Thomomys* spp.) and ground squirrels (*Spermophilus* spp.). The Marten and the Badger are not obligated to subsist on their normal prey; each would find adequate nutrition in the food of the other. Nevertheless, each is moderately specialized to pursue its normal food, and such specializations undoubtedly account for the presence of the Marten in coniferous forests and the Badger in dry, arid areas of loose soil. Clearly, these two carnivores must compete only rarely where their habitats meet or overlap, even though their dietary differences reflect their habitats more than their preferences for different foods.

Ruminants

Some grazing mammals have a four-chambered or ruminant stomach, which is a specialization for the digestion of certain plant tissues. Mammals lack enzymes necessary to digest cellulose, a major part of the cell walls of grasses, forbs, and other plant materials; this material is broken down by cellulose-fermenting bacteria and protozoa. The use of such a microbiota in the digestive tract of herbivores is most elaborately developed in ruminants, such as deer and sheep. In the rumen (one chamber of the four-chambered stomach) of these mammals, cellulose, a complex sugar, is digested to simple sugars; these, in turn, are utilized by the bacteria and protozoa of the gut. In many nonruminant herbivores this process goes on in the large intestine and is probably of little benefit to the mammal. Ruminants regurgitate partly digested plant foods from the rumen and, "chewing their cud," masticate food particles that have been chemically changed by the microbiota. When this material enters the small intestine of the ruminant, the mammal can absorb some of the products of microbial digestion.

Other herbivores, such as voles or the Porcupine, have cellulose-digesting bacteria and protozoa in the caecum or even in the large intestine. Presumably these mammals derive less benefit from the digestive activities of their microbiota than do the more specialized ruminants. Indeed, fermentation of cellulose in the large intestine may be of little or no benefit to the herbivorous mammal.

The palatability of plants to herbivores is partly a function of the toxicity of certain plant compounds to the microbiota of herbivores. Some plants contain oils and alkaloids that may immobilize or destroy the cellulose-digesting bacteria and protozoa.

Microbiota benefit ruminants by making available the protein contents of plant cells, and also through the animal's digestion of the microbes themselves. The sugar products from the digestion of cellulose provide energy for the maintenance of the microbiota.

Coprophagy

Several different kinds of mammals, but notably rabbits and hares, alternately produce soft feces and hard feces. Typically, soft feces are passed at night and promptly eaten as they leave the anus. The mammal benefits from reingesting its own fecal pellets by the assimilation of some materials digested by the intestinal microbiota; these materials contain nutrients that would otherwise be lost. The process in which a mammal reingests its own fecal material is called *coprophagy.*

Apart from the nutrition provided by coprophagy, the process also conserves water. In arid environments, loss of body water is always a threat, and desert mammals develop several means of reducing dissipation of body fluids. Coprophagy is reported from pocket gophers and also from shrews; quite possibly it also occurs in other groups. In a nonruminant herbivore, coprophagy makes available the digestive products released by the microbiota in the large intestine.

Generalized Diets

Some carnivores are bona fide generalists and do not even restrict their diet to meat. Although the Gray Fox (*Urocyon cinereoargenteus*) seems to relish mice and small birds, it readily feeds on berries and other plant parts. It is at home in trees and takes birds' eggs in season. The Gray Fox is structurally and phylogenetically a carnivore, but it is, in fact, a generalized feeder—an omnivore. Two other carnivores, the Raccoon (*Procyon lotor*) and the Coyote (*Canis latrans*), are notorious for their versatility in feeding, although the Coyote eats much less plant material than does the Raccoon. The Black Bear (*Ursus americanus*), another carnivore, actually feeds largely on plant materials.

Rodents are commonly considered to be plant feeders, and many species are said to subsist on seeds, but in fact most species are true generalists and highly versatile in their pursuit of food. Deer Mice (*Peromyscus maniculatus*) and their allies tend to eat whatever is abundant and nutritious. This adaptability can give the *appearance* of specialization, for if one type of food is abundant, Deer Mice may feed on it exclusively. In years when pines produce a very heavy seed crop, Deer Mice may subsist on pine seeds to the exclusion of other palatable foods. There are years, however, when many trees and shrubs fail to produce seeds, and there are no pine seeds, acorns, manzanita berries, or other energy-rich foods from plants. Such widespread seed shortages are not unusual, and seed scarcities occur among some plants in most years, perhaps caused by late frosts, insect attacks, or similar variables. At such times the generalist turns to other food items, such as insects and underground fungi.

When Deer Mice feed heavily on pine seeds, they have not really become specialists. A true generalist is a *facultative specialist*—an opportunist. Perhaps the generalist seeks the most nutritious food that can be obtained at the least expense of energy. We have, for example, known Deer Mice to feed entirely on maggots in a rotting deer carcass. Such food contained little waste and required no expense in foraging. Such specialization is frequently observed at times of abundance of a particular food item. In the summer of 1951 an outbreak of the California Tortoiseshell Butterfly (*Nymphalis californica*) provided an abundance of caterpillars and pupae. For a short time the caterpillars formed the main food of the Golden-mantled Ground Squirrel (*Spermophilus lateralis*), which pursued these insects on the ground. The caterpillars climbed into bushes to pupate, and chipmunks fed on the hanging chrysalids; the Yellow Pine Chipmunk (*Tamias amoenus*) ate little else while the chrysalids lasted.

The ability to shift attention from one sort of food to another enables the generalist to thrive at all times, in spite of variations in food supplies. Many small mice feed heavily on insects, and, in most environments the volume of insects eaten by mice probably far exceeds that taken by insectivorous birds; in addition, these birds are largely migratory and in mid-latitudes are present for only part of the year. The versatility of the generalist is obtained at the cost of occasional extensive foraging, an expense not usually shared by the specialist. Thus, there is a trade-off. The specialist has its food supply continuously at hand and never needs to search for it, but when its food supply fails (which it seldom does), the specialist is doomed. The generalist must forage for its food, and foraging takes energy, so the energy in the generalist's food must exceed the cost in obtaining it, but the generalist always has *something* to eat.

There are conditions intermediate between the restricted diet of the specialist and the broad spectrum of food taken in by the generalist. Two common California rodents eat plant fibers. The Porcupine (*Erethizon dorsatum*) spends most of its time sitting against the trunks of small conifers, especially pines, from which it strips the woody tissue beneath the bark. In the springtime, however, the Porcupine regularly repairs to mountain meadows and feeds on grass. On the other hand, voles (*Microtus* spp.) normally and by choice feed on the tender growth of forbs and grasses but have been known to eat the bark and roots of trees, even to the extent of killing orchard trees.

Normally, one studies the food of mammals either by direct observation of feeding in the field or by examining the contents of the stomach. Animals provided with fur-lined cheek pouches frequently fill them with seeds as they forage and in this way can reduce the number of trips needed in foraging. By examining the contents of cheek pouches, the mammalogist can obtain some notion of the seeds the animals are eating. Rodents such as pocket mice (*Perognathus* spp.) and kangaroo mice (*Microdipodops* spp.), however, may at times prey heavily on insects, and live insects are rarely placed in the pouches of these rodents. Thus, examination of cheek pouches indicates the plant foods eaten by these rodents but presents a biased picture of their total diet.

Insectivores, as indicated by their name, do feed largely on insects, but they, too, are somewhat flexible in their choice of food. Shrews (*Sorex* spp.) will take quantities of conifer seeds in years of heavy seed production, and the Water Shrew *(Sorex palustris)*, which preys mostly on nymphs and larvae of aquatic insects, will also capture small minnows and trout fry. Although moles (Talpidae) feed on insects to a large extent, they also take large numbers of annelid worms and have been known to eat underground parts of plants.

Seasonal changes in food are apparent to anyone who examines the feeding patterns of a generalized feeder in detail. Available food supplies change not only from year to year but from season to season. A Deer Mouse that eats pine seeds as they fall on bare ground in September and October may have to search for different foods when snow falls. Unlike chipmunks (*Tamias* spp.), which are notorious for laying up large amounts of food for later use, Deer Mice appear not to have such a well-developed habit. The winter diet of the Deer Mouse consists mostly of such food as insects and fungi that occur about the snowless margins of fallen logs or large boulders. With the approach of spring in the Sierra, seeds of the Big Leaf Maple (*Acer macrophyllum*) may fall and germinate on the surface of the snow. These seeds are eagerly eaten by Deer Mice and are probably extremely nutritious. Later, as the snow disappears and the sun warms the ground, Deer Mice eat sprouting seeds and tender shoots of grasses and forbs, and green plant material forms much of their fare in the spring. In this way the diet of a generalist changes with the seasons.

The diets of wild mammals are in fact poorly known, and for most species our knowledge comes from fragmentary anecdotal accounts. There are many opportunities to discover patterns of feeding and dietary similarities and differences among related species and sympatric species, as well as seasonal variations in food availability and use. There is much to be learned, for example, regarding dietary overlap among sympatric bats and the effect of foraging mice and rabbits on the plant cover.

Species Accounts

IDENTIFICATION OF MAMMALS

Classification and Names

In this book the taxonomic arrangement of orders, families, and genera of mammals and the generic-specific combinations are taken from *Mammal Species of the World* (Honacki et al., 1982). Most of these names are familiar to mammalogists, and the new combinations are based mostly on recent evidence of mammalian relationships. We have tried to indicate where such changes are controversial.

As knowledge of mammalian relationships accumulates, generic concepts are sometimes modified, and there may also be changes in the allocation of species to genera. One should bear in mind that, in contrast to a species, which is presumed to exist as a discrete entity in nature, a genus is a concept that exists only in our own minds. The western species of chipmunks are, in this volume, placed in the genus *Tamias*. This arrangement is not agreed upon by all mammalogists. Other changes account for unfamiliar combinations of generic and specific names. A very old and familiar name, *Citellus,* well known as the generic name for most of the ground squirrels in California, has been changed to *Spermophilus* owing to nomenclatorial technicalities.

These changes can be frustrating. We urge readers of this volume not to be hasty in adopting every proposed taxonomic change as soon as it appears. Many changes may be suggested for invalid reasons and are quickly forgotten.

Over the past thirty years there have been great efforts to standardize common names of animals, especially vertebrates.

We praise these efforts but cannot always condone the results. There are no rules for the use of common names. Generally we have followed the common names employed by Jones et al. (1982) and Laudenslayer and Grenfell (1983). We do, however, believe that it is desirable to adopt an attitude of flexibility with common names; when we thought it was appropriate, we have deviated from the lists used by these authors.

"Wild Boar" is a well-established name for wild populations of *Sus scrofa,* whether descendants of European Wild Boar or feral domestic strains or mixtures of the two. The name "Wild Boar" appears in the titles of many articles on this species, and it would be confusing to drop this name, as has been proposed, and adopt "Wild Pig."

Similar confusion results from the name change of our most common free-tailed bat, *Tadarida brasiliensis.* Previously the scientific name was *Tadarida mexicana,* and it was sometimes called the "Mexican Free-tailed Bat." Because it is acknowledged that *mexicana* is really a subspecies of *Tadarida brasiliensis,* the full scientific name is *Tadarida brasiliensis mexicana*—a name that has prompted some workers to adopt the inappropriate "Brazilian Free-tailed Bat," a procedure which creates confusion not clarity. We have employed "Guano Bat," a much older name with historical significance.

The thoughtless adaptation of a common name from a scientific name has resulted in many new names that are designed to replace older, more familiar common names. There are many mouselike rodents with a specific name derived from "California," and *Peromyscus californicus* has recently been called the "California Mouse." It had previously been known as the Parasitic Mouse, which is appropriate and has biological significance.

The common names of a great many species are well known but pathetically misleading. The Ornate Shrew (*Sorex ornatus*) is a dull-colored creature, for example, and the Broad-footed Mole (*Scapanus latimanus*) has forefeet no broader than those found in other members of the genus. We have in our state many other examples of mammals with misleading common names.

In this volume we have followed conventional practice in almost all common names of California mammals. We hope readers will have no difficulty in recognizing species referred to by a different, usually an older, name. Confusion can be kept to a minimum by learning the generic and specific name for each species. This is the name by which they are indexed in scientific literature, and it is our sincere wish that this volume will stimulate readers to develop a greater, more intense interest in the study of wild mammals.

References

Honacki, J. H.; Kinman, K. E.; and Koeppl, J. W. 1982. *Mammal species of the world.* Lawrence, Kansas: Association of Systematics Collections and Allen Press.

Jones, J. K., Jr.; Carter, D. C.; Genoways, H. H.; Hoffmann, R. S.; and Rice, D. W. 1982. *Revised checklist of North American mammals north of Mexico, 1982*. Occasional Papers of the Museum of Texas Technological University, No. 80.

Laudenslayer, W. F., Jr.; and Grenfell, W. E., Jr. 1983. A list of amphibians, reptiles, birds and mammals of California. *Outdoor California,* Jan.–Feb.: 1–14.

Keys

The identification of a mammal depends on the condition of the specimen and the parts available. The keys are intended to be used in conjunction with the descriptions, plates, and range maps. A skull from an owl pellet, for example, may be adequate material for mammals such as shrews or voles but insufficient for others, such as white-footed mice. To assist in specific recognition, the keys usually include details of dentition and skull features as well as aspects of color and body proportions.

The keys are arranged so that a single major group is divided into two lesser groups—dividing a family into genera, for example, or genera into species. A key first uses a couplet of general contrasting characters which separate the major category into two subgroups. The specimen should fit into only one half of the couplet, and the number at the right-hand margin leads to the next appropriate couplet. Eventually the process of going from one couplet to another should lead to the identification of the specimen at hand.

Keys are not flawless, however, and individual variation is sometimes difficult to include in the terse contrasting phrases of a couplet. The keys, therefore, should be used together with descriptions, plates, and range maps. After a specimen has been identified, it should be saved. It will enhance the identification of subsequently found individuals. Labeled with date and locality, moreover, it may expand the information contained in this volume.

Species

The following pages contain information on the wild mammals to be found in California today. There is an unfortunate paucity of knowledge on some kinds of mammals, especially among the species of bats, shrews, wild mice, and squirrels. Most of these forms are not rare, however, and our relative ignorance of their habits should challenge enterprising mammalogists to discover more about them.

Included are species which exist in California today or which approach our borders and might reasonably be expected to occur here. Not included are several introduced forms which are either very limited in their occurrence or exist in a state of semicaptivity.

MARSUPIALIA

This order contains the opossums of the New World as well as the various kangaroos, wallabies, and phalangers of Australia and New Guinea. These mammals produce young after a very brief gestation and little maternal nourishment during pregnancy. The newborn young are in a near embryonic condition; only the forearms are well formed. The minute infant crawls to its mother's pouch, or marsupium, which contains the nipples.

References

Hartman, C. G. 1952. *Possums.* Austin: University of Texas Press.
Hunsaker, D. (Ed.) 1977. *The biology of marsupials.* New York: Academic Press.

Opossums (Didelphidae)

Superficially ratlike but structurally unique in many ways. Skull with five upper and four lower incisors on each side. Braincase very small. Tail prehensile. Foot with an opposable thumb.

When mammals began to emerge as a distinct group, some 100 million years ago, North America was the center of marsupial evolution. Opossums represent a rather primitive form of marsupial and are the only type to exist in the United States today. Opossums are the basic marsupial type from which so many diverse modern forms have evolved.

Virginia Opossum (*Didelphis virginiana*) [fig. 7]

Description: A large, ratlike animal with many small teeth. Differs clearly from rodents by the presence of large canine

FIG. 7. Opossum (*Didelphis virginiana*).

teeth, the absence of a pair of chisellike incisors, and the presence of a thickened, prehensile tail. Fur rather long and loose, light gray. Hind foot with an opposable thumb. Female with a pouch in which the young are carried. The skull is distinctive, with a high dorsal ridge called the sagittal crest (fig. 8b). TL 700–900, T 290–400, HF 62–75, E 47–57. Weight: 1.5–3.1 kg; females are about two-thirds the size of males. Dentition: 5/4, 1/1, 3/3, 4/4.

Distribution: First introduced in California at San Jose in 1910. Now occurs widely in cultivated areas at lower elevations. Readily adjusts to agriculture, from which it seems to obtain much of its food and shelter. Seems to avoid mountains above 1000 m. From the eastern edges of the Sacramento Valley and San Joaquin Valley west to the coast and from San Diego north to Oregon. From most of the eastern half of the United States south to include most of Mexico.

Food: A very generalized feeder. Forages for almost anything edible, plant or animal. Most of its food consists of soil-

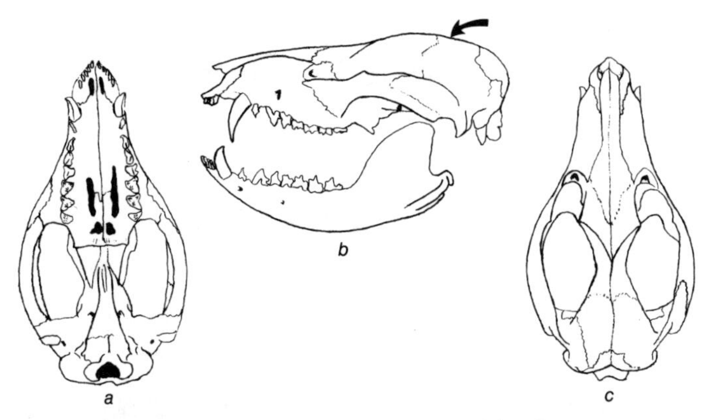

FIG. 8. Skull of the Opossum (*Didelphis virginiana*); *a*, ventral view; *b*, lateral view (arrow to sagittal crest); *c*, dorsal view.

dwelling insects but also eats small mice, birds' eggs, nuts, and berries.

Reproduction: The only mammal in California to reproduce without a well-formed placenta. Breeds any time, but most mating occurs from January to July. Gestation lasts only some twelve and a half days. Minute young, about the size of a bumblebee, crawl to the mother's pouch and virtually fuse to a nipple. Young remain in the pouch about two months. They nurse for approximately one month more during which period they make excursions from the pouch and also take some solid food. Sexually mature in the first year and may produce two litters of four to ten young annually.

The generic name (*Didelphis* = two-womb) refers to the complete separation of the uterus into two uterine horns that lead from the genital opening. The penis is divided at its apex, as well. When mating, each half of the bifid penis enters a uterine horn.

INSECTIVORA

Insectivores are mostly small, mouse-sized, rather primitive mammals with fine and sharply pointed teeth. These teeth serve well in the insectivore's probing search for insect food, and the molars have sharp cutting edges with which these little mammals crush the exoskeleton of their prey. In California there are two families of insectivores, the shrews (Soricidae) and the moles (Talpidae). These two families can be separated by the following key.

1. Eye orifice small but distinct, and eye clearly visible; ear pinna clearly present; forefeet somewhat narrower than hind feet; teeth with some red pigment; zygomatic arch absent . . . . Soricidae

   Eye orifice minute or grown over with skin; ear pinna absent; forefeet broader than hind feet; teeth white; zygomatic arch weak and delicate but present . . . . . . Talpidae

Shrews (Soricidae)

Shrews are rather delicate, mouselike insectivores of a dull brown or gray color. They are immediately distinguished from mice by the canine nature of the upper incisor teeth and by the presence of a wine-red pigment on most of the teeth (in California species). The snout is long and the skull lacks the zygomatic arch, which is present in moles.

Shrews have a distinctive life cycle. The young are born in the spring and rarely become sexually mature before late winter of the following year. Sexual maturity is indicated by a rapid increase in body weight and reproductive structures, and mating occurs in late winter or early spring. In many species, there is but a single litter. Shortly after the young are weaned, the adults die, so that by midsummer virtually the entire population consists of young born during the previous few weeks.

Reference

Junge, J. A., and Hoffmann, R. S. 1981. *An annotated key to the long-tailed shrews (Genus Sorex) of the United States and Canada, with notes on Middle American Sorex.* Occasional Papers of the Museum of Natural History, The University of Kansas, No. 94.

Key to Genera of Shrews (Soricidae) in California

1. Upper jaw with a row of three unicuspid teeth (see fig. 9b) . . . . *Notiosorex*

   Upper jaw with a row of five unicuspid teeth (see fig. 13a) . . . . *Sorex*

Desert Shrew (*Notiosorex crawfordi*) [pl. 1e]

Description: A small gray or brown shrew with a short tail; teeth faintly reddish. It is distinct from species of long-tailed shrews (*Sorex* spp.) in having only three unicuspid teeth in the upper jaw (fig. 9b); in *Sorex* there are four or five. Ears large for a shrew. TL 81–90, T 24–26, HF 9–11, E 8–9. Weight: 3.0–5.0 g.

Distribution: The Desert Shrew is a rarely encountered resident of the southernmost part of the state. Most likely to be found in areas of scattered scrub oaks, juniper, sycamore, and cottonwood; also in more arid habitats of sagebrush and associated shrubs. East to western Oklahoma and south to central Mexico. (See p. 343.)

Food: Presumably a generalized diet of insects in the wild, eating whatever large terrestrial insects it can find. Subsists on a broad variety of insects in captivity.

Reproduction: A litter of three to five young. Pregnancies from April to November. Possibly two or more litters a year.

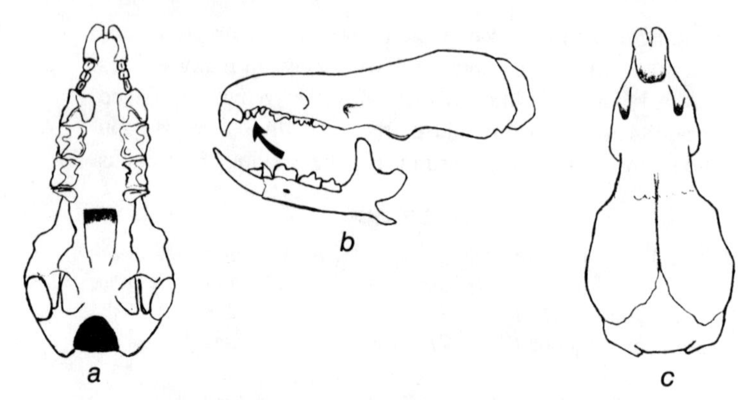

FIG. 9. Skull of the Desert Shrew (*Notiosorex crawfordi*): *a*, ventral view; *b*, lateral view (arrow to unicuspid teeth); *c*, dorsal view.

Long-tailed Shrews (*Sorex* spp.)

Long-tailed shrews (*Sorex* spp.) are a group of closely related and frequently very similar insectivores. Their identification rests largely on details of their dentition; body color and tail color are useful characters, but they are not always reliable in determining specific recognition. Some species are known from very few specimens and virtually nothing is known of their biology. The comments on these species are, therefore, regrettably skimpy. For some common species, details of the life history are rather well known.

Key to Species of Long-tailed Shrews (*Sorex* spp.) in California

1. Inner side of lower jaw with a postmandibular canal (fig. 10a); inner (median) surface of unicuspid teeth without a pigmented ridge (fig. 11a) . . . Subgenus *Sorex* (2)

   Inner side of lower jaw without a postmandibular canal (fig. 10b); inner (median) surface of unicuspid teeth with a pigmented ridge (fig. 11b) . . . Subgenus *Otisorex* (3)

2. Third upper unicuspid larger than fourth (fig. 12a); up-

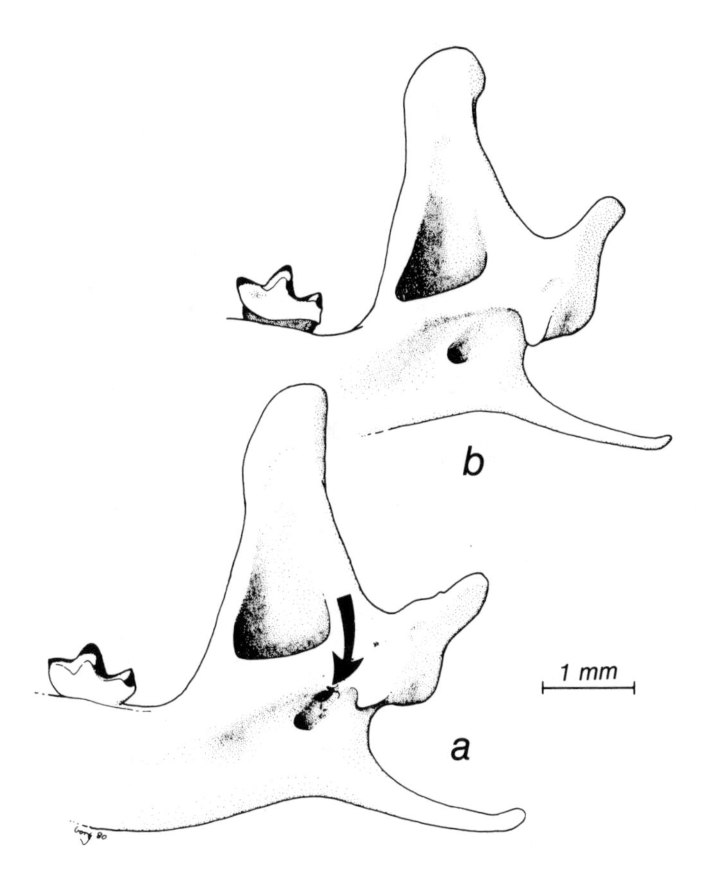

FIG. 10. Lower jaws of long-tailed shrews (genus *Sorex*); *a*, subgènus *Sorex*, showing the characteristic postmandibular foramen (arrow); *b*, subgenus *Otisorex*, which lacks a postmandibular foramen. (From Junge and Hoffmann, 1981.)

per incisor lacking a median tine (or minute lobe) when viewed anteriorly (fig. 12b); sagebrush and semiarid regions . . . . *merriami*

Third upper unicuspid smaller than fourth (fig. 13a); upper incisor with a distinct median tine when viewed anteriorly (fig. 13b); montane forested regions . . . . *trowbridgii*

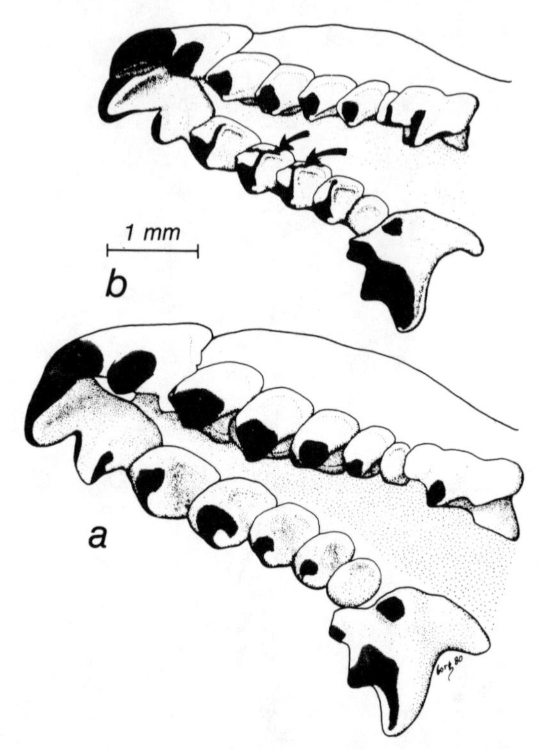

FIG. 11. Inner surfaces of upper tooth rows of long-tailed shrews (*Sorex*): *a,* subgenus *Sorex,* with unicuspid teeth lacking pigmented ridges; *b,* subgenus *Otisorex,* with pigmented ridges (arrows) on unicuspid teeth. (From Junge and Hoffmann, 1981.)

3. Third upper unicuspid smaller than fourth 4
   Third upper unicuspid larger than fourth . . . . . *lyelli*

4. Skull length greater than 19.0 mm 5
   Skull length less than 19.0 mm 7

5. First and second unicuspids equal or nearly equal; color brown dorsally 6
   First unicuspid smaller than second (fig. 14a); hind foot with distinct lateral fringe of stiff hairs; color gray or black dorsally . . . . *palustris*

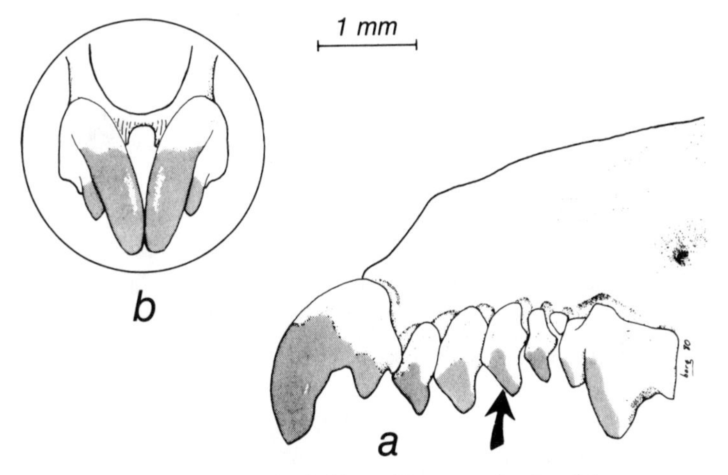

FIG. 12. Teeth of Merriam's Shrew (*Sorex merriami*): *a,* lateral view, showing the large third unicuspid (arrow); *b,* anterior view of upper incisors, showing lack of median tine (compare with fig. 13*b*). (From Junge and Hoffmann, 1981.)

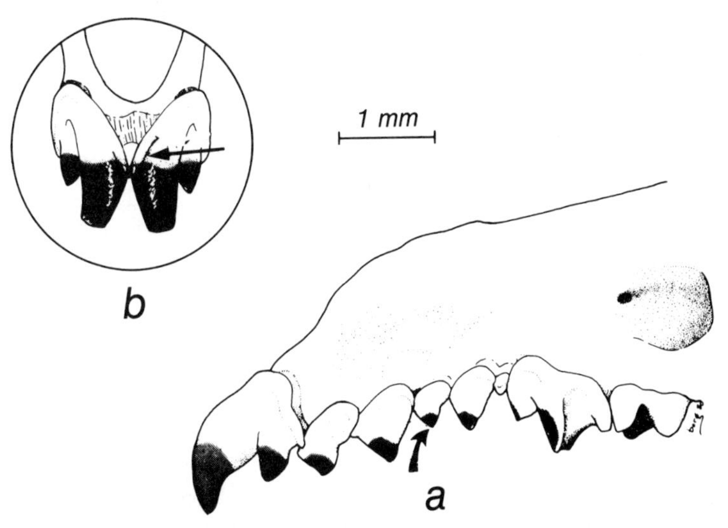

FIG. 13. Teeth of Trowbridge's Shrew (*Sorex trowbridgii*): *a,* lateral view, showing the small third unicuspid (arrow); *b,* anterior view of upper incisors, with median tine (arrow). (From Junge and Hoffmann, 1981.)

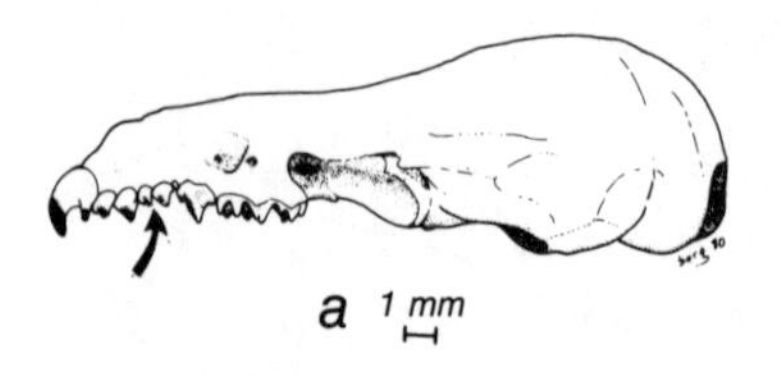

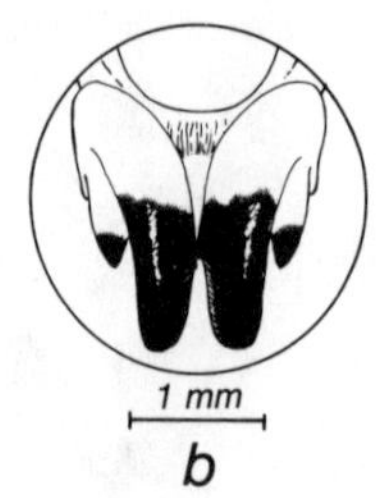

FIG. 14. Skull of the Water Shrew (*Sorex palustris*): *a,* the first unicuspid smaller than the second, and the third unicuspid (arrow) smaller than the fourth; *b,* anterior view of upper incisors. (From Junge and Hoffmann, 1981.)

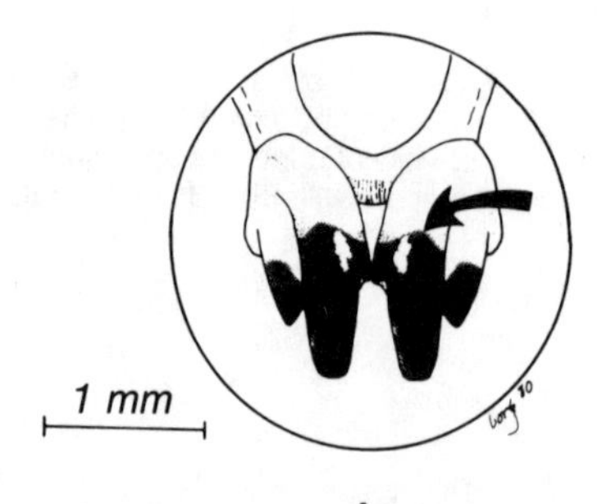

FIG. 15. Anterior views of upper incisors: *a,* Vagrant Shrew (*Sorex vagrans*); *b,* Montane Shrew (*Sorex monticolus*). Arrows show extent of pigment on incisors. (From Junge and Hoffmann, 1981.)

6. Upper incisor with a median tine or lobe; extreme northwestern California . . . . *bendirii*

   Upper incisor without a median lobe or tine; coastal forests north of San Francisco Bay . . . . *pacificus*

7. Pigment on anterior surface of upper incisor not extending dorsally to median tine or lobe; northern half of California, mostly in wet grasslands (fig. 15a) . . . . *vagrans*

   Pigment on anterior surface of upper incisor extending above median tine or lobe (fig. 15b) 8

8. Skull length 15.4 mm or more 9

   Skull length 15.3 mm or less; arid regions south of Lake Tahoe . . . . *tenellus*

9. Cranium rather flat (fig. 16a); skull length 15.4–17.0 mm; third unicuspid conspicuously smaller than fourth . . . 10

   Cranium convex (fig. 16b); third unicuspid very slightly smaller than fourth; Sierra Nevada, San Gabriel, and San Bernardino mountains . . . . *monticolus*

10. Color very dark brown or black; salt marshes at north end of San Pablo Bay, Suisun Bay, and Grizzly Island . . . . *sinuosus*

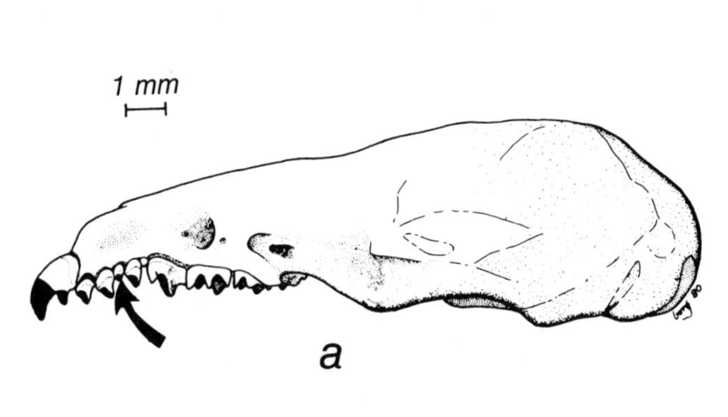

FIG. 16. Lateral views of skulls: *a,* Ornate Shrew (*Sorex ornatus*), with the third unicuspid (arrow) conspicuously smaller than the fourth (note the flatter skull); *b,* Montane Shrew (*Sorex monticolus*): with the third unicuspid (arrow) only slightly smaller than the fourth (note the convex skull). (From Junge and Hoffmann, 1981.)

Color light brown; Central Valley to coast, including Santa Catalina Island . . . . *ornatus*

Marsh Shrew (*Sorex bendirii*)

Description: A large, dark brown shrew, slightly modified for an aquatic life. It lacks the fringe of stiff hairs on the hind foot, which identifies *S. palustris;* pelage of *bendirii* is dull, but it is shiny in *palustris*. Upper incisor with a median tine; third unicuspid smaller than fourth. Ventral color nearly as dark as back. TL 145–170, T 60–80, HF 18–21, E 7–9. Average weight: 16 g.

Distribution: Northwestern corner of California, south to Gualala. Near standing water; sometimes along creeks. North to British Columbia, along the coast. (See p. 343.)

Reproduction: At least one litter of four to six young (usually five) from March to June.

The Marsh Shrew is not well known but presumably feeds on aquatic insects that occur in the lowland marshes.

Mt. Lyell Shrew (*Sorex lyelli*)

Distribution: Known only from high elevations in the southern Sierra Nevada. (See p. 344.)
HF 11–12, E 6.

Distribution: Known only from high elevations in the southern Sierra Nevada. (See p. 344.)

Reproduction: Little known. One record indicated four embryos in June.

This shrew has been collected few times, and very little is known other than region of occurrence.

Merriam's Shrew (*Sorex merriami*)

Description: A small species of *Sorex* with dark red teeth and no accessory tines. Tail conspicuously bicolored. Second unicuspid largest and fourth smaller than the third (see fig.

12a). TL 99–107, T 33–42, HF 11–13, E 8–9. Weight: 4.4–6.5 g.

Distribution: Essentially a Great Basin species; in California found in the high-elevation sagebrush and piñon–juniper along the eastern border. Throughout the Great Basin from eastern Washington to western North Dakota and Nebraska. (See p. 344.)

Food: Small spiders, orthopterans, and such soil-dwelling insects as beetle larvae and cutworms.

Reproduction: From five to seven young in a litter; probably only one litter a year. Pregnancies known to occur from late March to June.

Montane Shrew (*Sorex monticolus*)

Description: A rather small shrew with a brown summer coat and a darker, grayer color in the winter. It resembles *S. vagrans,* but in contrast to *vagrans* the pigment on the anterior surface of the upper incisor extends above the median tine; cranium convex (see fig. 15b). TL 111–120, T 46–55, HF 13–15, E 6–7.

Distribution: Sierra Nevada and San Bernardino Mountains. From central Mexico through the Rocky Mountains north to Alaska. (See p. 344.)

Food: Many small insects, especially soil-dwelling larvae. Little vegetable food.

Reproduction: From four to eight young in a litter. Pregnancies occur from April or May; possibly only one litter.

This shrew is well named, for it is truly a montane species.

Ornate Shrew (*Sorex ornatus*)

Description: A rather small, dull brown shrew with a faintly bicolored tail. Some populations around the north shore of San Pablo Bay are very dark brown. Most similar to *S. vagrans* and *S. monticolus* but separable on the basis of the characters in the key (see fig. 16a) and the geographic distribution of the three species. TL 89–108, T 32–44, HF 12–13, E 6–7. Weight: 3–7 g.

Distribution: Central Valley and Coast Ranges south to Baja California. Typically in rather open areas. (See p. 345.)

Food: Small soil-dwelling insects.

Reproduction: Breeds from February until the adults die in early summer; from three to five young in a litter.

This is the shrew commonly found in the Central Valley, and other species are not likely to occur with it. Along lower elevations of the western slope of the southern Sierra Nevada it may occur with *Sorex monticolus*.

Pacific Shrew (*Sorex pacificus*) [pl. 1b]

Description: A large shrew distinctive by its reddish, light brown color and also by the absence of a median tine on the upper incisor. First and second unicuspids equal or nearly so; third unicuspid smaller than fourth. Tail is almost uniformly colored. TL 120–158, T 45–65, HF 13–18, E 7–8. Weight: 14–18 g.

Distribution: Coniferous coastal forest and brushy area from Marin County north to the central Oregon coast. Mostly near creekside thickets. (See p. 344.)

Food: Known to feed heavily on centipedes and spiders; also eats slugs and snails.

Reproduction: Three to five young in a litter; probably more than a single litter a year.

This large, brightly colored shrew is rather common along the north coast in woodlands and streamside thickets but is, nevertheless, not very well known.

Water Shrew (*Sorex palustris*) [pl. 1d]

Description: A large black or dark brown shrew with a distinctive dense velvety fur that is water-repellent. Third unicuspid smaller than fourth; upper incisor lacking median tine. Hind feet with a fringe of rather stiff whitish hairs on sides. TL 144–158, T 73–78, HF 18–21, E 6. Weight: 8–14 g. (See fig. 14a.)

Distribution: Throughout the Sierra Nevada, Cascades, and North Coast Ranges, from about 1300 m and above, near water. Much of the forested regions of United States and Canada north to Alaska. (See p. 343.)

Food: Mostly nymphal and larval stages of aquatic insects. Commonly mayflies, stoneflies, caddisflies, and some small crustaceans and occasionally small fish.

Reproduction: From four to seven young in a litter; possibly two litters a year.

This is the most aquatic of North American shrews. Its water-repellent fur retains a bubble of air about its body and enables it to float on the surface like a duck. People who have been fortunate enough to observe this animal report that it swims and dives with the agility of an otter. It is rarely found more than a meter from water and is sometimes common along small mountain streams.

Suisun Shrew (*Sorex sinuosus*)

Description: A small, dark brown or blackish shrew similar to *ornatus*. Separable on basis of color and geographic occurrence. TL 98–106, T 35–44, HF 11–13, E 7–8. Weight: 4.5–6.8 g.

Distribution: Restricted to salt marshes about San Pablo Bay, Suisun Bay, and Grizzly Island.

Food: Insects and small crustaceans.

Reproduction: Breeds from late February to early June; lactating females have been found in July. From two to nine young in a litter, based on counts of embryos. One or two litters a year.

It has been suggested that this shrew represents a very local subspecies of *Sorex ornatus*.

Inyo Shrew (*Sorex tenellus*)

Description: A small brownish shrew with a faintly bicolored tail. Resembles *S. vagrans,* but has pigment on the upper incisor extending above the median tine or lobe. Third unicuspid smaller than fourth. Skull length 15.3 mm or less. TL 85–106, T 32–42, HF 10–12, E 6–7. Weight: 4–8 g.

Distribution: The high arid region immediately east of the southern Sierra Nevada, southeast to southern Nevada. (See p. 345.)

Very little is known of this interesting species.

Trowbridge's Shrew (*Sorex trowbridgii*) [pl. 1c]

Description: Brown in summer and gray in winter; tail bicolored. Upper incisor with a distinct median tine; third unicuspid smaller than fourth (see fig. 13b). TL 95–132, T 40–62, HF 12–15, E 6–8. Weight: 4.8 g.

Distribution: Found in coniferous and mixed forests in northern two-thirds of state; absent from Central Valley. North to British Columbia. (See p. 345.)

Food: A generalized feeder that takes many kinds of beetles, moths, heteropterans, spiders, and centipedes. Also earthworms in fall and winter, when soil moisture brings these invertebrates to the surface. At times they feed heavily on conifer seeds, especially those of Douglas Fir, but generally they do not take much vegetable food.

Reproduction: Sexual activity starts in February (in the northern Sierra Nevada) and continues until the latter half of May or gradually ceases as the adults die. Pregnancies as early as February; some females have at least two broods. From three to six young (most commonly five).

Like most shrews, this is an annual species, with a complete turnover of generations every year. The adults die rather rapidly at about one year of age, and by the end of the summer almost the entire population consists of young born that spring.

Vagrant Shrew (*Sorex vagrans*) [pl. 1a]

Description: A rather variable species that differs in appearance from place to place. Brown in summer and gray in winter; tail bicolored. Pigment on anterior surface of upper incisor not extending to median tine or lobe (see fig. 15a), third unicuspid smaller than fourth. TL 90–120, T 34–42, HF 11–12, E 6–8. Weight: 5–7 g.

Distribution: Found in Coast Ranges and Sierra Nevada, usually in grassy meadows and other moist open areas. East to Idaho and north to British Columbia. (See p. 346.)

Food: Opportunistic in feeding, taking small arthropods, earthworms, and slugs.

Reproduction: Breeds as early as January. From three to eight in a litter; perhaps two or more litters.

Like most other shrews, the Vagrant Shrew is an annual species; all adults die by the end of their second summer.

Moles (Talpidae)

These insectivores are fossorial (burrowing) mammals. They are most likely to be found in light and sandy soils and seem to avoid ground with a heavy clay content. Most species have greatly broadened forefeet for digging through the soil, although this specialization is less well marked in the Shrew-mole (*Neurotrichus gibbsii*), which tends to be shrewlike in habits. No external ear. Moles have three upper incisors per side, and all teeth are white. The skull has a zygomatic arch (fig. 17b).

Key to Species of Moles (Talpidae) in California

1. Forehand as wide or wider than long; tail less than 25 percent of total length . . . . *Scapanus* (2)

   Forehand longer than wide; tail length more than 25 percent of total length . . . . *Neurotrichus gibbsii*

2. Unicuspid teeth evenly spaced (figs 17b and 18b); color

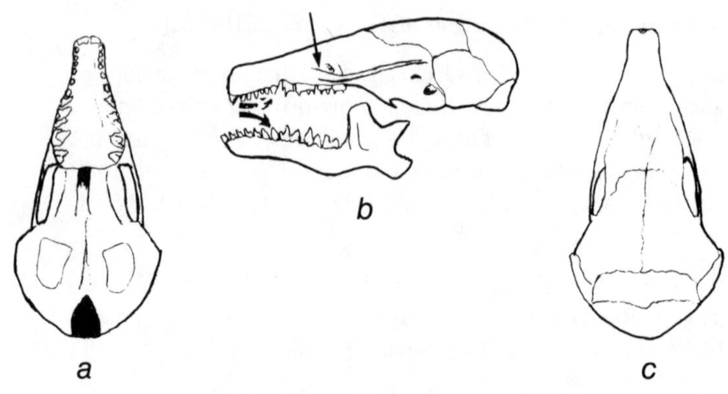

FIG. 17. Skull of Townsend's Mole (*Scapanus townsendii*): *a,* ventral view; *b,* lateral view (arrows to sublacrymal maxillary ridge and unicuspid teeth); *c,* dorsal view.

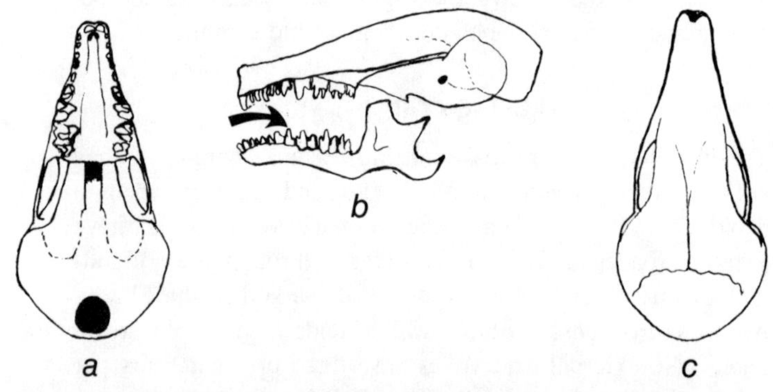

FIG. 18. Skull of the Coast Mole (*Scapanus orarius*): *a,* ventral view; *b,* lateral view (arrow to unicuspid tooth); *c,* dorsal view.

usually black; north coast north of San Francisco Bay 3

Unicuspid teeth unevenly spaced (fig. 19b); color usually light gray or brownish; throughout the state except in arid regions of the southeast and Central Valley . . . . *Scapanus latimanus*

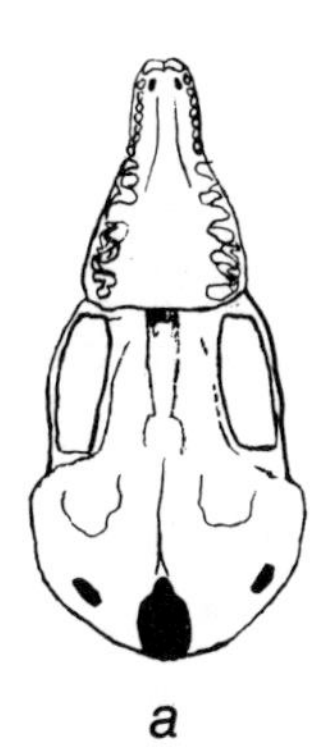

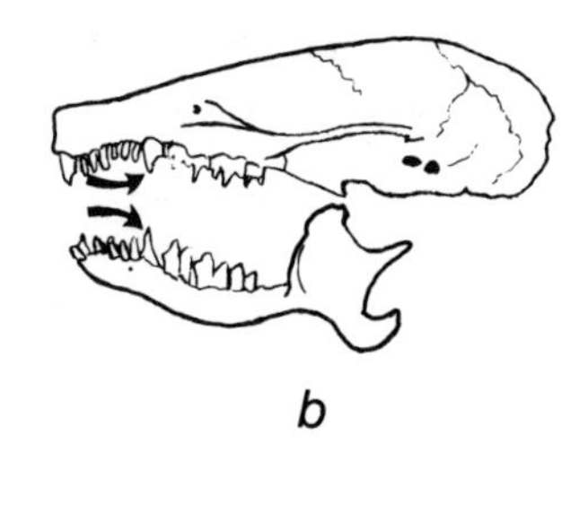

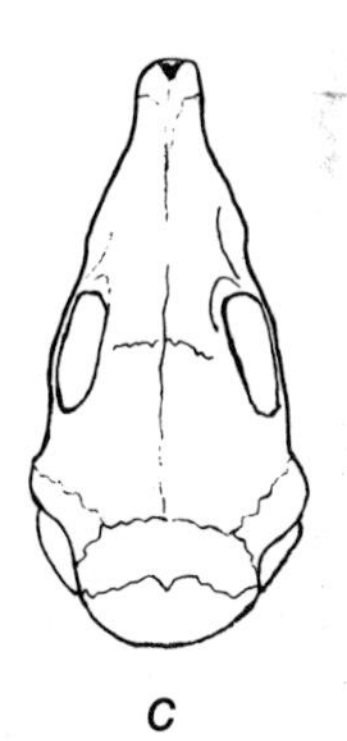

FIG. 19. Skull of the Broad-footed Mole (*Scapanus latimanus*): *a,* ventral view; *b,* lateral view (arrows to unicuspid teeth); *c,* dorsal view.

3. Total length usually more than 200 mm; skull with a distinct ridge (sublacrymal maxillary ridge) above maxillary region (see fig. 17b) . . . . *Scapanus townsendii*

   Total length less than 180 mm; skull lacking the distinct ridge (see fig. 18) . . . . *Scapanus orarius*

Shrew-mole (*Neurotrichus gibbsii*) [pl. 1f]

Description: A small, dark or black mole with forefeet only moderately expanded. The tail is distinctive, about one-third the total length, and constricted at the base. TL 110–125, T 33–45, HFD 15–18. Weight: 10–13 g.

Distribution: This little mole is found in many humid habitats; unlike moles of the genus *Scapanus,* it is not restricted to light and sandy soils. It occurs along the coastal redwood woodlands from Monterey Bay north to southern British Columbia and is found in the northern Sierra Nevada south at least to Plumas County. Unlike the other California moles, the Shrew-mole is frequently active aboveground and may forage either by day or by night. (See p. 346.)

Food: Like most insectivores, Shrew-moles seem to feed indiscriminately on a broad spectrum of soil-dwelling insects, pillbugs, and centipedes. At times they feed heavily on earthworms, and this may be their preferred food.

Reproduction: The Shrew-mole appears to breed in late winter, although there is little known about its reproductive patterns in California. There may be one to four young in a litter.

Broad-footed Mole (*Scapanus latimanus*) [pl. 1h]

Description: A light gray mole of medium size. Distinctive characters as in key and illustrations; unicuspid teeth unevenly spaced (see fig. 19b). TL 135–190, T 20–45, HF 18–25.

Distribution: Widely distributed in California up to at least 2000 m in the mountains. Favors light, sandy soils but is absent from heavily cultivated areas. Especially numerous on floodplains with a high soil moisture and a strong growth of forbs and soil invertebrates. Oregon south to Baja California and east to extreme western Nevada. (See p. 347.)

Food: Soil invertebrates, especially earthworms.

Reproduction: A single litter of two to four young, born in late winter.

This is the most widely distributed and commonly encountered mole in the state.

Coast Mole (*Scapanus orarius*)

Description: The smallest and most delicately built of California moles. Skull delicate; last upper unicuspid larger than that immediately in front of it; unicuspid teeth evenly spaced (see fig. 18b). Grayer than Townsend's Mole and sometimes brownish. TL 165–175, T 25–36, HF 20–23. Weight: 54–62 g.

Distribution: Coastal California north of San Francisco Bay; north to British Columbia and east to Idaho. (See p. 347.)

Food: Preys heavily on earthworms but probably takes most soil invertebrates.

Reproduction: A single litter of two to four young is born in late winter.

This mole may occur in gardens, but it is not the pest that Townsend's Mole can be.

Townsend's Mole (*Scapanus townsendii*) [pl. 1g]

Description: A large, black or dark gray mole. Last upper unicuspid about equal to that immediately anterior; unicuspid teeth evenly spaced (see fig. 17b). Distinguished from other California moles by its large size, dark color, and geographic distribution. TL 195–240, T 33–52, HF 24–28. Weight: 115–150 g.

Distribution: Extreme northwest corner of state. In rich, moist soil. Common in meadows and river valleys at low elevations. (See p. 347.)

Food: Soil invertebrates and sometimes underground parts of plants. It has been known to become a pest in commercial bulb plantings.

Reproduction: A single litter of two to four young is born in midwinter.

CHIROPTERA

Bats have been flying mammals for perhaps 60 million years or more and are highly specialized in both structure and behavior. The fingers are extremely long and slender, and they are connected by a thin, delicate membrane. The hind legs too are connected by a membrane in most species (and in all bats in California). This interfemoral membrane usually encloses the tail. Typically a small bone, the *calcar,* projects from the hind foot and extends the interfemoral membrane when the bat is in flight (see fig. 25). The eyes are greatly reduced, but there is evidence indicating that bats nevertheless do see well. The ears are large in many species; in some the ears are highly mobile and can be directed independently of one another. The inner ear of bats is sensitive to extremely high frequencies.

The pectoral muscles are large and strong, and the sternum is keeled, like the "breastbone" of a fowl, providing attachment for the pectoral muscles. Among the diverse kinds of bats there are different specializations for flight, and these are reflected in the wing structure. For this reason the length of the humerus, or forearm (labeled "F" in the keys), is of importance in identification of bats.

References

Allen, G. M. 1939. *Bats.* Cambridge: Harvard University Press.

Barbour, R. W., and Davis, W. H. 1969. *Bats of America.* Lexington: University Press of Kentucky.

Griffin, D. R. 1958. *Listening in the dark.* New Haven: Yale University Press.

Hill, J. E., and Smith, J. D. 1984. *Bats: A natural history.* London: British Museum (Natural History).
Kunz, T. H. (Ed.) 1982. *Ecology of bats.* 3 vols. New York: Academic Press.
Schober, W. 1984. *The lives of bats.* London: Croom Helm/Arco.
Slaughter, B. H., and Walton, D. W. 1970. *About bats: A chiropteran biology symposium.* Dallas: Southern Methodist University Press.
Wimsatt, W. A. (Ed.) 1970–1977. *Biology of bats.* 3 vols. New York: Academic Press.
Yalden, D. W., and Morris, P. A. 1975. *The lives of bats.* Newton Abbot: David & Charles.

Key to Families of Bats (Chiroptera) in California

1. Tail entirely (or almost entirely) contained within the interfemoral membrane; if tail tip projects beyond margin of interfemoral membrane, face has a leaflike appendage 2

   Tail projecting conspicuously beyond margin of interfemoral membrane; face without a leaflike appendage . . . . Molossidae

2. Face with a flat leaflike appendage . . . Phyllostomidae

   Face without a leaflike appendage . . . Vespertilionidae

Leaf-nosed Bats (Phyllostomidae)

A diverse group of bats, mostly in Central and South America. Distinguished by peculiarities of the skull and, in most species, by a distinctive "nose-leaf" on the tip of the snout. They are rather different from one another in wing structure, flight patterns, and diet. Some eat fruit while others specialize on nectar, pollen, or insects.

Key to Leaf-nosed Bats (Phyllostomidae) in California

1. Snout very long; upper molars with poorly defined W pattern (fig. 20a); tail short, not extending halfway to margin of interfemoral membrane . . . *Choeronycteris mexicana*

   Snout not elongate; upper molars with a distinct W pattern

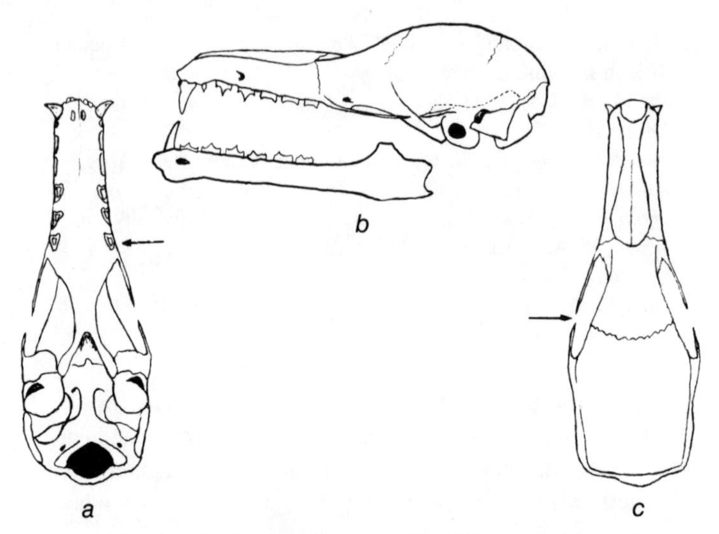

FIG. 20. Skull of the Hog-nosed Bat (*Choeronycteris mexicana*); *a,* ventral view (arrow to upper molar); *b,* lateral view; *c,* dorsal view (arrow to incomplete zygomatic arch).

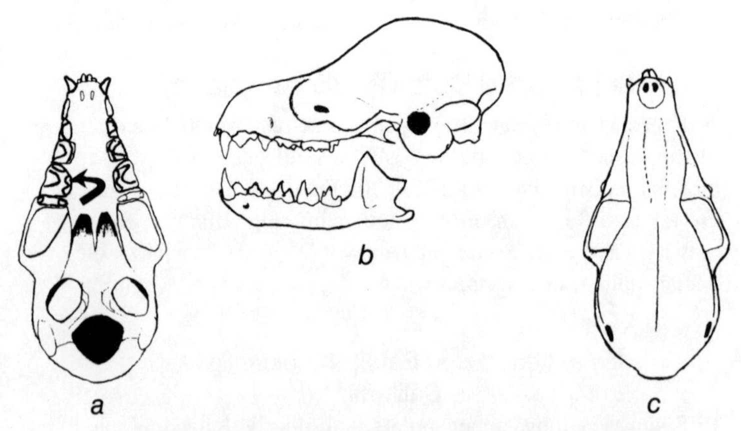

FIG. 21. Skull of the California Leaf-nosed Bat (*Macrotus californicus*): *a,* ventral view (arrow to molar); *b,* lateral view; *c,* dorsal view.

(fig. 21a); tail long, projecting slightly beyond interfemoral membrane . . . . *Macrotus californicus*

Hog-nosed Bat (*Choeronycteris mexicana*) [pl. 2l]

Description: A leaf-nosed bat with a very elongate snout bearing a horn-shaped appendage at the front. Skull with zygomatic arch incomplete (see fig. 20c). Color brownish gray. Tail reaching less than halfway to margin of interfemoral membrane. TL 81–103, T 5–10, HF 10–13, E 14–17, F 42–47. Weight: 12–21 g.

Distribution: Extreme southern California; from southern Arizona south through Mexico, including Baja California, to Guatemala. Sleeps in the darkest regions of caves and mines. (See p. 348.)

Food: The extremely long snout and extrudable tongue are adaptations for feeding on pollen and nectar of night-blooming flowers.

Reproduction: A single young is born in spring. Until the infant is rather large, it is carried by the mother when she forages.

California Leaf-nosed Bat (*Macrotus californicus*) [pl. 2i]

Description: Snout with a simple leaflike appendage at tip. Ears very large (more than 20 mm from crown to tip). Upper molars with a W pattern (fig. 21a). Color light chocolate brown, with ends of hairs conspicuously darker than underfur. Belly pale. Tail extends slightly beyond interfemoral membrane. TL 77–108, T 25–42, E 24–29, F 44–58. Weight: 10–14 g. (See fig. 21.)

Distribution: Extreme arid southern regions of the state. From southern Nevada and Arizona south through most of Mexico, including Baja California, and West Indies. Roosts in buildings and mines. (See p. 348.)

Food: Mostly large and heavy-bodied insects, such as noctuid moths, orthopterans, and scarabiid and carabid beetles.

Reproduction: A very unusual pattern of fertilization and developmental rate distinguishes this bat from other North American bats. Both mating and fertilization occur in the autumn (without either sperm storage or delayed fertilization). Embryonic development is slow for the first five months; but embryonic growth accelerates in the spring, and birth follows mating by approximately eight months. Ovulation is only from the right ovary and implantation is only in the right uterine horn.

Vesper Bats (Vespertilionidae)

This is the predominant family of bats in the Northern Hemisphere. It is a diverse family structurally, and most important features that characterize the group are skeletal details. The ears are large in some species and small in others, and the ears are joined or separate at the top of the head. The snout never has a leaflike appendage at the tip. The tail extends to, but never beyond, the margin of the interfemoral membrane. The anterior incisors are separated from each other by a toothless space medially, so that there is no anterior concavity on the roof of the mouth when viewed ventrally.

Almost all species are insectivorous. Two species (not present in California) capture small fish at the surface of quiet waters. A conspicuous peculiarity of their reproductive cycle is the phenomenon of sperm storage: most species mate in the autumn, and sperm remains viable in the upper oviduct of the female during hibernation. In the spring, shortly after arousal from hibernation, the females ovulate and the ova are fertilized by the stored sperm. In some species males hibernate separately from females. At least short migrations characterize most vespertilionid bats, and some kinds move hundreds of kilometers each year.

You are most likely to observe bats as they fly at dusk. Like many small mammals, bats are crepuscular—that is, they fly at dusk and again at dawn, and during the middle of the night they rest in a "nighttime roost." When they rest, they may pass scats from their evening meal. Nighttime roosts may be under the eaves of a porch roof, beneath which there may be pieces of insect wings and legs that the bats have removed as they feed.

Unless you encounter bats as they rest at night, the accumulation of scats and insect parts might remain a mystery.

Key to Genera and Species of Vesper Bats (Vespertilionidae) in California

1. With two pairs of upper incisors 5
   With one pair of upper incisors 2

2. Ears very long; interfemoral membrane nude; snout blunt with a ridge on side and top . . . . *Antrozous pallidus*
   Ears short; interfemoral membrane well furred dorsally; snout without a ridge on sides and top . . . *Lasiurus* (3)

3. Color reddish or brownish gray, tips of hairs white; two upper premolars, usually with an inner peglike upper premolar (fig. 22a) 4
   Color yellow or yellowish brown; one upper premolar; tips of hairs not white . . . . *Lasiurus ega*

4. Color brick-red; forearm less than 45 mm. . . . . *Lasiurus borealis*
   Color grayish brown; forearm more than 45 mm. . . . . *Lasiurus cinereus*

5. Color brown or blackish and more or less uniform; lower canine large and pointed 6

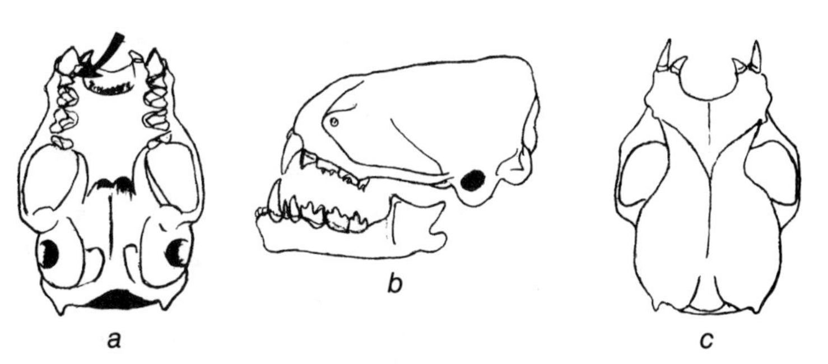

FIG. 22. Skull of the Hoary Bat (*Lasiurus cinereus*): *a*, ventral view (arrow to minute, inner premolar); *b*, lateral view; *c*, dorsal view.

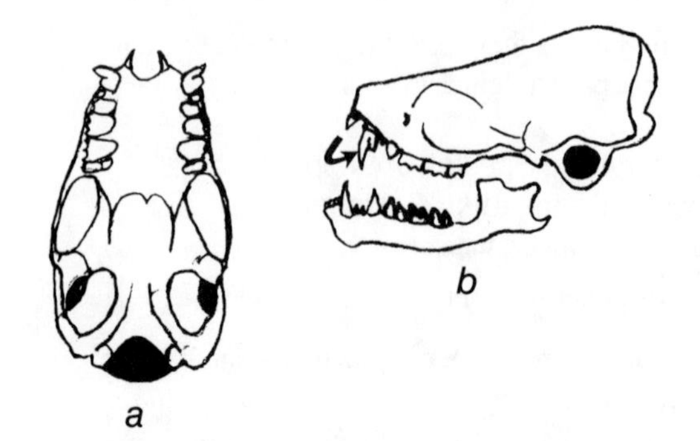

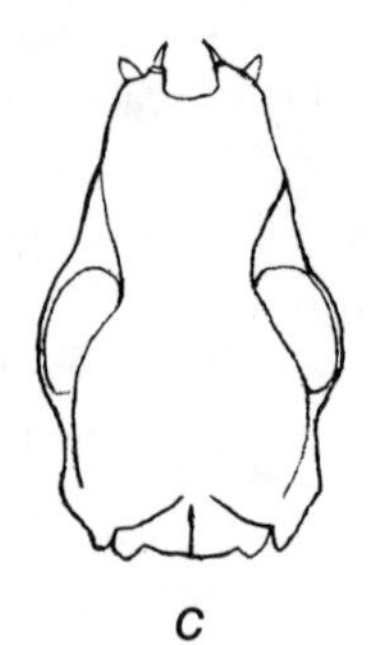

FIG. 23. Skull of the Silver-haired Bat (*Lasionycteris noctivagans*): *a*, ventral view; *b*, lateral view (arrow to bicuspid upper incisor); *c*, dorsal view.

Black with three large white spots on dorsum; lower canine small and appearing bifid laterally . . . . *Euderma maculatum*

6. With two lower premolars 9
   With three lower premolars 7

7. Ears short; color black, hairs tipped with white; interfemoral membrane furred; inner upper incisor bicuspidate (fig. 23a, b) . . . . *Lasionycteris noctivagans*
   Ears long; color brownish or olive; interfemoral membrane nude or scantily furred . . . . . . *Plecotus* (8)

8. With two large fleshy glands on side of muzzle . . . . *Plecotus townsendii*
   Muzzle without large glands . . . *Idionycteris phyllotis*

9. With two pairs of upper premolars 10
   With one pair of upper premolars (fig. 24b) . . . . *Eptesicus fuscus*

10. Tragus about one-third ear length; upper incisors about equal . . . . *Pipistrellus hesperus*
    Tragus more than one-half ear length; outer upper incisor clearly larger than the inner . . . . *Myotis* (11)

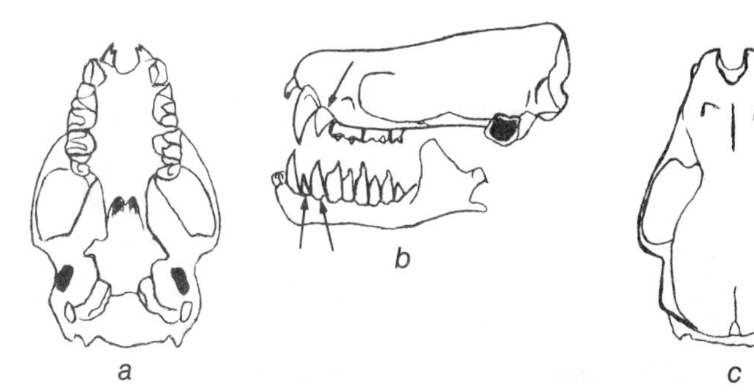

FIG. 24. Skull of the Big Brown Bat (*Eptesicus fuscus*): *a*, ventral view; *b*, lateral view (arrows to single upper premolar and two lower premolars); *c*, dorsal view.

11. Margin of interfemoral membrane with a fringe of fine hairs 12

    Margin of interfemoral membrane with fine hairs absent or extremely few 13

12. Forearm more than 40 mm; ear not extending more than 5 mm beyond muzzle when laid forward . . . . *Myotis thysanodes*

    Forearm less than 40 mm; ear long, extending more than 6 mm beyond muzzle when laid forward . . . . *Myotis evotis*

13. Ear short, not extending more than 2 mm beyond muzzle when laid forward 14

    Ear long, extending more than 6 mm beyond muzzle when laid forward . . . . *Myotis evotis*

14. Calcar keeled (fig. 25a) 15

    Calcar not keeled (fig. 25b) 17

15. Lower side of wing not furred to elbow; hind foot less than 8.5 mm; forearm usually less than 35 mm 16

    Lower side of wing with fur to elbow; hind foot more than 8.5 mm; forearm usually more than 35 mm . . . . *Myotis volans*

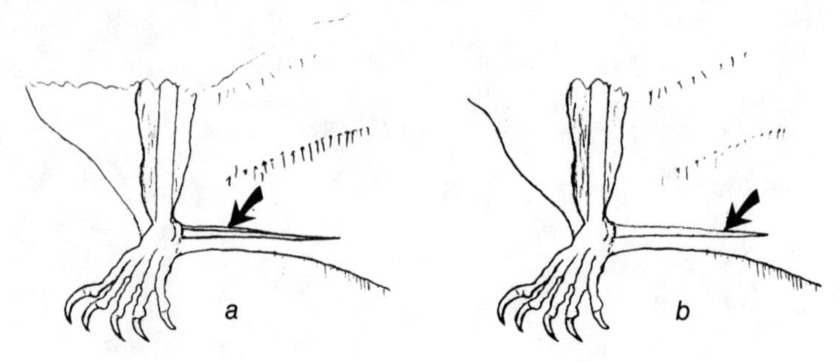

FIG. 25. Calcars of *Myotis* spp.: *a,* with a keel (arrow); *b,* without a keel (arrow). The presence of a keel is usually apparent in fresh specimens but may be difficult to see after the skin is dry.

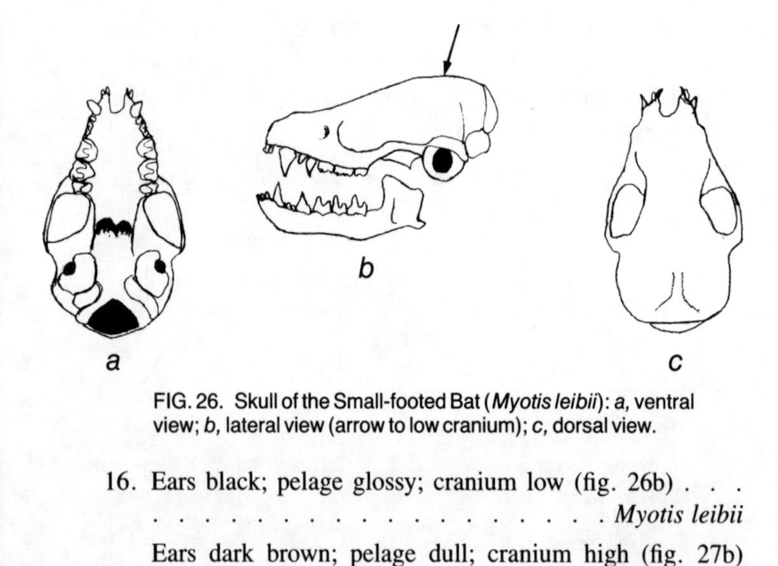

FIG. 26. Skull of the Small-footed Bat (*Myotis leibii*): *a,* ventral view; *b,* lateral view (arrow to low cranium); *c,* dorsal view.

16. Ears black; pelage glossy; cranium low (fig. 26b) . . . . *Myotis leibii*

    Ears dark brown; pelage dull; cranium high (fig. 27b) . . . . *Myotis californicus*

17. Forearm less than 40 mm; cranium without a distinct sagittal crest (figs. 28b and 29b) 18

    Forearm more than 40 mm; cranium with a sagittal crest (fig. 30b, c) . . . . *Myotis velifer*

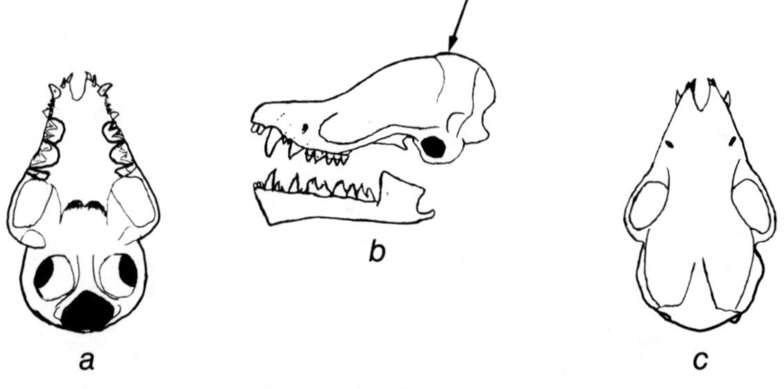

FIG. 27. Skull of the California Bat (*Myotis californicus*): *a*, ventral view; *b*, lateral view (arrow to high cranium); *c*, dorsal view.

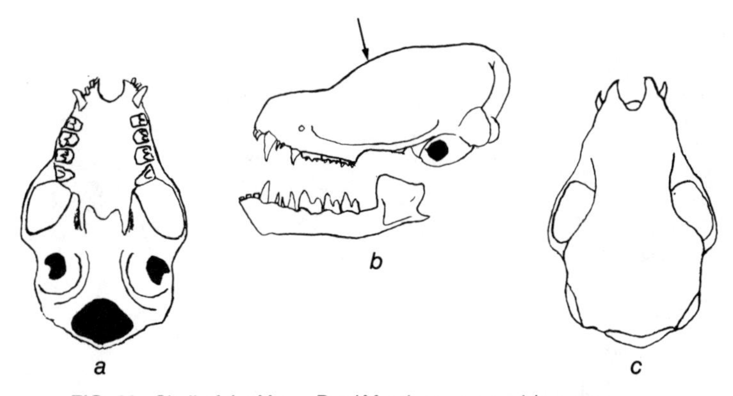

FIG. 28. Skull of the Yuma Bat (*Myotis yumanensis*): *a*, ventral view; *b*, lateral view (arrow to abrupt rise in braincase); *c*, dorsal view.

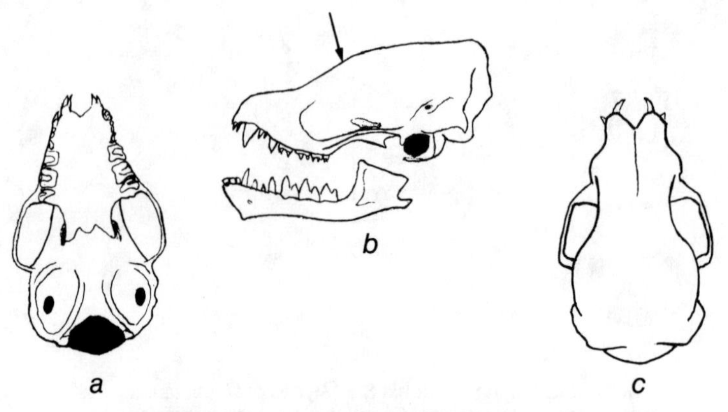

FIG. 29. Skull of the Little Brown Bat (*Myotis lucifugus*): *a,* ventral view; *b,* lateral view (arrow to gradual rise in braincase); *c,* dorsal view.

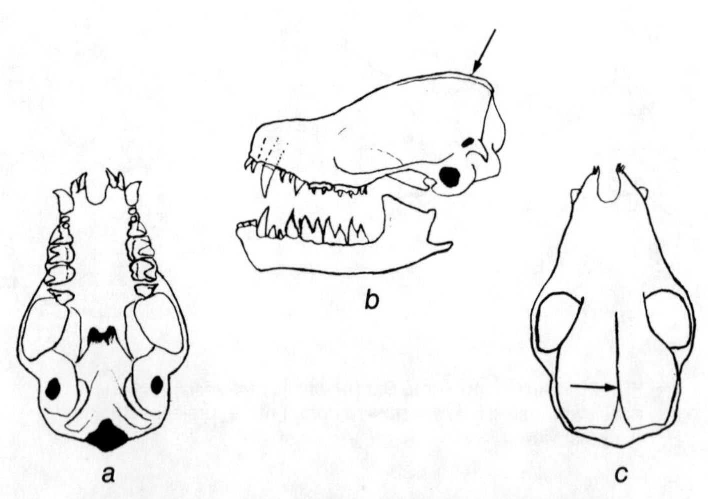

FIG. 30. Skull of the Cave Bat (*Myotis velifer*): *a,* ventral view; *b,* lateral view; *c,* dorsal view (arrows in *b* and *c* to sagittal crest). The sagittal crest may not be developed in young individuals.

18. Pelage dull; forearm 32–37 mm; braincase abruptly raised (see fig. 28b) . . . . *Myotis yumanensis*

    Pelage glossy; forearm 35–40 mm; braincase gradually elevated (see fig. 29b) . . . . *Myotis lucifugus*

Pallid Bat (*Antrozous pallidus*) [pl. 2c]

Description: A medium-sized bat with buff or sandy-colored fur. Ears very long and clearly separated at the base. (Ears are joined at base in Townsend's Long-eared Bat and Allen's Long-eared Bat; the California Leaf-nosed Bat has a triangular leaf-like appendage on the snout and the tail extends beyond the interfemoral membrane.) Hairs dark at the tips and light next to the skin. Only a single pair of upper incisors. TL 92–135, T 43–49, E 28–32, F 48–61. Weight: 16–19 g. Dentition: 1/2, 1/1, 1/2, 2/3.

Distribution: Most of California; especially common in open, lowland areas, generally below 2000 m. Much of western United States, from central Mexico to British Columbia. Makes local seasonal movements but apparently not migratory. (See p. 349.)

Food: Feeds largely on flightless insects, which it captures by foraging on the ground. Jerusalem crickets, scorpions, and June beetles figure largely in the diet of this bat.

Reproduction: Mating occurs in autumn; sperm stored in reproductive tract of female. Frequently collects in small nursery colonies. From one to three embryos have been found, but litters consist of two young (less commonly a single young). Birth occurs in late June. The female hangs upside down during labor, and the young drop into a sack formed from the interfemoral membrane.

Big Brown Bat (*Eptesicus fuscus*) [pl. 2h]

Description: Larger than the species of *Myotis* in California. Distinguished by a combination of features: ears and wing membranes are nude and darkly pigmented, color brownish and rather glossy; two lower premolars and a single upper premolar (see fig. 24b). TL 105–120, T 39–51, E 13–19, F 39–45. Weight: 14–23 g.

Distribution: A widespread and common species. Frequently enters buildings, sometimes invading attics and abandoned buildings in numbers. Enters caves, especially in winter, and tends to remain near the entrance. From Colombia and Venezuela and north to Canada, including the Greater Antilles, and east to Atlantic coast.

Food: An early-flying species, usually appearing long before the sky is dark. Captures beetles and dipterans.

Reproduction: Mating occurs in August or September; sperm stored in reproductive tract of female. Ovulation in April, and a single young is born in June.

The Big Brown Bat hibernates in cold weather, but it is much hardier than most species and not infrequently forages in winter evenings.

Spotted Bat (*Euderma maculatum*) [pl. 2k]

Description: A black bat with huge ears and three large white spots dorsally. Throat with a large, apparently nonglandular, nude area. Quite unlike any other California bat. TL 107–115, T 47–50, E 44–50, F 48–51. Weight: 13–18 g. Dentition: 2/2, 1/1, 2/2, 3/3.

Distribution: Known from southern California, from both montane open coniferous forests and low deserts. Primarily a cave-dwelling species. From central Mexico north to southern Montana and east to Texas. (See p. 349.)

Food: Seems to favor noctuid moths; known also to take terrestrial insects.

Reproduction: A single young born in late May or early June; nursing continues into August.

Allen's Long-Eared Bat (*Idionycteris phyllotis*) [fig. 31]

Description: Medium-sized bats with very long ears, joined at the top of the head. A pair of flaps from ears project over forehead and snout. Color pale tawny, with base of hairs dark brown. Muzzle with only faintly developed dermal glands. Calcar with a strong keel (see fig. 25a). TL 103–108, T 46–55, E 38–43, F 39–49. Weight: 7–17 g. Dentition: 2/3, 1/1, 2/3, 3/3.

Distribution: Montane forests of oaks and pines. Roosts in caves and mineshafts. From central Mexico to southern Nevada, Arizona, and New Mexico. Probably enters adjoining area of California, but not yet known to occur in our state.

Food: Emerges late in the evening and forages for flying insects. Calls loudly while hunting.

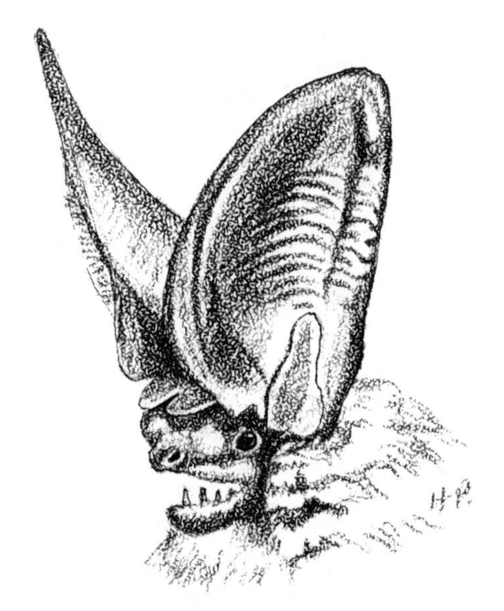

FIG. 31. Allen's Long-eared Bat (*Idionycteris phyllotis*).

Reproduction: A single young born in June or July. Known to form nursery colonies.

This species has long been placed in the genus *Plecotus*. Primarily on the basis of chromosomal morphology, the species *phyllotis* is considered to be more closely related to *Euderma maculatum* than to those species currently placed in *Plecotus*. Its placement in the genus *Idionycteris,* and as the sole member of that genus, is not universally accepted.

Silver-haired Bat (*Lasionycteris noctivagans*) [pl. 2d]

Description: Dorsal fur dark brown or black with hair conspicuously white-tipped, giving a frosted appearance. Ear short and broad. Interfemoral membrane well furred on basal half. Inner incisor (of upper jaw) strongly bicuspidate (see fig. 23b). TL 92–115, T 35–45, E 11–14, F 38–44. Weight: 6–12 g. Dentition: 2/3, 1/1, 2/3, 3/3.

Distribution: Common in forested areas in most of the northern half of the state. A tree-dwelling species but sometimes enters buildings. From extreme northeastern Mexico north to Alaska and east to Atlantic coast. (See p. 350.)

Food: Flying insects of many orders; fond of moths.

Reproduction: Mating occurs in September or October; sperm stored in oviducts until ovulation the following spring. After a gestation of fifty to sixty days, a litter of two young is born. Young fledged in about three weeks and weaned shortly thereafter.

Genus *Lasiurus*

These colorful bats have sometimes been called "hairy-tailed bats," for most or all of the dorsal surface of the interfemoral membrane is well furred. The underside of the wings is also densely covered with hair along the bones and over muscles and tendons. Ears are short and rounded; the margin is generally nude and darkly pigmented. Some species of this genus are highly migratory and make extensive flights over the oceans.

They typically roost in dense foliage of trees and are rarely found in concentrations. Dentition: 1/3, 1/1, 1–2/2, 2/2.

Unlike most bats, these species have two pairs of mammae and have litters of two to four (usually three or four) young.

Key to Species of *Lasiurus* in California

1. Usually with a small, peglike premolar next to the inner margin of upper incisor (fig. 22a); dorsal pelage russet or brownish, but hair white-tipped 2

   Without a small, peglike premolar; dorsal pelage an even yellow-brown, hairs not white-tipped . . . *Lasiurus ega*

2. Color reddish or russet; forearm 37–44 mm . . . . *Lasiurus borealis*

   Color brownish; forearm 50–56 mm. . . . . *Lasiurus cinereus*

Red Bat (*Lasiurus borealis*) [pl. 2f]

Description: Dorsal color basically bright russet with tips of hairs white. TL 98–110, T 38–43, E 8–11, F 37–44. Weight: 9–15 g.

Distribution: Statewide. Most commonly encountered in August and September when migrating. Found in and near deciduous trees. North to Canada, east to Atlantic coast, and south to Chile and Argentina; absent from most of Great Basin and Rocky Mountain region.

Food: Flies early in the evening, well before dark, foraging at decreasing heights as the sky becomes darker. Feeds frequently on moths and is also known to take terrestrial insects.

Reproduction: Mating occurs in late summer (August–September); sperm stored in oviducts until ovulation the following spring, as in many other vespertilionid bats. Usually three or four young born in June. Young become independent at about four to five weeks of age.

Hoary Bat (*Lasiurus cinereus*) [pl. 2g]

Description: A dull, chocolate brown with dorsal hairs white-tipped, giving a frosted appearance. With upper peglike premolar (see fig. 22). Easily separated from the Red Bat by its color and larger size. Quite unlike any other bat in our state. TL 126–143, T 48–63, E 9–14, F 50–56. Weight: 23–27 g.

Distribution: Generally distributed in wooded areas of California. Throughout the United States and Canada, migrating north to Hudson Bay in summer; south to Chile and Argentina. Populations also on Hawaiian Islands and Galápagos. Visits the Farallon Islands on migration.

Food: Begins to forage early in the evening long before dark. Captures many moths and beetles.

Reproduction: Usually two young born in June or July. Newborn young may cling to a twig while the mother forages in the evening, but the mother is capable of carrying both infants when moving them to a new roosting site.

This is a relatively common species, but it does not occur in aggregations. It has been seen flying in the daytime in September. In winter, it sometimes roosts on the vertical trunks of trees, even 3 or 4 ft (1 m) from the ground, and then closely resembles a protruding part of the tree.

When picked up alive, individuals of this species produce a most startling rattling hiss accompanied by an impressive show of teeth.

Western Yellow Bat (*Lasiurus ega*) [fig. 32]

Description: A medium-sized insectivorous bat with light yellow fur and dorsal surface of interfemoral membrane furred on anterior half. Ear short. TL 109–124, T 45–50, HF 8–11, E 11–13, F 43–52. Weight: 12–19 g.

Distribution: Extreme southwestern deserts. Sometimes found roosting in dense foliage of palms. Arizona south to Argentina and Uruguay. (See p. 350.)

FIG. 32. Western Yellow Bat (*Lasiurus ega*).

Food: Emerges late in the evening; feeds on flying insects.

Reproduction: Two or three young born in late June.

Genus *Myotis*

This is one of the more difficult groups of small mammals to identify. Their distinguishing features are most significant when used in combination—that is, in making an identification you must consider not only general color and form (and the species are remarkably similar) but also cranial features. In most species the tail vertebrae extend to the margin of the interfemoral membrane (or uropatagium), but in some it may project slightly beyond. The calcar is a small bone extending from the hind foot and supporting the interfemoral membrane; in some species of *Myotis* the calcar may be clearly keeled while in others it is rounded in cross section (see fig. 25). As in many bats, there is a small projection at the base of the ear, the tragus, and this may be an important diagnostic feature.

There is no single common name for bats of this genus, and they are sometimes referred to simply by their generic name, such as the Long-legged Myotis (for *Myotis volans*). For many years, however, *Myotis lucifugus* has been called the Little Brown Bat; although this name is familiar to generations of American mammalogists, the genus consists of many species of "little brown bats." We have tried to adopt the most familiar and descriptive common name for each species, but the realistic solution to this problem is to learn the scientific names.

California Bat (*Myotis californicus*) [pl. 2b]

Description: Close to *M. leibii* and definitely separable only on the basis of the outlines of their skulls (see figs. 26b and 27b). A small, buff-colored species with medium-sized ears, which barely extend beyond the tip of the snout when laid forward. Ears and face darkly pigmented. TL 78–87, T 35–40, E 11–15, F 29–36. Weight: 3–5 g.

Distribution: Occurs statewide; roosts in buildings, caves, and mine shafts. In California associated with oak woodlands and also juniper–piñon areas of the desert; more characteristic of lower elevations, not commonly entering mountains or inhabiting coniferous forests. From central Mexico to British Columbia east to the Rocky Mountains.

Food: Forages for flying insects early in the evening, often hunting only 5 or 10 ft (2 or 3 m) above the ground.

Reproduction: A single young born between late May and early July.

Recovery of banded individuals of this species reveals that it may live up to fifteen years in the wild.

Long-eared Bat (*Myotis evotis*) [pl. 2a]

Description: Dorsum a pale golden color. Ears long, extending 7 mm beyond tip of snout when laid forward. Ears tend to be darkly pigmented. Interfemoral membrane sometimes with a few marginal hairs, but much less conspicuous than in the Fringed Bat. Sagittal crest weak or absent. TL 84–95, T 38–41, E 18–23, F 35–42. Weight: 4–8 g.

Distribution: Ranges over the entire state, often in montane forests, seldom in large numbers. Western North America from Mexico to Canada.

Food: A late-flying species, emerging after dark. Forages

low, from 4 to 6 ft (1 to 2 m) above the ground. Captures flying insects.

Reproduction: A single young is born in June. Apparently does not form nursery colonies.

Small-footed Bat (*Myotis leibii*)

Description: Light or golden brown dorsally; tips of hairs tend to be glossy. Ears and face frequently black. Similar to *M. californicus,* but separable on the outlines of their skulls (see figs. 26 and 27). Ear scarcely extending beyond tip of snout when laid forward. Calcar with a keel (see fig. 25a). TL 60–90, T 32–45, E 12–15, F 28–36. Weight: 3–6 g.

Distribution: Most of California except coastal redwood region. Usually solitary; not known to be colonial. From central Mexico north to southern Canada and east to Rocky Mountains; also eastern United States.

Food: Emerges rather early in the evening. Captures small moths, beetles, and other flying insects.

Reproduction: A single young is born in June or July. Nursery colonies apparently do not occur in this species.

Little Brown Bat (*Myotis lucifugus*)

Description: This is perhaps the best-known member of the genus in North America and may serve as a basis of comparison for the identification of other species. The rather glossy fur is dark brown or blackish. The ears are not large and extend about to the tip of the snout when laid forward. The calcar is rounded and without a keel (see fig. 25b). The skull lacks a sagittal crest (see fig. 29b). TL 80–93, T 31–39, HF 8–10, F 35–41. Weight: 7–10 g.

Distribution: This bat is found throughout California except for the southwest region of the state. It ranges over much of North America from northern Mexico to far into Alaska and east to Newfoundland. It is frequently abundant in buildings.

Food: As in many species of the genus, the diet depends

largely on availability, and the Little Brown Bat feeds on a variety of flying insects. Some workers report that it has a penchant for adult forms of flying insects, especially aquatic species, such as midges.

Reproduction: This bat, like virtually all local species of the family Vespertilionidae, mates in the fall, usually September or October. Ovulation and fertilization occur the following spring, for the sperm are stored in the upper region of the uterine tract if not in the oviduct itself; the single young is born about sixty days later. Bats normally roost with the head directed downward; during the birth of the young this position is reversed, however, and the infant is dropped into a "basket" the mother forms with her interfemoral membrane. When very small the young bat clings to its mother while the latter forages at night. At about three weeks of age the young is capable of flight and is weaned shortly thereafter.

Much of the research on bats has been conducted on this species. Movements and mortality have been studied by placing number bands on individuals. By this technique individuals have been found to live up to thirty-one years.

Fringed Bat (*Myotis thysanodes*)

Description: Dorsum brown with a distinct cinnamon aspect. Interfemoral membrane with a delicate fringe of fine hairs. Ears of medium length extending some 3 to 5 mm beyond tip of snout when laid forward. Skull with a sagittal crest. Calcar without a keel (see fig. 25b). TL 80–91, T 34–42, E 16–19, F 40–66. Weight: 3–6 g.

Distribution: Throughout California; most frequent in coastal and montane forests and about mountain meadows. British Columbia east to the Rocky Mountain states and south to Mexico.

Food: Presumably a broad variety of flying insects. Beetles and moths seem to be favored.

Reproduction: A single young is born in early summer (late

June to early July). Forms nursery colonies in caves or old buildings.

Cave Bat (*Myotis velifer*)

Description: A large *Myotis* with small ears; dorsum brown or black and dull (not glossy). Forearm unusually long. Calcar not keeled (see fig. 25b). The mature skull has a distinct sagittal crest (see fig. 30b, c). TL 90–104, T 39–47, HF 9–10, E 13–16, F 35–47. Weight: 9–13 g.

Distribution: In caves and buildings along the Colorado River; occasionally in large colonies. East to Oklahoma and Kansas and south to Honduras. (See p. 350.)

Food: Feeds frequently over streams and ponds; seems to favor beetles and small moths. Emerges rather late in the evening, commonly after dark.

Reproduction: As in other species of the genus, mating occurs in late summer or autumn and sperm are stored in the reproductive tract of the female until ovulation the following spring. A single young is born in mid-June or July. Capable of flight at about three weeks of age; weaning follows a few days later.

Long-legged Bat (*Myotis volans*)

Description: Dorsal fur tawny to dark brown, hairs rather long. Base of interfemoral membrane and wing membrane furred. Ears short, reaching about to tip of snout when laid forward. Cranium rather highly elevated with a poorly developed sagittal crest. Calcar is keeled (see fig. 25a). TL 89–98, T 34–45, E 10–14, F 35–42. Weight: 5–7 g.

Distribution: Throughout California, in both forested regions and brushy areas, up to 2500 m in deserts and 3300 m in the mountains. Roosts in buildings, trees, and crevices in cliffs. Western United States from northern British Columbia south to Central Mexico.

Food: Emerges early, long before the sky is dark, and presumably specializes on early-flying nocturnal insects.

Reproduction: A single young is born in very late spring or early summer.

Yuma Bat (*Myotis yumanensis*)

Description: Dull brown or buff dorsally. Skull with a conspicuous rise but without a sagittal crest (see fig. 23b). Calcar lacks a keel (see fig. 25b). TL 82–88, T 32–38, E 11–15, F 32–38. Weight: 6–8 g.

Distribution: Throughout California; especially common along wooded canyon bottoms. A colonial species roosting in caves and old buildings in aggregations up to 2000 individuals. Common in deserts of southeastern California. Central Mexico (including Baja California) north to British Columbia and east to Colorado and Texas.

Food: Flying insects, especially small moths, beetles, and midges.

Reproduction: A single young born in spring or early summer. Prior to birth of the young, females segregate into nursery colonies, occasionally of large size.

Western Pipistrelle (*Pipistrellus hesperus*)

Description: A very small bat with yellowish or dull pelage. Ears short and darkly pigmented. Calcar keeled (see fig. 25a). TL 69–81, T 25–31, E 10–11, F 29–32. Weight: 3–5 g. Dentition: 2/3, 1/1, 2/2, 3/3.

Distribution: Found throughout California, often common in open arid areas at lower elevations. Solitary or in small colonies. Hibernates in caves but may emerge sporadically in winter. Central Mexico north to Washington and east to Oklahoma.

Food: Emerges early in the evening, long before the sky is dark, and also flies at dawn. Flies close to the ground and takes small moths, beetles, and dipterans.

Reproduction: Usually two young born in late June. Nursery colonies have been found.

Because of its early emergence in the evening, this bat is commonly seen. It can be recognized by its rather erratic flight.

Townsend's Long-eared Bat (*Plecotus townsendii*)

Description: Medium-sized light brown bat with very long ears. Ears joined at top of the head. Large glandular swellings on side of the snout. Belly brown. (See also Pallid Bat, *Antrozous pallidus.*) TL 89–112, T 35–54, E 30–41, F 39–47. Weight: 7–12 g. Dentition: 2/3, 1/1, 2/3, 3/3.

Distribution: A cave-dwelling species but also found in buildings. Desert scrub and piñon–juniper associations. Females often in small groups, but males tend to be solitary. Ranges over all of the state, including the Channel Islands. From Mexico to British Columbia and the Rocky Mountain states; some parts of the central Appalachians. (See p. 351.)

Food: Emerges late in the evening; favors small moths.

Reproduction: Mates in autumn and a single young is born in May or June in California. Nursery colonies have been found in caves, mine shafts, and buildings.

Free-tailed Bats (Molossidae)

This family is mostly a tropical group of bats which has several species that enter temperate regions. The species in California have short, dense fur of a dark chocolate color. The interfemoral membrane is naked (or only thinly provided with hair) and distinctly thick and leathery. The tail lies within the interfemoral membrane, but the tip of the tail projects beyond the margin of the membrane—hence the name "free-tailed."

Key to Species of Free-tailed Bats (Molossidae) in California

1. Size not large (total length less than 140 mm); anterior margin of bony palate with a marked indentation between upper incisors 2

   Size large (total length greater than 140 mm); upper incisors virtually in contact with each other . . . . *Eumops perotis*

2. Ears long, extending conspicuously beyond muzzle when laid forward; inner bases of ears connected to each other . . . . *Nyctinomops*

   Ears shorter, not extending much beyond muzzle when laid forward; inner bases of ears not connected to each other . . . . *Tadarida brasiliensis*

3. Forearm longer than 57 mm . . . *Nyctinomops macrotis*

   Forearm shorter than 50 mm . . . . *Nyctinomops femorosacca*

Western Mastiff Bat (*Eumops perotis*) [pl. 2m]

Description: A large bat with broad, truncate ears that are joined across the top of the head. Fur short and dull; gray or dark brown, hairs white or very light at base. TL 155–185, T 35–45, E 27–32, F 73–82. Dentition: 1/2, 1/1, 2/2, 3/3. Weight: 80–100 g.

Distribution: In caves and buildings in southern half of state; usually open, arid areas with high cliffs. Roosts in crevices in small colonies. Makes local seasonal movements but remains in California throughout the year. Southern Arizona to Texas and south into Argentina; Cuba; distribution discontinuous. (See p. 351.)

Food: Emerges when the evening light has nearly disappeared. Forages very high (600–700 m), usually over mesquite, where it catches strong flying insects such as dragonflies, moths, beetles, and hymenopterans.

Reproduction: Unlike many species of bats, both sexes remain together throughout the year, including the period of birth of the young. Mating in early spring. From one to two young born anytime between late June and September.

This is a very vocal bat that calls continuously while hunting.

Pocketed Free-tailed Bat (*Nyctinomops femorosaccus*) [fig. 33]

Description: A medium-sized free-tailed bat, dark brown or gray in color; base of hairs distinctly lighter than tips. Ears

FIG. 33. Pocketed Free-tailed Bat (*Nyctinomops femorosaccus*).

joined at the base. TL 100–108, T 38–43, E 21–24, F 45–49. Dentition: 1/2, 1/1, 2/2, 3/3. Weight: 9–13 g.

Distribution: Southern part of state in desert areas. Roosts in buildings and also in crevices in high, vertical rock outcrops. From southern Arizona east to Texas and south to central Mexico, including Baja California. (See p. 352.)

Food: Flies late in the evening; captures flying insects, especially moths and beetles.

Reproduction: A single young is born in late June or July. Nursing may continue into September.

Big Free-tailed Bat (*Nyctinomops macrotis*) [fig. 34]

Description: Recognized by its large ears, which extend far beyond tip of snout when laid forward; ears united at the base. Fur glossy, reddish brown to black; hairs distinctly lighter at the base. TL 108–141, T 48–61, E 26–29, F 57–62. Dentition: 1/2, 1/1, 2/2, 3/3. Weight: 12–19 g.

Distribution: Apparently an uncommon resident in piñon–juniper regions of the arid parts of the state. Associated with high cliffs and rocky outcrops, where it roosts in crevices. East

FIG. 34. Big Free-tailed Bat (*Nyctinomops macrotis*).

to Kansas and south to Uruguay. Known to move northward to British Columbia in late summer.

Food: Captures moths and beetles.

Reproduction: A single young born in June or July; nursing may continue to August. Aggregate in nursing colonies.

This bat is known to be very vocal in flight.

Guano Bat (*Tadarida brasiliensis*) [fig. 35; pl. 2n]

Description: A small, chocolate brown free-tailed bat; fur is dull and hairs are not usually much (if any) lighter at the base. Bases of ears close together but definitely not joined. Ears extend no more than slightly beyond tip of snout when laid forward. Anterior margin of bony palate concave (fig. 35a). TL 88–112, T 31–41, E 14–20, F 36–46. Dentition: 1/3, 1/1, 2/2, 3/3. Weight: 8–14 g.

Distribution: The most common free-tailed bat in California. Assembles in large numbers on occasion, in buildings, caves, or mine shafts, from sea level to 1500 m or more in the mountains. Migratory but some numbers present in California throughout the year. Across southern United States to Atlantic coast, south through northern South America, including the Greater Antilles.

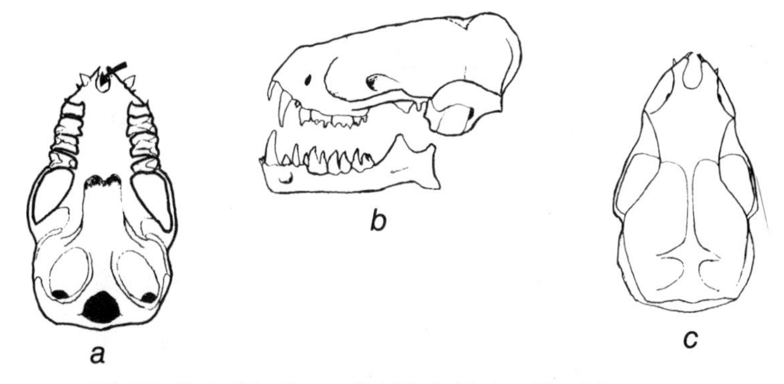

FIG. 35. Skull of the Guano Bat (*Tadarida brasiliensis*): *a,* ventral view (arrow to indentation of bony palate); *b,* lateral view; *c,* dorsal view.

Food: Emerges early in the evening and is a very rapid flyer. Feeds largely on small moths and is said to consume a gram per night.

Reproduction: Breeds in late winter and gestation takes about 100 days. Development occurs almost exclusively in the right uterine horn; rarely does implantation occur in the left side. A single young is born from late June to early July. Large nursery colonies are formed, and the mother–young relationship seems not to be firm; a lactating female allows any of the young to take milk and seems unable to distinguish her own young from others.

The Guano Bat is so named because of the excavation and sale of the guano in the Carlsbad Cave of New Mexico. These deposits were up to 30 m in depth and sustained a commercial guano mine for the first twenty years of this century.

This species is also known as the Brazilian Free-tailed Bat, referring to its specific name. When its specific name was *Tadarida mexicana,* it was called the Mexican Free-tailed Bat. Such name changes serve no useful purpose.

CARNIVORA

Most carnivores are clearly modified for a diet of meat and the capture of living quarry. Their dentition is distinctive in having the incisor teeth rather small and the canine teeth conical and long for holding and tearing meat (fig. 4). The molars are variously shaped and usually serve to shear or crush bones. As in meat eaters in general, carnivorous mammals have a rather short, simple gastrointestinal tract.

Major groups of carnivores are classified on the basis of the structure of the foot. Bears and Raccoons, for example, walk on the palms or plantar surface of the feet and are therefore called *plantigrade*. In contrast, dogs and cats walk on their toes and are called *digitigrade*. Weasels (Mustelidae) are somewhat intermediate, for some species are plantigrade and others digitigrade. Seals, sea lions, and walruses are appropriately designated *pinnipeds*—literally fin-feet—although they do not constitute a natural group.

Carnivores tend to be solitary except during the mating season. In the dogs (Canidae), however, both parents tend to care for the growing young. Even when mature, there is a strong bond in a dog family, which sometimes remains as a hunting group.

References

Andersen, H. T. (Ed.) 1969. *The biology of marine mammals*. New York: Academic Press.

Barnes, C. T. 1960. *The cougar or mountain lion*. Salt Lake City: Ralton Co.

Bekoff, M. 1978. *Coyotes: Biology, behavior and management.* New York: Academic Press.
Dobie, J. F. 1950. *The voice of the coyote.* Boston: Little, Brown & Co.
Ewer, R. F. 1973. *The carnivores.* Ithaca: Cornell University Press.
Grinnell, J.; Dixon, J. S.; and Linsdale, J. M. 1937. *Fur-bearing mammals of California: Their natural history, systematic status and relation to man.* Berkeley: University of California Press.
Harrison, R. J.; Hubbard, R. C.; Peterson, R. S.; Rice, C. E.; and Schusterman, R. J. 1968. *The behavior and physiology of pinnipeds.* New York: Appleton, Century, Crofts.
Kenyon, K. W. 1969. *The sea otter in the eastern Pacific Ocean.* North American Fauna 68. Washington, D.C.: U.S. Government Printing Office.
King, J. E. 1983. *Seals of the world.* London: British Museum (Natural History); Ithaca: Cornell University Press.
Leydet, F. 1977. *The coyote: Defiant songdog of the west.* San Francisco: Chronicle Books.
Powell, R. A. 1982. *The fisher: Life history, ecology and behavior.* Minneapolis: University of Minnesota Press.
Ridgway, S. H., and Harrison, R. J. 1981. *Handbook of marine mammals.* 2 vols. London: Academic Press.
Scheffer, V. B. 1958. *Seals, sea lions and walruses.* Stanford: Stanford University Press.
Verts, B. J. 1967. *The biology of the striped skunk.* Urbana: University of Illinois Press.
Young, S. P., and Goldman, E. A. 1946. *The puma: Mysterious American cat.* New York: Dover Publications Inc.

Dogs, Foxes, and Allies (Canidae)

Dogs and their relatives are somewhat generalized carnivores that are adapted for running. They are light-bodied and have rather slender legs. They are digitigrade (they walk on their toes), and their claws are nonretractile. The snout is elongate, eyes rather large, and ears erect. Their molar teeth are formed for both shearing and crushing bones, indicating a dietary position somewhat between cats and bears (figs. 36–38).

Coyote (*Canis latrans*) [pl. 3a]

Description: A medium-sized, rangy, doglike carnivore. Color gray, sandy, or brown. With the approximate form of a German shepherd. Most Coyotes weigh from 8 to 20 kg. Some

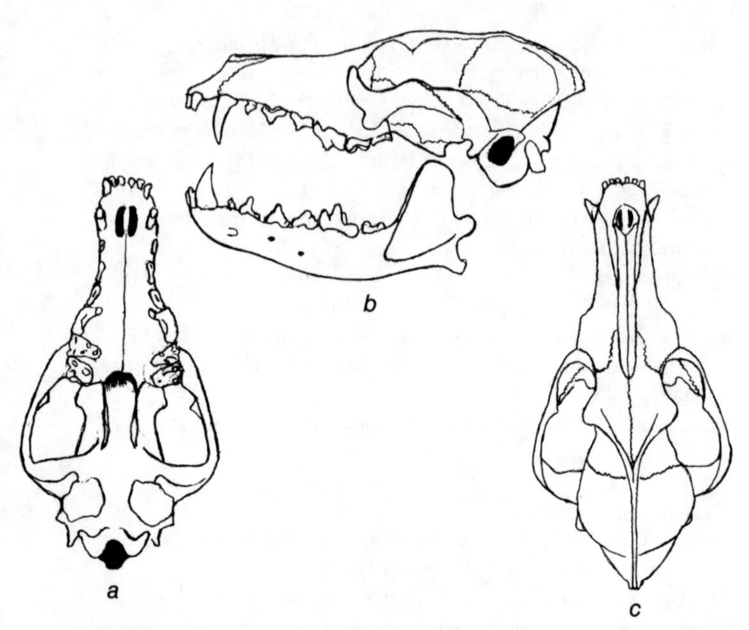

FIG. 36. Skull of the Coyote (*Canis latrans*): *a*, ventral view; *b*, lateral view; *c*, dorsal view.

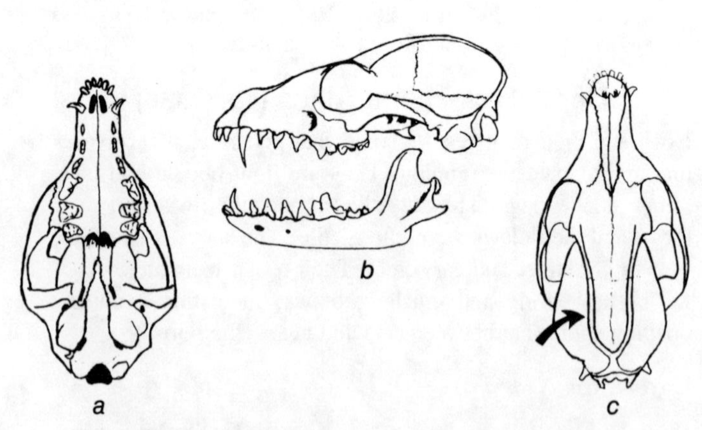

FIG. 37. Skull of the Gray Fox (*Urocyon cinereoargenteus*): *a*, ventral view; *b*, lateral view; *c*, dorsal view (arrow to sagittal crest).

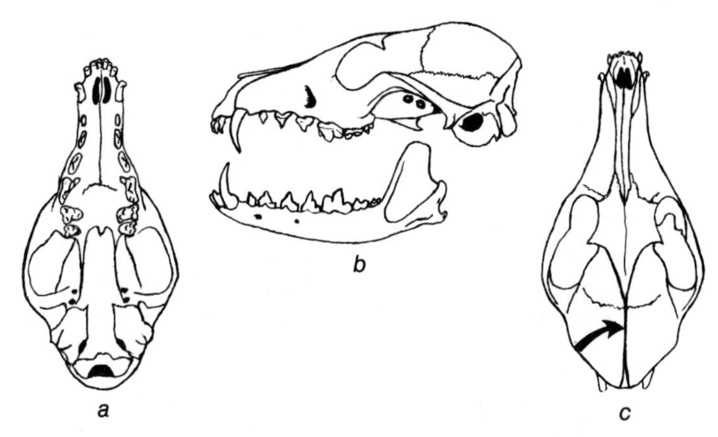

FIG. 38. Skull of the Red Fox (*Vulpes vulpes*): *a,* ventral view; *b,* lateral view; *c,* dorsal view (arrow to sagittal crest).

extremely large individuals, weighing from 25 to 33 kg, could be mistaken for Wolves. This species has been known to mate with domestic dogs; the hybrid, called a "coydog," is sometimes found in the wild. TL 1300–1700, T 300–400. (See fig. 36.)

Distribution: Throughout the state, but typically in open country. It can occur almost anywhere and even invades the outskirts of Los Angeles. Common in the sagebrush plains of the Great Basin. Most common west of the Mississippi River but in the last forty years has spread to many regions to the east, being found on the Atlantic coast in some areas. From Alaska and central Canada south to Panama.

Food: Preys extensively on both jackrabbits and cottontails. This fare is supplemented with small mice and ground squirrels. The Coyote also eats fruits, berries, insects, and carrion.

Reproduction: Mates in February in California; from five to ten pups born some two months later. The Coyote is very prolific and well able to sustain its numbers in the face of many decades of persistent hunting and poisoning. Both parents remain with the young until autumn. Family groups tend to disperse in early winter when food becomes scarce. In some montane populations there is a tendency to move downhill in autumn.

Versatility is the key to the Coyote's success. Its major conflict with human society stems from the penchant some individuals develop for domestic stock, especially sheep. Once such a habit is formed, it is not lost, and such individuals do much harm.

Coyotes are sometimes seen "mousing," at which time they may be quite oblivious to human observers. One may encounter pairs, or even groups of five animals, representing perhaps parents and grown young.

Gray Fox (*Urocyon cinereoargenteus*) [pl. 4a]

Description: A silvery gray fox with conspicuous patches of yellow, brown, or white on the throat and belly. Tail with a mid-dorsal mane of stiff hairs not present on other foxes (except the Island Fox). The Gray Fox has rather short legs, perhaps an adaptation for its tree-climbing tendencies. Skull with distinct sagittal crests which end in a lyre-shaped flare posteriorly (see fig. 37b, c). TL 800–1125, T 275–443, HF 100–150. Weight: 3–5 kg, occasionally to 7 kg.

Distribution: By far the most common and widespread fox in California. Survives well in cultivated land, in chaparral, and in forested areas. Sometimes dens under farm outbuildings or even under suburban homes. Throughout most of the United States except the northwest and Rocky Mountains; south through most of South America.

Food: Small rodents, birds, and berries; also insects and fungi. Their climbing ability allows them to obtain a greater variety of foods than is available to other foxes.

Reproduction: Mating in late winter is followed by birth of three to five young in April or May. Larger litters may represent combined broods of two females that have denned together. The male remains with the female while the young are dependent.

This is a very beautiful fox that has a rather good quality fur. Because of its attractive colors, abundance, and wide distribution, the Gray Fox is an important furbearer.

Like others of the dog family, the Gray Fox is playful at times. One was observed tossing a dried cow-pie into the air, chasing after it, and flinging it up again, whether by mouth or paw was not clear.

Island Fox (*Urocyon littoralis*)

Description: A diminutive replica of the Gray Fox, which it closely resembles in every way but size. TL 590–780, T 110–290, HF 98–157. Weight: 2.0–2.2 kg.

Distribution: On six of the Channel Islands, where they are generally abundant.

Food: Omnivorous, taking more plant and insect food than does the Gray Fox. It feeds extensively on berries such as manzanita, toyon, saltbush, prickly pear, and ice plant. Mice are also eaten.

Reproduction: Mates in February and March; a litter, usually of two kits, is born in late April and May. Dens in a hole in the ground or a hollow tree. Both parents care for the young, and the family forages as a group.

This little fox does not conflict with human activity and apparently does not suffer much from the presence of humans. Its main competitor is the domestic house cat. The Island Fox, although presumably nocturnal, is not infrequently seen in the daytime.

Kit Fox (*Vulpes macrotis*) [pl. 4b]

Description: A small gray fox with exceptionally large ears. Color rather uniform except for a black tip to the tail. Appreciably smaller than the Red Fox, which has small ears, is reddish, and has a white tip to the tail. The Kit Fox differs also from the Gray Fox, which has much shorter legs and is usually a mixture of red and gray. The Kit Fox lacks the sagittal crests of the Red and the Gray. TL 730–840, T 260–325, HF 113–137, E 78–94. Weight: 1.7–2.5 kg.

Distribution: Open, arid regions of southern part of the state. The range much restricted in recent years, for the Kit Fox picks up poisoned baits left out for Coyotes. Intensive agricultural use of much of the San Joaquin Valley renders the habitat unsuitable for the Kit Fox. One or more relict populations persist in Contra Costa County. From south central Oregon east to west Texas and New Mexico and south to central Mexico, including Baja California. (See p. 352.)

Food: Various small rodents, but especially kangaroo rats. Also mice and small squirrels, lizards, insects, and berries of wild shrubs. Brush Rabbits sometimes form an important part of the diet of this little carnivore.

Reproduction: Mates in winter; three to five young born in February or March. The female spends much of her time in the den when the young are small. Some family groups consist of one male with two females and their offspring. The young disperse in autumn.

This delicate little fox does not conflict with human activities, but it has declined with the expansion of agriculture and intensification of predator control. The Kit Fox is believed to increase in areas where the Coyote is reduced in numbers. This species has been confused in the literature with *Vulpes velox,* which does not occur in California.

Red Fox (*Vulpes vulpes*) [pl. 4c]

Description: A rather bright reddish fox with a white tip to the tail. Color variations include an all black or melanistic "silver fox" and a brown and gray "cross fox," but they are all, nevertheless, Red Foxes. The skull is distinctive in having the sagittal crests coming to a point posteriorly (see fig. 38c) in contrast to the lyre-shaped sagittal crests of the Gray Fox (see fig. 37c). TL 875–1000, T 340–390, HF 140–165. Weight: 5–8 kg.

Distribution: Two populations of the Red Fox are widely separated in California, one in the higher elevations of the Sierra Nevada and another in some areas of the Sacramento Valley. Only the montane Red Fox is native to California. In the nineteenth century a population of the eastern Red Fox was introduced into the lowlands of the state. These animals are believed to have descended from foxes that were either released or escaped from fur farms, and they most closely resemble Red

Foxes from the northern Central Plains states. The Sierran population lives at 1500 m and above. This species is found throughout North America and also Eurasia. (See p. 352.)

Food: Small rodents, birds, berries, and insects for the bulk of the diet of this species. It can capture birds up to the size of a Pheasant or Mallard.

Reproduction: Little is known of the breeding of the Red Fox in California. A litter of five to ten kits is born in early spring. As in other members of the dog family, the male assists the female in providing food for the young.

Bears (Ursidae)

Bears are large, heavy-bodied, almost tailless carnivores. Limbs are stout and claws are heavy, blunt, and nonretractable. Bears are plantigrade; that is, they walk on the palmar and plantar surfaces of their feet. Despite their bulk and short limbs, Black Bears climb well.

Their molar teeth are flat-crowned—shaped for crushing as are the teeth of swine and humans, not for cutting as in cats. The crushing molars of bears reflect their omnivorous diet (fig. 39).

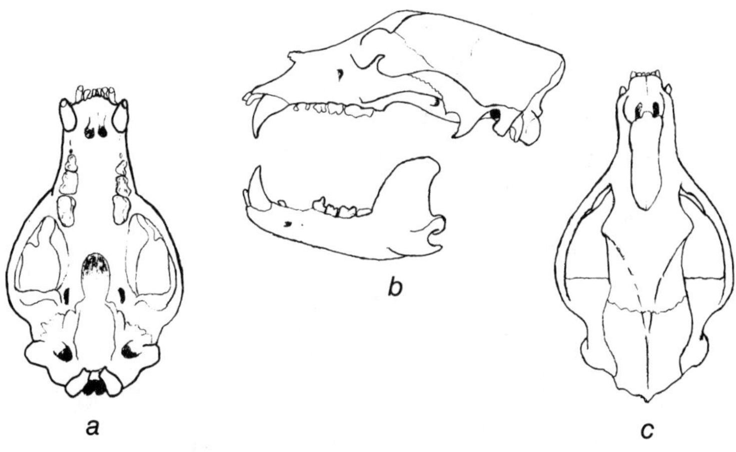

FIG. 39. Skull of the Black Bear (*Ursus americanus*): *a*, ventral view; *b*, lateral view; *c*, dorsal view.

There are three species of bears in North America. The Polar Bear (*Ursus maritimus*) is an arctic species; the Grizzly Bear (*Ursus arctos*) no longer exists in California; but the Black Bear (*Ursus americanus*) is common in many parts of our state. The Grizzly Bear is discussed in the first part of the book under Mammals Recently Extinct in California.

Black Bear (*Ursus americanus*) [fig. 40]

Description: A large, stout bear with coarse black or brown fur and a white or pale patch on the throat or belly. Sometimes more than 190 kg but usually less. Claws on forefeet about the same length as those on hind feet; in the Grizzly Bear the claws of the forefeet are much the longer. The tail is minute. (See fig. 39.)

Distribution: Most forested regions of the state; a pest in some of the montane parks. In the Cascades and Sierra Nevada the Black Bear occurs from the upper edge of the forested elevations down to about 1000 m or less. In the northwest coastal forests they may occur at sea level and may even venture out on the beaches. Frequently they forage in garbage dumps in mountain communities. They may invade apple orchards in autumn and also cause great damage to beehives. From the Canadian coniferous forests to Mexico. (See p. 353.)

Food: True omnivores, finding nutrition in almost any organic food. Bears are fond of berries, nuts, and other vegetable foods, and in autumn they often subsist on manzanita berries and acorns. Like other forest-dwellers, bears are fond of underground fungi or "truffles." Most of their animal food consists of insects, especially ant larvae and beetle larvae, but they also eat mice and ground squirrels and occasionally a ground-nesting bird.

Reproduction: The Black Bear mates in summer, and implantation is delayed for several months (see Delays in Birth). This schedule provides for birth of one or two small young in midwinter, when the female has retreated for a winter rest. The cubs nurse and grow during the winter, while the mother remains in a somewhat lethargic state. Her fat stores, accumulated during the previous summer and autumn, contribute to

FIG. 40. Black Bear (*Ursus americanus*).

the manufacture of milk, which is the sole nourishment of the rapidly growing young. This species breeds only every other year. Apparently this schedule is necessary because of the tremendous drain of stored energy of the female during an extended period of fasting.

Raccoons and Ringtails (Procyonidae)

Raccoons and Ringtails, together with their Central American relatives, the Coati and Kinkajou, are related to bears and pandas. They are plantigrade, walking on the soles or plantar sur-

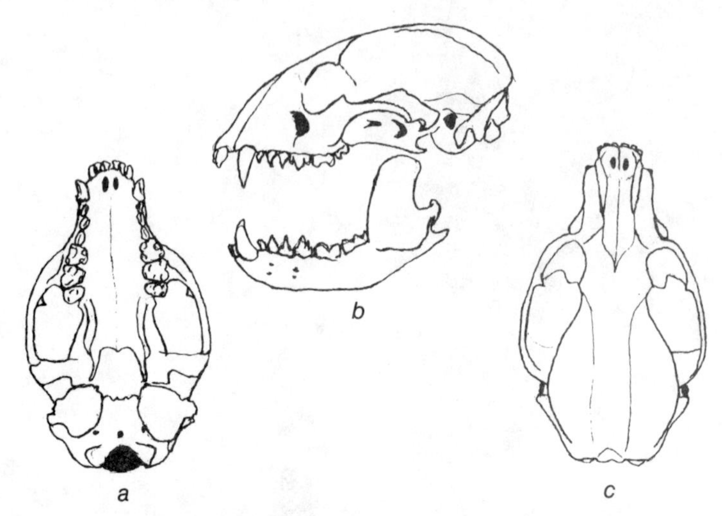

FIG. 41. Skull of the Raccoon (*Procyon lotor*): *a,* ventral view; *b,* lateral view; *c,* dorsal view.

face of the feet; soles bare or partly furred. Molar teeth for crushing and not for shearing as in cats (fig. 41).

Ringtail (*Bassariscus astutus*) [fig. 42]

Description: Somewhat like a small Raccoon but of slender build and with an extremely long tail. The huge, dark eyes of the Ringtail replace the black facial mask of the Raccoon, but the relationship is still obvious. Soles mostly or partly furred; claws partly retractable. TL 620–800, T 315–504, HF 55–75, E 45–50. Weight: 900–1150 g. Dentition: 3/3, 1/1, 4/4, 2/2.

Distribution: Brushy and wooded areas at lower and middle elevations; especially common in foothill canyons and sometimes in the Sacramento Valley. It is less common in the high mountains, but is known to live up to 2600 m. There is a population in the Sierra Buttes. The Ringtail, like the Raccoon, seems to prefer to live along watercourses. From Oregon and Colorado south into Central America. (See p. 354.)

Food: Preys on mice and wood rats and takes berries and soft fruits such as cherries, raspberries, and the fruits of the

FIG. 42. Ringtail (*Bassariscus astutus*).

madrone. Does not forage in water, and aquatic organisms are not included in its diet.

Reproduction: Little is known of the breeding pattern of this animal. Mating occurs in late winter, and three to four kits are born in May or June. Dens are secreted among large boulders near canyon bottoms.

Like the Raccoon, the Ringtail is nocturnal. The Ringtail shuns urban regions but is known to enter cabins in the mountains. According to early stories, gold miners welcomed the Ringtail because it kept their cabins free of mice and wood rats. This little carnivore is friendly and unafraid, and its presence should be encouraged.

Raccoon (*Procyon lotor*) [fig. 43]

Description: The Raccoon is perhaps the most familiar carnivore in North America. Its black mask and ringed tail distinguish it from all other carnivores but the Ringtail (*Bassariscus astutus*). TL 780–930, T 300–390, HF 100–130, E 50–60. Weight: 4–8 kg; females tend to be smaller than males. Dentition: 3/3, 1/1, 4/4, 2/2. (See fig. 41.)

Distribution: Found almost everywhere within the state; persists along creeks that course through urban areas. It is both a

FIG. 43. Raccoon (*Procyon lotor*).

forest-dweller and a creature of swamps and marshes, but it avoids extremely dry regions. It may make its den in woodlands far from water and also occurs in cattail marshes far from trees. This adaptability doubtless accounts, in part, for its abundance. Ranges over most of North America and far into Central America. (See p. 353.)

Food: Vertebrates and invertebrates, fruits, nuts, and berries. Commonly forages along watercourses and catches crayfish and frogs. Also mice and small birds, including birds' nests and their contents. The name *lotor* means "one who washes." When captives are provided with water, they usually wash their food before ingesting it; when deprived of a pan of water, they usually rub their food with their dry paws. The meaning of this activity is not known.

Reproduction: Mating in late winter. A litter of three to six young is born in a hollow log or tree. Mother and offspring remain together until the end of summer, at which time the young are nearly full grown. Raccoons are sociable, and family groups may remain in a unit through the winter. Dispersal occurs at about one year of age.

Apart from its value as a furbearer, the Raccoon is a prime quarry for those who run hounds. Nevertheless, the Raccoon survives where other carnivores disappear. Its persistence may be partly due to its tree-climbing ability and also to its extreme ferocity when encountered by dogs.

In some cities Raccoons have learned to live in storm drains, where, presumably, they prey upon rats.

Weasels, Marten, and Allies (Mustelidae)

A diverse group of small to medium-sized carnivores with an underlying similarity. Skull with a short snout, an elongate braincase, and rather large auditory capsules (*auditory bullae*). Eyes small and ears broad and rounded. Body long and legs short; in skunks, body and leg proportions concealed by long, loose fur. Anal scent glands present in most and well developed in some.

Mustelids vary from the tiny weasels to the formidable Wolverine. Although most species are terrestrial, the Marten and the Fisher are expert tree-climbers, the Badger is greatly modified for digging, the Mink is slightly adapted for an aquatic life, and the otters are clearly modified in many ways for swimming and foraging in the water. The skunks are unique in many features but retain the basic mustelid body pattern.

Reproduction in many species of mustelids includes delayed implantation (embryonic diapause)—mating and fertilization of the egg with a subsequent prolonged quiescent stage during which no growth and development occur. Later implantation follows and cell division resumes. In such species, mating occurs in summer or autumn and implantation follows in late December, apparently under the influence of day length (see Delays in Birth).

Key to Genera of Mustelidae in California

1. Premolars 4/4 2

   Premolars fewer than 4/4 3

2. Body with a lateral stripe more lightly pigmented than the rest of the dorsal and lateral fur . . . . *Gulo*

   Body without a lateral stripe . . . . *Martes*

3. Fur conspicuously black and white dorsally 4

   Fur dorsally an even brown or mottled gray; head may be striped 5

4. With two broad white stripes on back . . . . . *Mephitis*

   With two indistinct rows of white patches or spots . . . . *Spilogale*

5. Tail at base not greatly thickened (but sometimes bushy); premolars 3/3; feet variable but not broad and webbed 6

   Tail thick and muscular at base; premolars 4/3 or 3/3; feet broad and webbed 7

6. Dorsal fur of an even brown or black; tail slender . . . . *Mustela*

   Dorsal fur mottled gray; head with a white stripe, sometimes extending onto shoulders; tail bushy . . . . *Taxidea*

7. Premolars 4/3 . . . . *Lutra*

   Premolars 3/3 . . . . *Enhydra*

Sea Otter (*Enhydra lutris*) [pl. 5d]

Description: The most aquatic member of the weasel family and clearly the largest of living mustelids. Tail thickened at base. Hind feet compressed into flippers, sparsely furred. Foreclaws retractile. Molars broad and crushing (not for shearing and cutting). Color dark brown or black. Adult male almost 2 m long, female somewhat smaller. Weight: 21–45 kg (males), 14–33 kg (females). Dentition: 3/2, 1/1, 3/3, 1/2.

Distribution: An estimated population of 2000 along the coast of California, with concentrations in Monterey County. Associated with kelp beds in our area but may occur independent of kelp elsewhere. Found sporadically in Oregon and reestablished (by transplanting individuals from Alaska) in Washington and British Columbia. Also on Aleutians and on Commander Islands and Kuriles. Rarely south to the coast of Baja California. Coastal, to depths of 100 m, the Sea Otter does not enter fresh water nor does it migrate.

Food: Feeds on a variety of bottom-dwelling invertebrates. A large lung capacity allows it to dive to about 100 m, where it gathers bivalves, abalone, sea urchins, crabs, and a few slow-moving fish. It sometimes feeds on mussels and scallops.

Reproduction: Births and mating are not seasonal in California, but young are born in summer in the Aleutians. A single young is born after a gestation of six to nine months, a part of which time includes a delay in implantation. In California birth usually takes place while the mother is offshore, in kelp beds, but in Alaska the pup is most commonly born on land. The young is precocial, fully furred and active at birth. Nursing may continue until it is full grown. Mating and birth probably occur every other year.

There is a complex relationship between densities of the Sea Otter, sea urchins, abalone, and shallow-water fish. Whereas moderate levels of these various organisms do not conflict with each other, excessive removal of abalone, clams, sea urchins, or other organisms may affect densities of the other members of the community. When Sea Otters abound, they seem to curb populations of sea urchins, which, in turn, feed on kelp. Shallow-water fish find protective cover in kelp; consequently, a healthy growth of kelp allows larger populations of some fishes. Thus the Sea Otter promotes an increase of kelp—and therefore fishes—whereas a scarcity of Sea Otters allows an increase of sea urchins, which tends to depress the dense growth of kelp and thereby the fish populations. Although abalone are taken by Sea Otters, this large snail is far less important in its diet than are sea urchins.

The Sea Otter is one of the few mammals to use a tool. It is often seen to come to the surface with a rock and a shellfish. Lying on its back, the shellfish on its chest, the mammal breaks it with the rock. It is also reported to use rocks to dislodge abalone attached to bedrock.

The demand for Sea Otter pelts first led the Russians to Alaska and eventually to the California coast in the mid-eighteenth century. Early in the nineteenth century the Sea Otter was abundant along the coast of Oregon, where the Indians not only used the pelts for their own clothing but also traded them to merchants for beads.

Wolverine (*Gulo gulo*) [fig. 44]

Description: Exceeds all other terrestrial members of the weasel family in size and ferocity. Heavy-bodied and short-legged. Nearly equal to a large Fisher in length, but much

FIG. 44. Wolverine (*Gulo gulo*).

heavier. Fur dark brown, head whitish between eyes and ears. A wide, light band on side of body, with considerable individual variation. TL 900–1125 (males), 880–970 (females); T 190–260 (males), 170–195 (females); HF 180–200 (males), 170–185 (females). Weight: 13.6–16.0 kg (males), 9.0–11.5 kg (females). Dentition: 3/3, 1/1, 4/4, 1/2.

Distribution: In high montane forests; rather rare, seldom seen. Mostly High Sierra south of Lake Tahoe; also northwest coast counties (Humboldt, Del Norte, Trinity). North to Oregon and Washington and across much of the coniferous forests of northern North America. Also Eurasia. (See p. 354.)

Food: The most powerful predator of the Mustelidae. Hunters tell of bears and Mountain Lions that retreat from their meals at the approach of a Wolverine. Apparently not a threat to livestock or humans. Takes various squirrels up to size of Marmot; also Porcupines, the quills of which may bring death to a Wolverine. At times a carrion-feeder.

Reproduction: Known dens of Wolverines are on the ground, in crevices under rock ledges at 3000 m or above, well above timberline. A litter of one to four young is born in spring from mating in the winter.

Legends of the ferocity of this animal may be partly based on the damage it can to do mountain cabins. One cabin in

Plumas County was broken into and ransacked, and the distinctive hairs of a Wolverine were found where the window had been forced open. It is completely protected today and apparently increasing in numbers, though nowhere common.

River Otter (*Lutra canadensis*) [fig. 45]

Description: One of the largest of the weasel family. With the thick, dense fur characteristic of aquatic mammals. Color dark brown, appearing black at a distance or in poor light. Toes connected by webbing. TL 889–1300, T 300–500, HF 100–145. Weight: 5–10 kg. Dentition: 3/3, 1/1, 4/3, 1/2.

Distribution: Along margins of rivers and larger streams in the Cascades and Sierra Nevada down to the Central Valley and Delta; also in major drainages in Coast Ranges north of San Francisco. Tracks often seen on sandbars of larger rivers, and otter slides are frequent in the Delta. Occurs over much of North America. (See p. 355.)

Food: Crayfish, frogs, fish, and shellfish. Sometimes a nuisance near fish hatcheries, where control is simple. Most fish eaten are rough fish (minnows and suckers). Small mice, birds,

FIG. 45. River Otter (*Lutra canadensis*).

and birds' eggs are also eaten, but the River Otter seldom forages far from water. Also mussels along the coast.

Reproduction: Breeds at two years of age. Two to four young born in a streamside burrow in April or May. Delayed implantation accounts for a very long gestation after a fall mating.

The otter was one of the great incentives for the exploration of California prior to the gold rush. Commercial trapping ceased in 1961 in California; like the Marten and Fisher, the River Otter has greatly increased since then. In this century the River Otter has never been taken in large numbers: from 1921 to 1961, from 14 to 163 were taken annually; over this forty-year period there was no apparent trend of increase or decrease in abundance. Intensity of trapping is probably greatly influenced by abundance as well as by price, but this species holds its numbers with controlled commercial trapping.

River Otters are usually playful and commonly construct slides on muddy streambanks. These are worn smooth and slippery by constant use. In the mountains in the winter they make similar slides in the snow.

Marten and Fisher (*Martes*)

In North America the Marten and the Fisher represent the genus *Martes,* but in Eurasia there are a number of species, more or less Marten-like in general aspects. They are arboreal mustelids and feed mostly on tree squirrels. They live in temperate or cool areas and are active all winter. Perhaps as a result of these habits, they grow a rich fur which has always been highly valued. The fur known as sable comes from one or more Old World species of *Martes*.

Species of *Martes* differ from all other North American mustelids except the Wolverine in having four upper and four lower premolars. (They are fewer in other genera.) The Marten and the Fisher are also distinctive in their rather bushy tails and size (larger than *Mustela* spp.).

Key to Species of *Martes* in California

1. Tail more than 290 mm; color grayish brown to black on rump; tail entirely black . . . . *Martes pennanti*

Tail less than 290 mm; color chocolate or yellow-brown, to black on tip of tail . . . . *Martes americana*

Marten (*Martes americana*) [pl. 6b]

Description: Chocolate brown dorsally, sometimes yellow-brown with much white ventrally. Ears long and tail bushy. TL 570–1030 (males), 540–597 (females); T 170–210 (males), 170–206 (females); HF 80–92 (males), 82–84 (females). Weight: 1.2–1.5 kg (males), 0.9–1.1 kg (females). Dentition: 3/3, 1/1, 3/3, 1/2.

Distribution: In northwestern California as well as at high elevations in Cascade–Sierra ranges. In northwestern counties seems to favor redwood forests. In Cascades and Sierra Nevada in forests of pine, fir, and hemlock above 1200 m. Also on talus slopes and open rocky areas. Southern Canada, northern Rocky Mountains, and northeastern United States. (See p. 355.)

Food: Takes a great variety of vertebrates, especially tree squirrels and chipmunks. May prey heavily on Pikas, rabbits, and wood rats. Sometimes insects and many kinds of fruits and berries. Mountain ash (*Sorbus* spp.) is a favorite food, and scats of the Marten may consist almost entirely of the seeds of this small tree.

Reproduction: From two to four young born in April or May. Mating, with delayed implantation, shortly after birth of young.

The Marten is the most frequently encountered of the larger weasels and is frequently seen by persistent enthusiasts. Under total protection it has increased in numbers. A skillful climber, it is as likely to be seen in the trees as on the ground. This species is sometimes considered to be the same as the Old World marten (*Martes martes*).

Fisher (*Martes pennanti*) [pl. 6a]

Description: A rather large, dark brown or blackish weasel characterized by a heavy body and a long, thickly furred tail. Patches of white on throat or undersides. As in mustelids generally, males much larger than females. TL 900–1200 (males),

750–950 (females); T 381–422 (males), 340–380 (females); HF 113–128 (males), 89–115 (females). Weight: 3.5–5.5 kg (males), 2.0–2.5 kg (females). Dentition: 3/3, 1/1, 3/3, 1/2.

Distribution: Northwestern California, at rather low elevations, and Cascade–Sierra ranges, 1000 m and above. In stands of pine, Douglas Fir, and true fir, avoiding redwood forests. From southern Canada, northern Rocky Mountains, and northeastern United States. (See p. 356.)

Food: Many sorts of small mammals from small mice to rabbits; tree squirrels are a favorite item. A very formidable predator, the Fisher has been seen to capture and kill a Gray Fox. In northwestern California the false truffle (a hypogeous fungus) is an important food item. The Fisher does not capture fish but is known to eat them. Under duress it will attack and kill Porcupines.

Reproduction: A single brood of one to five young in April or May. Mating occurs in spring or summer, followed by an extremely long gestation of some 330 to 360 days.

Although the Fisher was rare or uncommon early in the century, sightings and evidence of its presence have increased from the 1960s. It may no longer be regarded as rare in some parts of Humboldt and Trinity counties.

Skunks

Skunks comprise a group of conspicuously black and white carnivores, infamous for their offensive odors. The color pattern is assumed to be a warning to predators. Certainly skunks have few enemies, which undoubtedly accounts for their lack of fear. One can easily approach a skunk closely as it feeds in the early evening, but proximity is dangerous, for they can fire the contents of the anal glands some 3 to 4 m with accuracy.

The smell is secreted into anal glands which are surrounded by voluntary muscles, and it can be ejected at will. Unlike the scents used for social communication by many kinds of mammals, the secretions of skunks are reserved to repel enemies. Prior to discharging its scent, a skunk turns its rear and elevates its large bushy tail. This display is sufficient to deter all but the naive or hungriest predator.

There are two species of skunks in California.

Striped Skunk (*Mephitis mephitis*) [fig. 46]

Description: Instantly recognized by almost everyone. Occasionally assumed to be just a "black and white woodpussy." Its long fur and bushy tail do give it a vaguely catlike aspect. Typically two broad stripes down the back but variable in width; frequently a white stripe on head. Scent glands extremely powerful. TL 575–800 (males), 600–725 (females); T 185–390 (males), 240–270 (females); HF 60–90 (males), 60–80 (females). Weight: 1800–2700 g, but large males may approach 4 kg.

Distribution: Throughout most of California except in the extremely arid southeastern deserts. Frequently in well-settled areas and often found in gardens that are not tightly fenced. Across the southern half of Canada, virtually all of the United States, and south into Mexico. (See p. 356.)

FIG. 46. Striped Skunk (*Mephitis mephitis*).

Food: Unlike most of the weasel family, a true omnivore. In well-watered lawns and gardens, it forages deliberately, searching every depression and crevice for beetle larvae, cutworms, mice, and earthworms. Does not shun plant food. Takes berries on low-growing bushes and eats underground parts of plants, such as bulbs and corms.

Reproduction: Four to seven (or more) young in June or May in a hollow log or underground chamber. Young remain with mother most of the summer. Mating in late winter, without delayed implantation. Remains in estrus for prolonged periods in the spring, as ovulation occurs only after mating (induced ovulation). Gestation sixty to seventy-seven days.

One occasionally sees a mother skunk setting forth in the evening with the young following in single file, a charming sight. In spite of their friendliness and beauty, skunks are a dangerous source of rabies. In fact, the incidence of rabies in the Striped Skunk exceeds that in the domestic dog. Although it brings a very low price, the fur has long been used extensively for trim as well as for full-length coats. The strong and foul odor ejected from its scent glands accounts for the scarcity of the enemies of both this and the Spotted Skunk.

Skunks foraging on open ground often move with a smooth, flowing motion, their long tails trailing behind. In the failing light of dusk, they seem like lifeless wraiths as they appear to glide over the ground. As they forage, their digging leaves numerous small, cup-shaped depressions where they found grubs, earthworms, and other comestibles.

Spotted Skunk (*Spilogale putorius*) [fig. 47]

Description: Considerably smaller than the Striped Skunk. Fur much softer, glossier than in the Striped Skunk, and pattern broken up into spots of white on black. About half the length of the Striped Skunk, and one-third or less in weight. TL 310–610 (males), 270–544 (females); T 80–280 (males), 85–210 (females); HF 32–59 (males), 30–59 (females). Weight: 535–800 g (males), 200–280 g (females). Dentition: 3/3, 1/1, 3/3, 1/2.

FIG. 47. Spotted Skunk (*Spilogale putorius*).

Distribution: Throughout most of the state, apparently avoiding high mountains. Santa Cruz and Santa Rosa Islands. Found over most of the United States (except the northeast); from Vancouver to central Mexico. (See p. 357.)

Food: Omnivorous, like the Striped Skunk. Forages both day and night. Searches for soil-dwelling insects, worms, mice. Also preys on small ground-nesting birds.

Reproduction: Dens underground, usually in the burrow of some digging mammal, such as a ground squirrel. Mating occurs in autumn or winter with delayed implantation. A litter of two to six young is born in June.

Like the larger Striped Skunk, the Spotted Skunk is fearless. Tameness, which might indicate rabies in a fox, is typical of skunks even in good health. The odor of the Spotted Skunk is just as offensive as that of its larger relative. The smaller species is known for various procedures that may indicate nervousness or display to induce fear in the observer. Some observers have noted that the Spotted Skunk is prone to stamp its forefeet. Better known are its handstands—it sometimes stands erect on its forefeet, with the tail and hind legs held high over the body, perhaps to alarm its potential attacker.

Its search for soil insects can tear up a lawn. Its fondness for insects sometimes leads it to beehives, where it can be destructive to domestic honeybees.

Because eastern populations of the Spotted Skunk mate in the spring and do not experience delayed implantation, it has been suggested that the form in California and elsewhere in the west is a separate species. There appears to be some variability in the time of implantation, especially in the southern part of its geographic range, and the distinctions between the eastern and western populations of these skunks may remain equivocal for some time.

Weasels and Minks (*Mustela*)

These little carnivores perhaps best illustrate the stereotype of the weasel family: the head is narrow, the neck long and very muscular, the legs short, and the body elongate. The *Mustela* skull (fig. 48) is easily recognized by the large ear capsules (auditory bullae). The dentition is 3/3, 1/1, 2–3/3–2, 1/2.

Key to Species of *Mustela* in California

1. Color brown dorsally; much lighter (even white) ventrally; may be white in winter 2

   Color dark brown or black; scarcely any lighter ventrally; no seasonal color change . . . . *Mustela vison*

2. Tail usually not more than 40 percent as long as head and body; hind foot less than 35 mm (males) or 28 mm (females) . . . . *Mustela erminea*

   Tail usually more than 40 percent as long as head and body; hind foot more than 40 mm (males) or 30 mm (females) . . . . *Mustela frenata*

Short-tailed Weasel or Ermine (*Mustela erminea*) [pl. 6e]

Description: The smallest California weasel, with a rather short tail. Usually brown or yellow-brown. Tail black on the distal third. A white winter coat in the mountains. Generally much smaller than the Long-tailed Weasel. TL 225–275 (males), 190–230 (females); T 55–75 (males), 50–63 (fe-

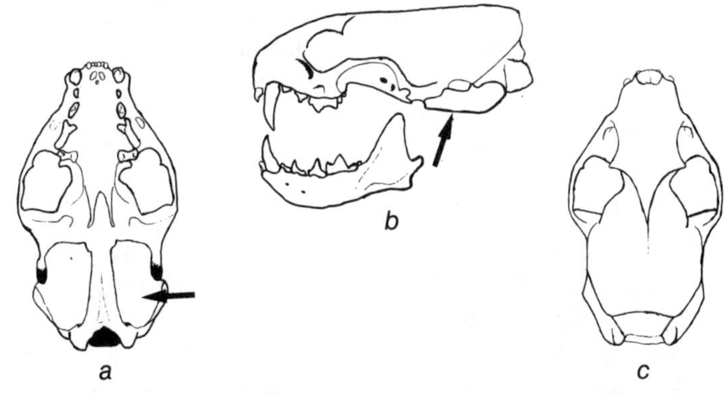

FIG. 48. Skull of the Long-tailed Weasel (*Mustela frenata*): *a,* ventral view; *b,* lateral view; *c,* dorsal view (arrows to auditory bullae.)

males); HF 27–34 (males), 23–27 (females). Weight: 50–60 g (males), 28–35 g (females); specimens in the mountains average smaller.

Distribution: Throughout the foothills and mountains of both Cascades and Sierra Nevada in northern half of state; Coast Ranges south to Marin County. Preferring coniferous forests. Throughout North America from Greenland to southwestern United States. Also Eurasia; in England it is called the Stoat.

Food: Mostly mice and small ground-nesting birds. From white-footed mice and voles up to rodents the size of a chipmunk or wood rat. Some reptiles and amphibians. Rarely fruits and berries. Climbs well and takes eggs and young of small birds. Active day and night, foraging over 50 to 100 acres.

Reproduction: Mating in late summer followed by delayed implantation. Four to eight young born in spring, about 1.7 g. Females may mate in first summer; males sexually mature the following summer.

Although this species is, in its white winter coat, the Ermine of commerce, the pelts bring only a dollar or two. An Ermine cape becomes expensive because of the large number of skins used in a single garment. The skin is thin and the fur quality is vastly inferior to that of the Mink. The development of the white winter pelage may be partly influenced by genetic factors, but experimentally it can be induced by low temperature. Molt itself is induced by day length.

Long-tailed Weasel (*Mustela frenata*) [pl. 6c]

Description: Both longer and heavier than the Short-tailed Weasel and with a relatively short tail. Color chocolate brown with the tail black-tipped. Frequently distinctive in having a white or light-colored facial mask, sometimes in the form of a whitish patch between the eyes. Winter pelage white in montane populations but brown at lower elevations where there is not normally snow on the ground. TL 350–450 (males), 335–395 (females); T 125–180 (males), 120–145 (females); HF 42–50 (males), 32–41 (females). Weight: 225–345 g (males), 115–220 g (females). (See fig. 48.)

Distribution: Statewide except in extremely arid areas. Occurs almost anywhere but avoids the streamside habitat apparently preempted by the Mink. Favors rockpiles and stacks of firewood. Throughout North America; from southern Canada to southern Bolivia. Up to 3300 m elevation in the United States. (See p. 357.)

Food: Not especially selective but clearly prefers warm-blooded vertebrates. Its small head and long neck permit the exploration of crevices and hiding places of mice. A variety of small mice, pocket gophers, ground squirrels, chipmunks, and small birds.

Reproduction: Mates in July with a prolonged delayed implantation (embryonic diapause). Implantation and continued development late the following winter; birth of four to eight young in June. Embryonic growth of four weeks or slightly longer. Females sexually mature the first summer, when older males are still sexually active.

Weasels forage at any hour of the day and are opportunists in their choice of prey. The predatory prowess of the Long-tailed Weasel is the result of its method of hunting and killing and certainly not its size. Typically a weasel succeeds in killing relatively large prey by clutching its back or neck and then biting the soft parts of the skull, especially about the ears. Such a technique enables these small carnivores to kill rabbits several times their own weight. However, weasels prey mostly upon mice and other mammals of approximately that size.

Although the Long-tailed Weasel sometimes kills grown chickens, it more commonly feeds upon rats that infest chicken coops. In view of the well-established rat-killing habits of weasels generally, it is foolish to destroy them on sight on the assumption that they are about to prey on domestic fowl.

The Long-tailed Weasel has a prolonged gestation due to the delay in implantation, which is characteristic of many species of this family. Copulation normally occurs after the young are weaned, at the end of summer. Mating in this weasel resembles the prolonged act of sex characteristic of the Mink: a pair of Long-tailed Weasels may remain clasped together for two hours or more and may repeat the performance the same day. The tiny, nearly naked young are born some 225 to 300 days later and weight approximately 3 g at birth. The young take milk for five weeks or so. The nest can be almost any concealed shelter, hollow log or squirrel nest, and is frequently lined with the fur of prey.

Like other members of this family, the Long-tailed Weasel is playful. A captive enjoyed chasing dead Starlings swung before it on a string; when finally allowed to catch its "prey," it performed a little dance, jumping high into the air while doing twists and aerial somersaults.

Mink (*Mustela vison*) [pl. 6d]

Description: Fur dark brown or black, dense and glossy. Small white spots about chin or throat. Weasellike in general form. Tail rather bushy for a weasel. TL 491–720 (males), 473–560 (females); T 160–211 (males), 157–203 (females); HF 60–75 (males), 58–64 (females). Weight: 880–1300 g (males), 540–750 g (females).

Distribution: Watercourses and marshes from San Joaquin Valley north to the Delta and throughout most of the northern half of California. Along tidal margins in the Delta and mudflats of San Francisco Bay. Streamsides to elevations of 2000 m or higher. North America to Alaska; introduced into Eurasia. (See p. 358.)

Food: Streamside invertebrates, such as crayfish, and vertebrates such as frogs and muskrats. Carrion. Ducks and coots sometimes captured.

Reproduction: Mates in late winter; three to ten young born in June or July. Den usually near water, under a log or beneath the ground.

As a predator the Mink is of neutral value. It not only preys on mice and carrion but takes some some game birds and rabbits and may occasionally molest domestic fowl and even domestic trout. The fur crop of the Mink is important, and this species survives well under sustained trapping. In 1978–1979, Mink pelts brought the trapper from $10 to $20, for a total of more than $7000 for the state. Most trappers are teen-aged boys earning a little extra money on weekends or retired people trying to survive on a limited fixed income.

Badger (*Taxidea taxus*) [fig. 49]

Description: A large mustelid rather obviously modified for a semifossorial life. Powerful forefeet for digging; body stout and flattened. Color silver gray; head dark with a white stripe. Tail short and moderately furred. TL 600–730, T 100–135, HF 92–126. Weight: up to 11.4 kg (large males), 4.5 kg (females). Dentition: 3/3, 1/1, 3/3, 1/2.

Distribution: A creature of open areas and sandy soils, including deserts. Statewide in open areas. Sporadically common in Sacramento Valley, fluctuating with populations of squirrels or pocket gophers. From south central Canada, over western and central United States, to central Mexico.

Food: Mostly ground squirrels and pocket gophers.

Reproduction: One to four young born in an extensive burrow system. Mating in late summer, followed by delayed im-

FIG. 49. Badger (*Taxidea taxus*).

plantation (embryonic diapause); implantation in December or January; young born in March or April.

The Badger is sometimes an important furbearer. The pelage is soft, durable, and very beautiful. Before the advent of the electric razor and brushless shaving cream, the fur of the Badger provided half the adult population with shaving brushes.

In its almost constant pursuit of quarry (ground squirrels and pocket gophers), a Badger tears up a great deal of ground and may damage levees infested by rodents. On open range, ranchers object to the burrows of the Badger because of the danger to horses. In the 1950s, Badgers were quite common in alfalfa fields which were heavily infested with pocket gophers—so common that their diggings sometimes damaged the blades of cutting machines.

Foraging Badgers may be watched by Red-tailed Hawks, which have been known to stoop down and snatch ground squirrels fleeing the mammalian predator.

Cats (Felidae)

Throughout the world there is little basic diversity among cats, and there are many similarities. Our two native species, the Mountain Lion and the Bobcat, have typical feline structure and resemble most cats from other parts of the world.

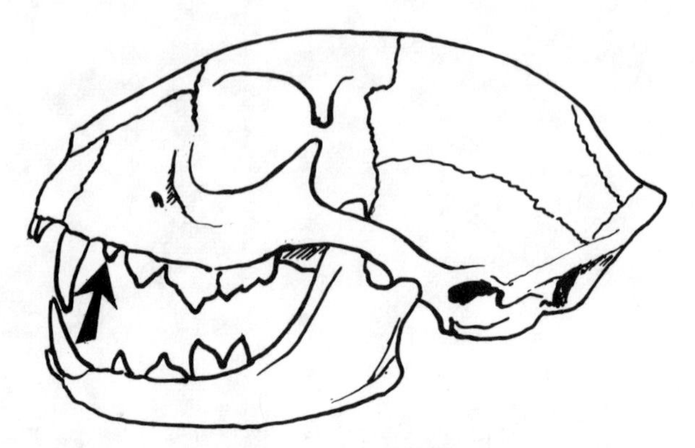

FIG. 50. Skull of the Feral House Cat (*Felis catus*). Note small upper premolar (arrow) behind the canine tooth, characteristic for this species.

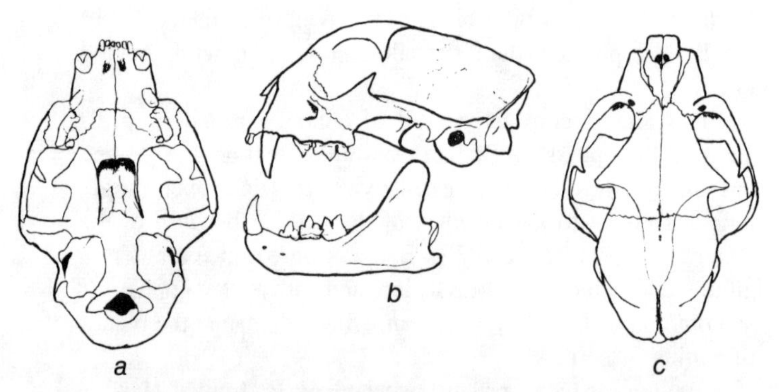

FIG. 51. Skull of the Bobcat (*Lynx rufus*): *a*, ventral view; *b*, lateral view; *c*, dorsal view.

The skull of cats reveals their diet, for the upper and lower molariform teeth meet so that their adjacent lateral surfaces form a shearing tool. This shape is in contrast to the molar surfaces in bears, which form a crushing or grinding surface. The snout of cats is short, in contrast to the pointed snout of dogs, and the eyes are large (figs. 50 and 51). The canine teeth are long and pointed. Cats walk on their toes (digitigrade), and in most species the claws are retractable: they can be withdrawn to prevent their becoming dulled when walking or extended to grapple with prey.

Feral House Cat (*Felis catus*)

In most agricultural areas there is a population of feral house cats. These feral populations are augmented annually when people abandon unwanted pets. These animals are frequently able to survive in the wild and seem to subsist mostly on wild rodents taken at some distance from human dwellings. Probably they take some carrion after the shooting of upland game birds. They are mentioned here to remind readers that the "wild mammal" they glimpse may be only a feral house cat.

The specific name *catus* was applied by Linnaeus to the domestic cat. The domestic cat presumably has been derived from *Felis sylvestris,* a wild species in Eurasia and North Africa, but to call our feral populations *sylvestris* would be very misleading. (See fig. 50.)

Mountain Lion (*Felis concolor*) [pl. 7]

Description: The Mountain Lion (also called Puma or Cougar) is the largest cat in our fauna. Distinctive in its sandy color and its long (1 m) and scantily furred tail. The kits are spotted; these indistinct brownish spots are lost in the first six months of life. TL 1500–2500, T 550–775, HF 220–275. Weight: 40–100 kg; males usually larger than females. Dentition: 3/3, 1/1, 3/2, 1/1.

Distribution: Throughout most of the forested and brushy regions of the state; avoids open areas. Tends to be shy, and its presence is known from tracks rather than sightings. The Mountain Lion is today not uncommon and is among the many wild mammals that are found as road kills on mountain highways. They survive within the city limits of Berkeley, Hayward, and Richmond. The most widely ranging mammal in the New World; found from the Canadian coniferous forest south to Patagonia. (See p. 358.)

Food: Preys heavily on deer and other mammals. One of the few predators to take skunks and Porcupines. Mountain Lions have been known to molest domestic stock, but this cat does not constitute the problem presented by the Coyote. They also take squirrels, rabbits, and even mice.
Reproduction: A litter of one to six kits born blind and helpless, usually in the spring. Sexually mature at 25 to 35 kg. Gestation about ninety days.

This magnificent cat survives close to civilization for two reasons: it is very shy and seldom seen, and it seldom conflicts with human activities. It is an extremely valuable predator, for it is the major check on numbers of deer. For many years it was assumed to have disappeared from eastern North America except for a population in Florida. In recent years, however, the unmistakable tracks have been reported from Georgia, Pennsylvania, New York, Maine, and New Brunswick. It has survived in these areas, undetected, for as long as seventy-five years.

Bobcat (*Lynx rufus*) [pl. 3b]

Description: A spotted cat with a short tail (which is white-tipped), tufted ears, and broad "whiskers." Rather long-legged for a cat. TL 700–1000, T 95–150, E (from crown) 60–75. Weight: 5–15 kg; males average larger than females; in eastern North America, Bobcats may weigh more than 30 kg. (See fig. 51.) Dentition: 3/3, 1/1, 2/2, 1/1.

Distribution: Statewide from Death Valley to high mountains. Equally at home in brushland, foothill chaparral, sagebrush, and forests. It remains relatively common despite heavy trapping and despite control on sheep pastures. From southern Canada to central Mexico.

Food: An opportunist whose diet varies with availability more than any apparent preference. Takes many rabbits and small squirrels; also small reptiles and birds. Occasionally molests sheep, but not nearly as troublesome as domestic dogs and the Coyote. In eastern North America it is a major predator of the White-tailed Deer; but eastern Bobcats are larger than

California specimens, and the White-tailed Deer is smaller than California deer.

Reproduction: A litter of one to six kits born in the spring or summer after a gestation of some fifty days. If a fertile mating does not occur at the first estrous period, estrus recurs one or more times. This accounts for the extended period over which young are born.

In contrast to the secretive Mountain Lion, the Bobcat is sometimes quite bold and may not run away at the first sight of human observers. It may occur close to buildings and has been known to crouch next to water troughs and strike down bats as they fly low to drink. Its spotted fur and erect pointed ears make it one of the most beautiful mammals in our fauna.

PINNIPEDIA

The following two families, the seals (Phocidae) and the sea lions or eared seals (Otariidae), have been placed together (with the walrus, Odobenidae) in a group called Pinnipedia; this major category used to be considered a separate order and was sometimes placed as a suborder of carnivores. Because walruses do not occur in California, we consider here only the seals and the sea lions. Fossil evidence suggests that these two families have had distinct evolutionary histories and that many of their similarities can be attributed to convergence. Sea lions may have diverged from a carnivore ancestral to bears; seals may have descended from a weasellike ancestor. If these two families have, in fact, had separate and distinct origins, a long history (15 million years or more) of aquatic existence could certainly account for many physical and physiological resemblances. Thus while it is convenient and reasonable to refer to these mammals collectively as "pinnipeds," it is not correct to place them in a discrete group quite separate from other carnivores. Seals and sea lions do differ from each other in a number of features, some of which are obvious even when these animals are viewed from a distance.

Pinnipeds are properly considered to be semiaquatic or amphibious, for they return to land to give birth to young. The infant, moreover, may not enter water immediately. On the other hand, locomotion on land is difficult and awkward, and pinnipeds do not obtain food on land. In the water they are swift and graceful, and all of their hunting is done beneath the surface.

As do many species of mustelids, sea lions and seals experience a period of delayed implantation during which the newly fertilized egg divides a few times and then enters a period of rest. This delay represents an adaptation to the temporary presence of both sexes together on land and provides for birth of the young when they again return to land almost exactly a year later. Actual embryonic development takes much less than one year.

Their senses also reflect their long history in a marine environment. Although their external ears are reduced, as in the sea lions, or absent, as in the seals, the hearing of pinnipeds is quite good underwater. An ability to echolocate is reported but disputed; in any case pinnipeds very likely recognize sounds produced by potential food items. Their eyes are also modified for vision in water. The eye is much larger than in land mammals of comparable body size: an increased size allows for the entrance of more light into the eye when foraging in murky depths. Most pinnipeds, moreover, seem to hunt at night. When light is bright, the pupil can be greatly reduced for aerial vision. Both the lens and the cornea are especially thickened—an adjustment to the increased refractive index of water—but this adaptation probably results in some degree of myopia when on land. Apparently the senses of smell and taste are not nearly as acute as in terrestrial carnivores.

The diving ability of pinnipeds results not only from a streamlined body form but also from less obvious physiological adaptations. Just before diving, a pinniped exhales and the heartbeat slows. Circulating blood is concentrated to the brain and heart, for the muscles accumulate oxygen while the animal is breathing. The Harbor Seal can remain submerged for nearly a half hour and forage to about 100 m. After surfacing, heavy breathing and rapid circulation restore oxygen to the tissues.

There are several conspicuous reproductive adaptations to a marine life. Although pinnipeds must come ashore to give birth, many species remain with the young until weaning. Such intensive parental care prevents the adults from feeding, but lactation is shortened by the provision of extremely fat-rich milk which enables the pup to grow rapidly. Lactation is more prolonged in most sea lions and fur seals, in contrast to the rather brief nursing period in seals; the milk of most sea lions is not as rich as that of most seals. Delayed implantation (discussed under Reproduction) times birth to occur eleven or eleven and a half months after mating, the time at which the females will return to their rookeries.

Sea Lions or Eared Seals (Otariidae)

Forelimbs and hind limbs rather long, and hind limbs or flippers capable of being rotated anteriorly. Flippers nude at their tips and with fewer than five toenails on each foot. Ears small but clearly visible, pointed and protruding from the outline of the head. Forelimbs used for propulsion both on land and in the water. Males much larger than females. Outer incisor canine-like; molars and premolars somewhat triangular, expanded in middle (fig. 52b).

Sea lions, including the fur seals, are better able than seals to progress on land. Because their hind limbs can be rotated into an anteriorly oriented position, they provide some thrust

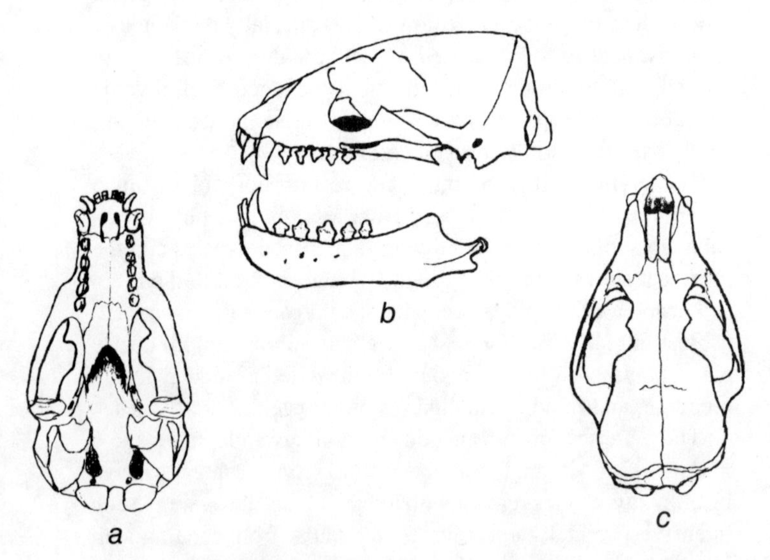

FIG. 52. Skull of the California Sea Lion (*Zalophus californianus*), female: *a*, ventral view; *b*, lateral view; *c*, dorsal view.

in movement on land and also assist, in a small way, in raising the rear of the animal off the ground.

The four California species are rather similar in many respects. In all of them the male greatly exceeds the female in size. The species do differ in shape, color, size, and distribution, but these differences are not always conspicuous, and field identification may not always be certain.

Guadalupe Fur Seal (*Arctocephalus townsendi*) [fig. 53]

Description: Smallest of the sea lions known to occur in California waters. Additionally distinguished by very long front flippers and a rather long, slender snout. Males reach 2 m in length; females typically reach 1 1/2 m. Color dark brown when wet, somewhat grayish when dry. Dense underfur. Old males tend to become a light yellow or gold on the chest. Weight: 150 kg (males), 45 kg (females). Dentition: 3/2, 1/1, 4/4, 2/1.

Distribution: Today known to breed only on Guadalupe Island, off Baja California, but in early times it bred on islands at least north to the Channel Islands or perhaps the Farallons. Sometimes seen off San Miguel and San Nicholas islands off southern California; may be expected to occur more frequently as populations increase.

Food: Unknown but probably medium-sized fish and shellfish.

Reproduction: On Guadalupe Island this fur seal repairs to caves for birth of the pups and breeding. One bull and a small number of cows form the breeding unit; pups are born in the spring. Mating follows within a week or so. Presumably the pup nurses for a prolonged period, at sea, for the population tends to disperse at the end of summer.

This species belongs to a group which is otherwise confined to the Southern Hemisphere. It has sometimes been placed in the same species as a fur seal (*Arctocephalus philippi*) known from Juan Fernández Islands, off the coast of Chile, but the

FIG. 53. Guadalupe Fur Seal (*Arctocephalus townsendi*).

Chilean population may now be extinct. Until the 1950s the Guadalupe Island population was considered to be extinct; since its rediscovery, it has been increasing steadily, and an estimated 1000 or more exist today.

Northern Fur Seal (*Callorhinus ursinus*) [fig. 54]

Description: A dark brown sea lion of medium size. Males more than 2 m; females 1 1/2 m. Flippers, especially hind flippers, quite long; nails only on dorsal surface. Weight: up to 225 kg (males), 50 kg (females). Dentition: 3/2, 1/1, 4/4, 2/1.

Distribution: Breeding in Pribilof Islands, west to Commander Islands, and breeding colonies on islands off the coast of Asia. During the nonbreeding season, populations are dispersed. Although bulls remain mostly in the Gulf of Alaska, cows and pups move far to the south, commonly as far south as our coast, although seldom near shore. A recently established colony on San Miguel Island greatly expanded the breeding range of this sea lion.

FIG. 54. Northern Fur Seal (*Callorhinus ursinus*).

Food: Because of the great commercial importance of this sea lion, its habits have been rather intensively studied. Foraging at night, the Northern Fur Seal may take advantage of the upward nocturnal movement of many fish and squid, their staple. In addition to several kinds of squid and cuttlefish, this pinniped takes many species of rockfish, herring, and mackerel.

Reproduction: The habit of breeding in groups of females within the territory of a solitary bull results in a large surplus of nonbreeding bulls, the "bachelor bulls." These nonbreeding bulls form the basis of the commercial sealing industry today. Because these individuals do not participate in the reproductive activity of the population, many can be removed without affecting the annual productivity. As in other sea lions, the Northern Fur Seal bulls arrive on the breeding grounds, the rookeries, and set up territories. They announce their presence to the cows, which arrive just at the time the pups are scheduled for birth. By autumn the cows and pups have begun a long migration to more southerly waters, but the bulls spend the winter in the arctic seas. Because mating occurs shortly after the birth of the pups, the cows are pregnant long before they depart for the autumnal migration.

The Northern Fur Seal, together with the Sea Otter, was in large measure responsible for early exploration of the North Pacific coast. In the eighteenth and nineteenth centuries these furbearers were taken in a totally indiscriminate manner, which included the killing of both sexes throughout the year. Consequently there were drastic declines in both species. Today a treaty between Canada, Japan, the Soviet Union, and the United States not only limits the numbers of Northern Fur Seals that may be killed but confines the take to the bachelor bulls. As a result, an important industry has been preserved and the populations of this pinniped have greatly increased.

Steller's Sea Lion (*Eumetopias jubatus*) [pl. 5c]

Description: Size large: adult males more than 3 m; females much smaller, about 2 m. Pups black, but adults straw or yellow-brown or even whitish when submerged. Ear small but distinct. Bulls lack the white raised forehead of the California Sea Lion. Weight: up to 1000 kg (males), 250 kg (females). Dentition: 3/2, 1/1, 4/4, 1/1.

Distribution: Breeds from the northern Channel Islands north to the Aleutians and Pribilofs. Sometimes seen on the California coast but less common inshore than the California Sea Lion. A breeding colony is on Año Nuevo Island. After breeding, southern populations move to the north and northerly populations tend to move south. Breeding populations extend westward to Kamchatka, the islands in the Sea of Okhotsk, north of Hokkaido, and on intervening islands.

Food: When breeding, bulls do not eat for some six weeks, but cows apparently feed at night. Little specific knowledge on the food of this offshore species; apparently takes a variety of medium-sized fish.

Reproduction: Females collect into groups of twenty to

thirty within the territory of a large bull. Bulls defend their territories, and battles may be vigorous and bloody. Cows apparently have no loyalty either toward a particular bull or his plot of rock but move about from one "harem" to another. A single pup is born in late spring or early summer; mating follows within about a week. Nursing may last for about a year. Females may breed in their third year but bulls not until several years later.

This species has also been called the Northern Sea Lion.

California Sea Lion (*Zalophus californianus*) [pl. 5e]

Description: Rather dark brown or blackish; color darker in males than in females and also darker when wet. Males 2 1/2 m; females some 2 m. Ear short, pointed. Weight: 250 kg (males), 100 kg (females). Dentition: 3/2, 1/1, 4/4, 2–1/1; molars and premolars nearly triangular (see fig. 52b).

Distribution: Breeds from the Channel Islands south to Baja California and occurs in the Gulf of California. Nonbreeding individuals may move north as far as British Columbia and are found in many areas along our coast. Separate populations (subspecies) also occur on the Galápagos and in the Sea of Japan. Known also to ascend coastal rivers.

Food: Takes squid and a variety of fish. Known to damage salmon in gill nets and to become entangled in nets themselves.

Reproduction: Known for its large and conspicuous aggregations during the breeding season. Females assemble on offshore islands and bulls move in. Bulls are territorial, but females move from one harem to another. A single pup is born in early summer; mating follows within ten days or two weeks. Thus the female may be pregnant for much of the period during which the infant is taking milk. Gradually, during the summer, the pups venture more and more into the sea. From the end of the summer until the following spring, mother and young may move in small or large groups, and the young may still nurse for the better part of a year.

This is the sea lion famous as the "trained seal." Only the females are used for this purpose. This sea lion is most commonly observed during the nonbreeding period when it moves up and down the coast. It habitually seeks certain beaches where it is not disturbed and frequently can be heard before it is seen. Its loud barking is a familiar sound that aids in distinguishing it from the Steller's Sea Lion.

While its predation on commercial fisheries may be negligible, it does damage gill nets.

Seals (Phocidae)

Limbs reduced in size and flexibility. Forelimbs furred. Limbs with some toenails or claws (in *Phoca*) or claws rudimentary or absent (*Mirounga*). Ear opening present but external ear (ear pinna) lacking. Sexes similar in size in most species.

Seals are perhaps more adapted to an aquatic life than are the various sea lions and fur seals. The body form of seals is more compact, with the neck not apparent, and their limbs are reduced in size and strength. Consequently, seals can do little more than wriggle along the rocks. Although awkward on land, seals are graceful and swift in the water. Their forelimbs lie appressed to the side of the body when swimming, and propulsion results from movement of the hind flippers and tail region. Sea lions, on the other hand, may use their fore flippers to assist in forward movement through the water. Probably reflecting the more thorough adaptation of seals to life in water, their testes are housed permanently within the body cavity, as is also the case with cetaceans, whereas in sea lions the testes lie in a scrotum as in most groups of land mammals.

Northern Elephant Seal (*Mirounga angustirostris*) [pl. 5a]

Description: A very large seal: males of more than 5 m; females very much smaller, 3 m long. Males also distinctive for its pendulus but inflatable snout. Color gray or brown; young black. The immense size alone, together with the seal body form, identifies this species. Weight: up to 1800 kg (males), 450 kg (females). Dentition: 2/1, 1/1, 4/4, 1/1.

Plates

PLATE 1 Shrews and Moles

	Map	Text
a. Vagrant Shrew (*Sorex vagrans*)	346	105
b. Pacific Shrew (*Sorex pacificus*)	344	102
c. Trowbridge's Shrew (*Sorex trowbridgii*)	345	104
d. Water Shrew (*Sorex palustris*)	343	102
e. Desert Shrew (*Notiosorex crawfordi*)	343	93
f. Shrew-mole (*Neurotrichus gibbsii*)	346	107
g. Townsend's Mole (*Scapanus townsendii*)	347	109
h. Broad-footed Mole (*Scapanus latimanus*)	347	108

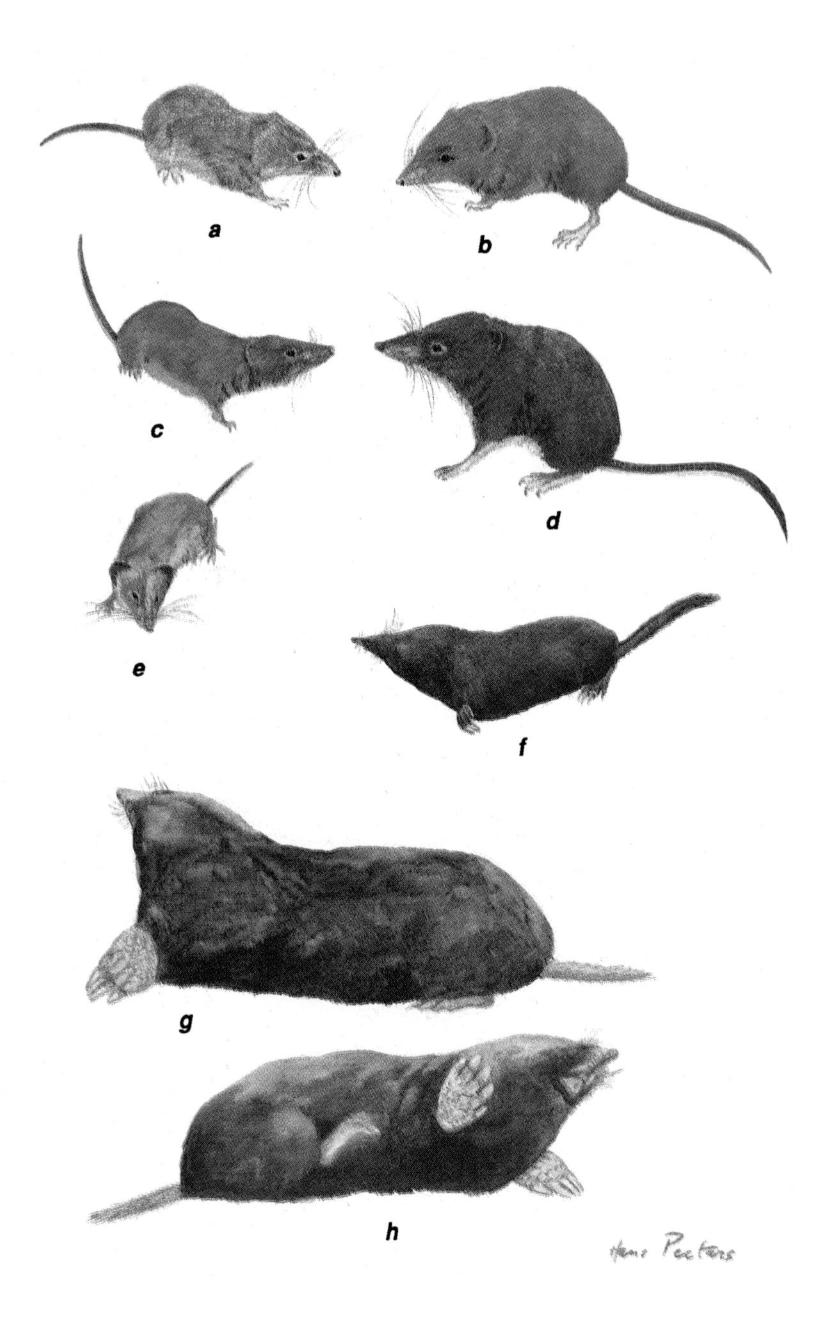
a
b
c
d
e
f
g
h
Hans Peeters

PLATE 2 Bats

Most *Myotis* species are extremely similar externally, and only two are shown. See species accounts.

		Map	Text
a.	Long-eared Bat (*Myotis evotis*)		128
b.	California Bat (*Myotis californicus*)		128
c.	Pallid Bat (*Antrozous pallidus*)	349	121
d.	Silver-haired Bat (*Lasionycteris noctivagans*)	350	124
e.	Western Pipistrelle (*Pipistrellus hesperus*)		132
f.	Red Bat (*Lasiurus borealis*)		125
g.	Hoary Bat (*Lasiurus cinereus*)		126
h.	Big Brown Bat (*Eptesicus fuscus*)		121
i.	California Leaf-nosed Bat (*Macrotus californicus*)	348	113
j.	Townsend's Long-eared Bat (*Plecotus townsendii*)	351	133
k.	Spotted Bat (*Euderma maculatum*)	349	122
l.	Hog-nosed Bat (*Choeronycteris mexicana*)	348	113
m.	Western Mastiff Bat (*Eumops perotis*)	351	134
n.	Guano Bat (*Tadarida brasiliensis*)		136

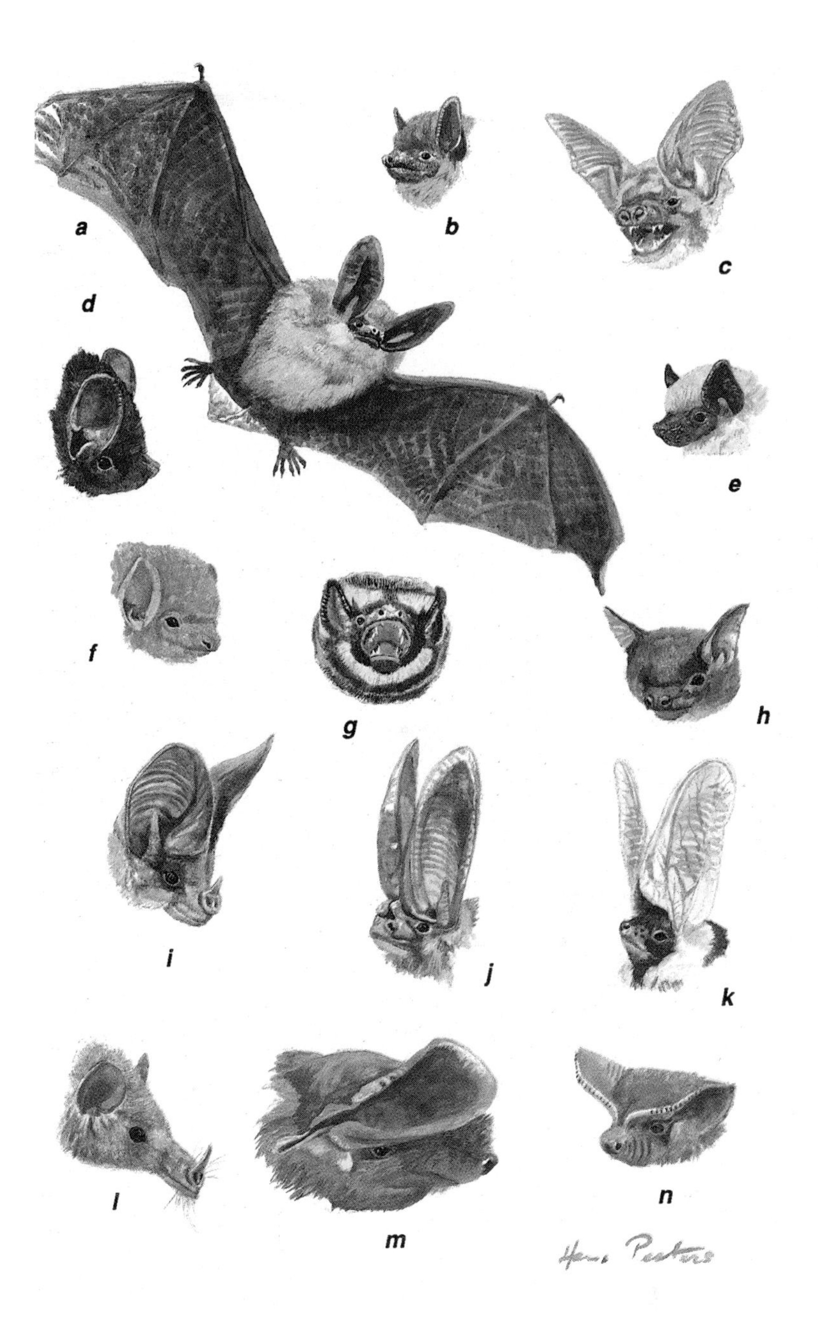
a
b
c
d
e
f
g
h
i
j
k
l
m
n

PLATE 3 Coyote and Bobcat

	Text
a. Coyotes (*Canis latrans*)	139
b. Bobcat (*Lynx rufus*)	170

PLATE 4 Foxes

	Map	Text
a. Gray Fox (*Urocyon cinereoargenteus*)		142
b. Kit Fox (*Vulpes macrotis*)	352	143
c. Red Fox (*Vulpes vulpes*)	352	144

PLATE 5 Seals, Sea Lions, and Sea Otter

a
b
c
d
e

PLATE 6 Weasels and Allies

	Map	Text
a. Fisher (*Martes pennanti*)	356	157
b. Marten (*Martes americana*)	355	157
c. Long-tailed Weasel (*Mustela frenata*). The distinctive face pattern is absent in montane populations (where the face is brown); these also molt into a white winter pelage in the mountains.	357	164
d. Mink (*Mustela vison*) with a favorite prey, a Muskrat (*Ondatra zibethicus*)	358	165
e. Short-tailed Weasel or Ermine (*Mustela erminea*) in winter pelage. The summer coat of this species is brown.		162

a
b
c
d
e

PLATE 7 Mountain Lion and Black-tailed Deer

Mountain Lion (*Felis concolor*) (map 358; text 169) chasing Black-tailed Deer (*Odocoileus hemionus*) (map 386; text 221).

PLATE 8 Ground Squirrels and Marmot

Townsend's Ground Squirrel (*Spermophilus townsendii*), which enters California in the Great Basin areas, looks like *S. tereticaudus* (with which it does not occur).

	Map	Text
a. Antelope Ground Squirrel (*Ammospermophilus leucurus*)	359	234
b. Nelson's Antelope Ground Squirrel (*Ammospermophilus nelsoni*)	359	235
c. Golden-mantled Ground Squirrel (*Spermophilus lateralis*)	361	237
d. California Ground Squirrel (*Spermophilus beecheyi*). This widespread species is highly variable in color and pattern. In addition, the winter coat is coarser and browner.	360	235
e. Round-tailed Ground Squirrel (*Spermophilus tereticaudus*)	360	238
f. Mojave Ground Squirrel (*Spermophilus mohavensis*)	361	238
g. Belding's Ground Squirrel (*Spermophilus beldingi*)	360	236
h. Rock Squirrel (*Spermophilus variegatus*)	361	240
i. Yellow-bellied Marmot (*Marmota flaviventris*)	362	242

a
b
c
d
e
f
g
h
i

PLATE 9 Tree Squirrels

	Map	Text
a. Northern Flying Squirrel (*Glaucomys sabrinus*)	362	241
b. Western Gray Squirrel (*Sciurus griseus*)	363	243
c. Douglas' Squirrel (*Tamiasciurus douglasii*)	363	245
d. Eastern Gray Squirrel (*Sciurus carolinensis*)		243
e. Fox Squirrel (*Sciurus niger*)		244

a
b
c
d
e

PLATE 10 Chipmunks (See Table 2)

In identifying chipmunks, note the locality, presence or absence of lateral dark stripes, overall color tone, contrast, and, finally, size and color of the patch behind the ear.

Three species which cannot be told apart from some of the following in the field are not shown. These are: (1) the Chaparral Chipmunk (*Tamias obscurus*), which looks like *merriami;* (2) the Redwood Chipmunk (*T. ochrogenys*), which looks like *senex;* and (3) the Siskiyou Chipmunk (*T. siskiyou*), which also looks like *senex.* See accounts of these species, and see Table 2 for differences in distribution.

	Map	Text
a. Yellow-pine Chipmunk (*Tamias amoenus*)	364	251
b. Merriam's Chipmunk (*Tamias merriami*)	364	251
c. Lodgepole Chipmunk (*Tamias speciosus*)	366	256
d. Long-eared Chipmunk (*Tamias quadrimaculatus*)	365	253
e. Sonoma Chipmunk (*Tamias sonomae*)	364	255
f. Least Chipmunk (*Tamias minimus*)	364	252
g. Shadow Chipmunk (*Tamias senex*)	365	254
h. Panamint Chipmunk (*Tamias panamintinus*)	365	253
i. Uinta Chipmunk (*Tamias umbrinus*)	366	256
j. Alpine Chipmunk (*Tamias alpinus*)	364	251

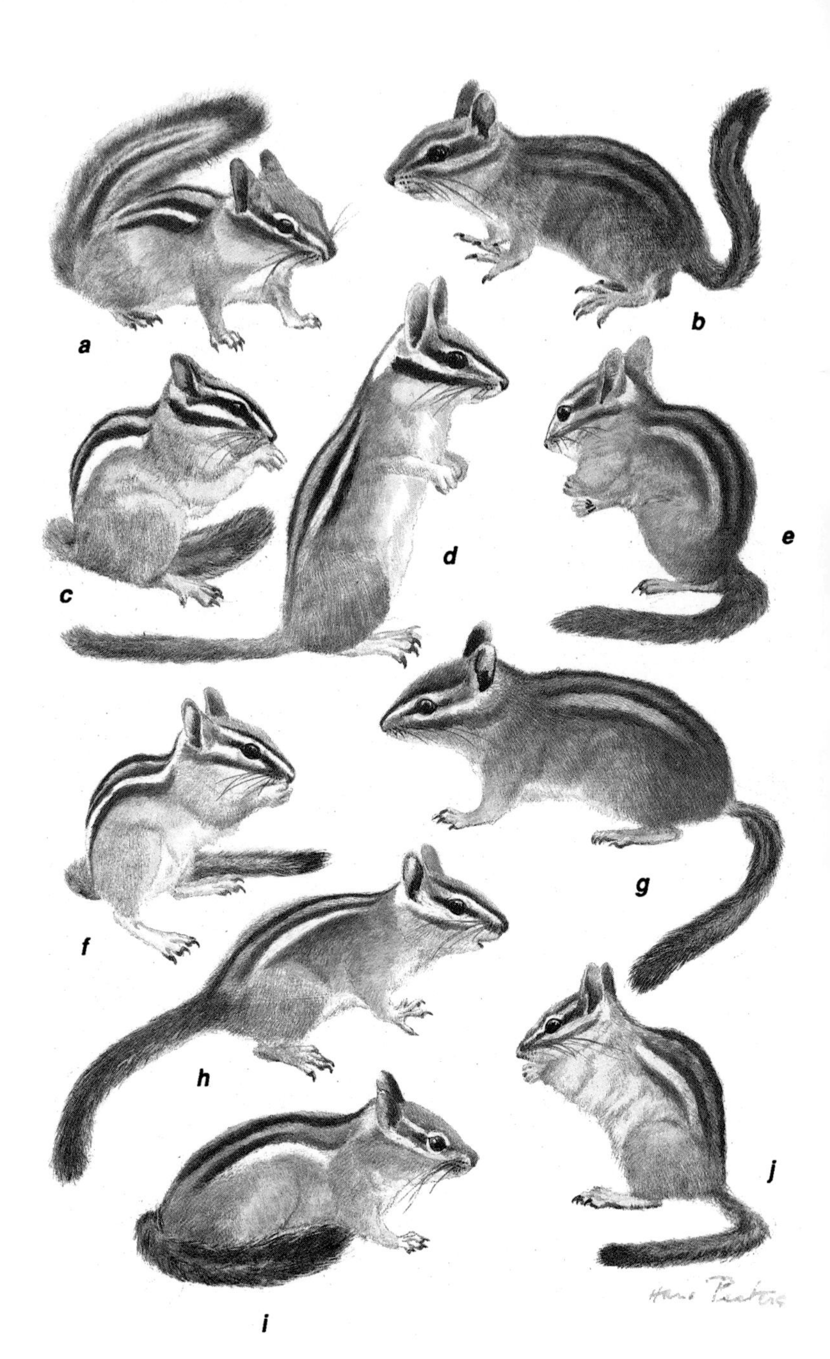
a
b
c
d
e
f
g
h
i
j

PLATE 11 Voles and Pocket Gophers

	Map	Text
a. Red Tree Vole (*Arborimus longicaudus*)	378	311
b. California Red-backed Vole (*Clethrionomys californicus*)	379	312
c. Sagebrush Vole (*Lagurus curtatus*)	379	312
d. Creeping Vole (*Microtus oregoni*)	381	319
e. California Meadow Vole (*Microtus californicus*)	380	314
f. Long-tailed Vole (*Microtus longicaudus*)	380	318
g. Botta's Pocket Gopher (*Thomomys bottae*)	366	258
h. Mazama Pocket Gopher (*Thomomys mazama*)	367	261
i. Mountain Pocket Gopher (*Thomomys monticola*)	367	262

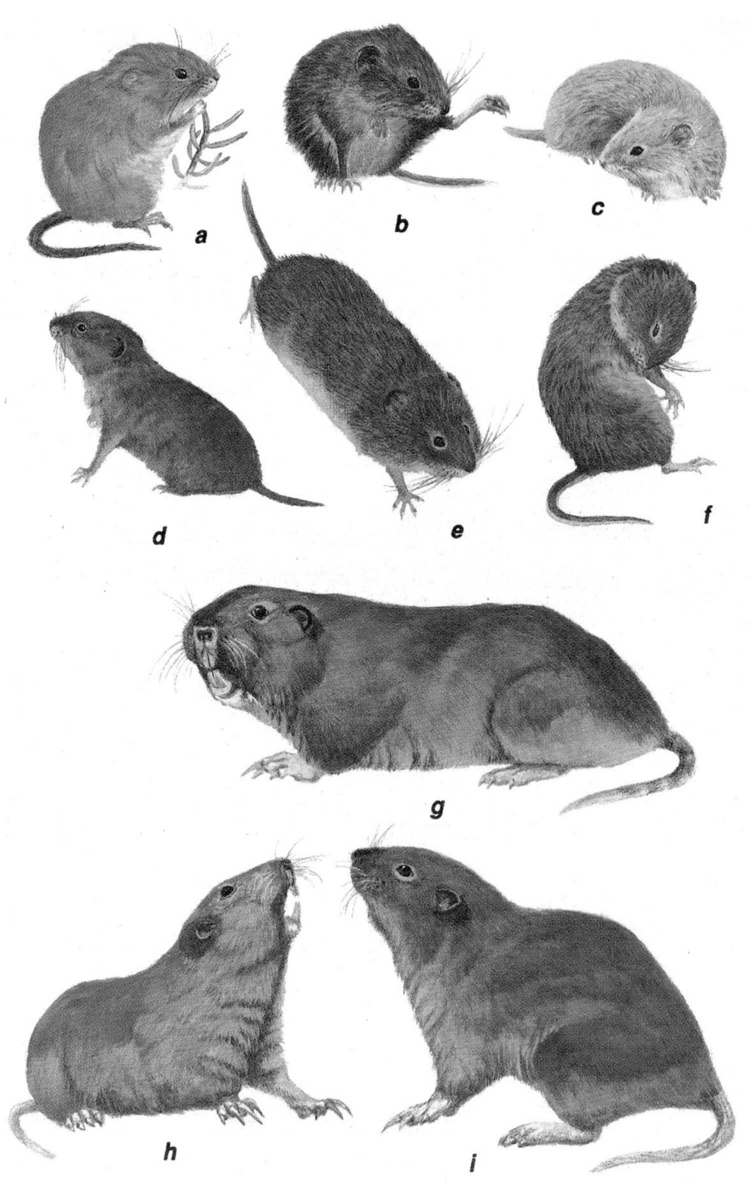
a
b
c
d
e
f
g
h
i
H Peeters

PLATE 12 Kangaroo Rats and Kangaroo Mice

Identification of these rodents is aided by noting the locality and, in the case of the rats, the relative size of the animal. Counting the number of hind toes is also helpful. Some of the species, such as *Dipodomys panamintinus,* vary in color with the color of the substrate through their ranges.

	Map	Text
a. Narrow-faced Kangaroo Rat (*Dipodomys venustus*)	369	276
b. Heermann's Kangaroo Rat (*Dipodomys heermanni*)	369	271
c. Panamint Kangaroo Rat (*Dipodomys panamintinus*)	368	275
d. San Joaquin Kangaroo Rat (*Dipodomys nitratoides*)	367	274
e. Desert Kangaroo Rat (*Dipodomys deserti*)	368	270
f. Merriam's Kangaroo Rat (*Dipodomys merriami*)	369	272
g. Dark Kangaroo Mouse (*Microdipodops megacephalus*)	370	278
h. Pale Kangaroo Mouse (*Microdipodops pallidus*)	370	278

a
b
d
c
f
e
h
g

PLATE 13 Pocket Mice

When identifying these rodents, note the locality and the presence or absence of spiny hairs on the rump.

	Map	Text
a. San Joaquin Pocket Mouse (*Perognathus inornatus*)	371	286
b. Desert Pocket Mouse (*Perognathus penicillatus*)	373	288
c. Little Pocket Mouse (*Perognathus longimembris*)	372	286
d. Great Basin Pocket Mouse (*Perognathus parvus*)	372	287
e. California Pocket Mouse (*Perognathus californicus*)	370	283
f. San Diego Pocket Mouse (*Perognathus fallax*)	371	284
g. Long-tailed Pocket Mouse (*Perognathus formosus*)	371	285
h. Bailey's Pocket Mouse (*Perognathus baileyi*)	370	283
i. Spiny Pocket Mouse (*Perognathus spinatus*)	372	288

a
b
c
d
e
f
g
h
i

PLATE 14 Rats

	Map	Text
a. Desert Wood Rat (*Neotoma lepida*)	375	297
b. White-throated Wood Rat (*Neotoma albigula*)	374	295
c. Bushy-tailed Wood Rat (*Neotoma cinerea*)	374	295
d. Brown or Norway Rat (*Rattus norvegicus*)		323
e. Roof Rat or Black Rat (*Rattus rattus*)		324
f. Black form, *R. rattus*		
g. Hispid Cotton Rat (*Sigmodon hispidus*)	380	307
h. Muskrat (*Ondatra zibethicus*); when dry, this species has very fluffy, thick fur.		320

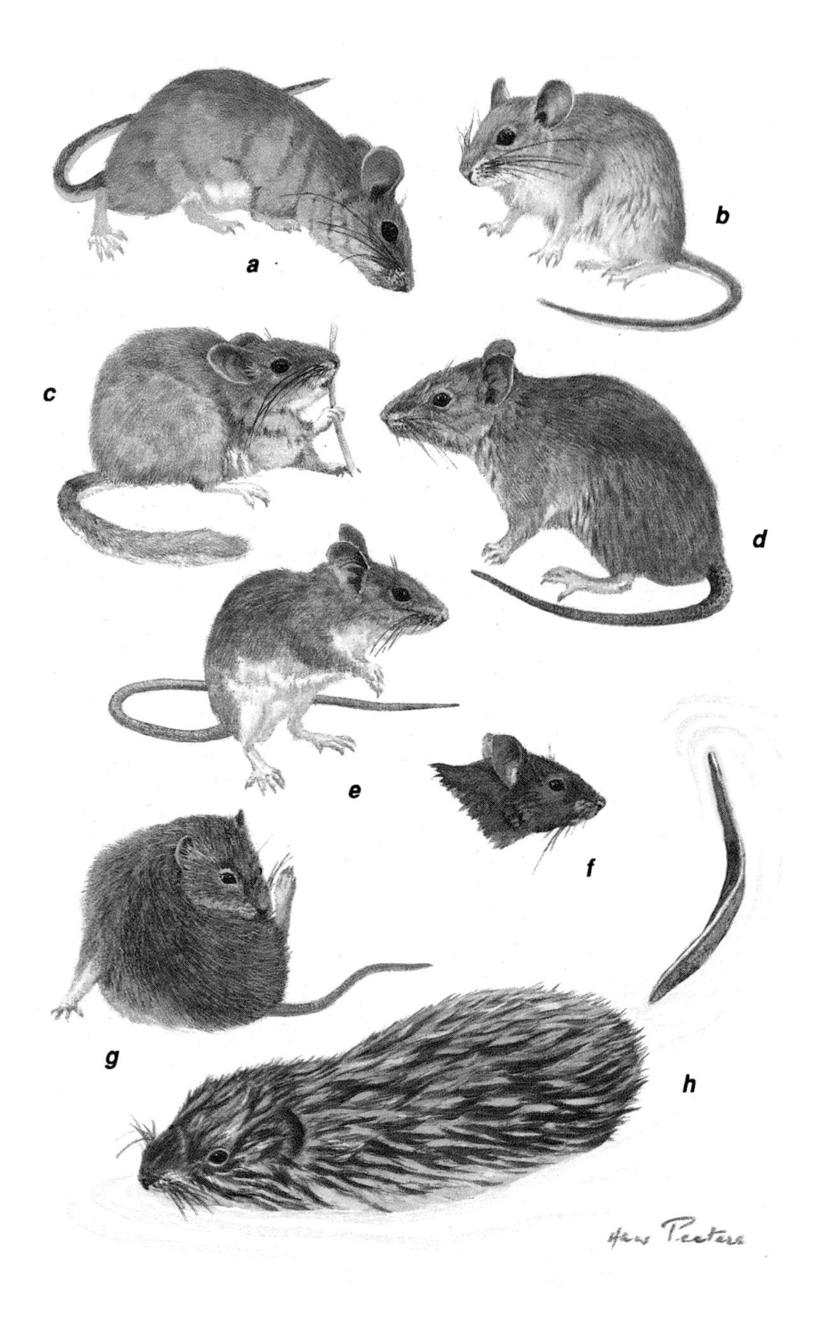
a
b
c
d
e
f
g
h

PLATE 15 Mice

	Map	Text
a. House Mouse (*Mus musculus*)		322
b. Pacific Jumping Mouse (*Zapus trinotatus*)	382	327
c. Harvest Mouse (*Reithrodontomys megalotis*)		306
d. Deer Mouse (*Peromyscus maniculatus*)		304
e. Brown form, *P. maniculatus*		
f. Cactus Mouse (*Peromyscus eremicus*)	377	303
g. Brush Mouse (*Peromyscus boylii*)	376	301
h. Piñon Mouse (*Peromyscus truei*); note very large ears.	377	305
i. Cañon Mouse (*Peromyscus crinitus*); note long hairs at tip of tail.	376	303
j. Parasitic Mouse (*Peromyscus californicus*)	376	302
k. Northern Grasshopper Mouse (*Onychomys leucogaster*); note extremely short tail.	375	299
l. Southern Grasshopper Mouse (*Onychomys torridus*); note short tail.	375	299

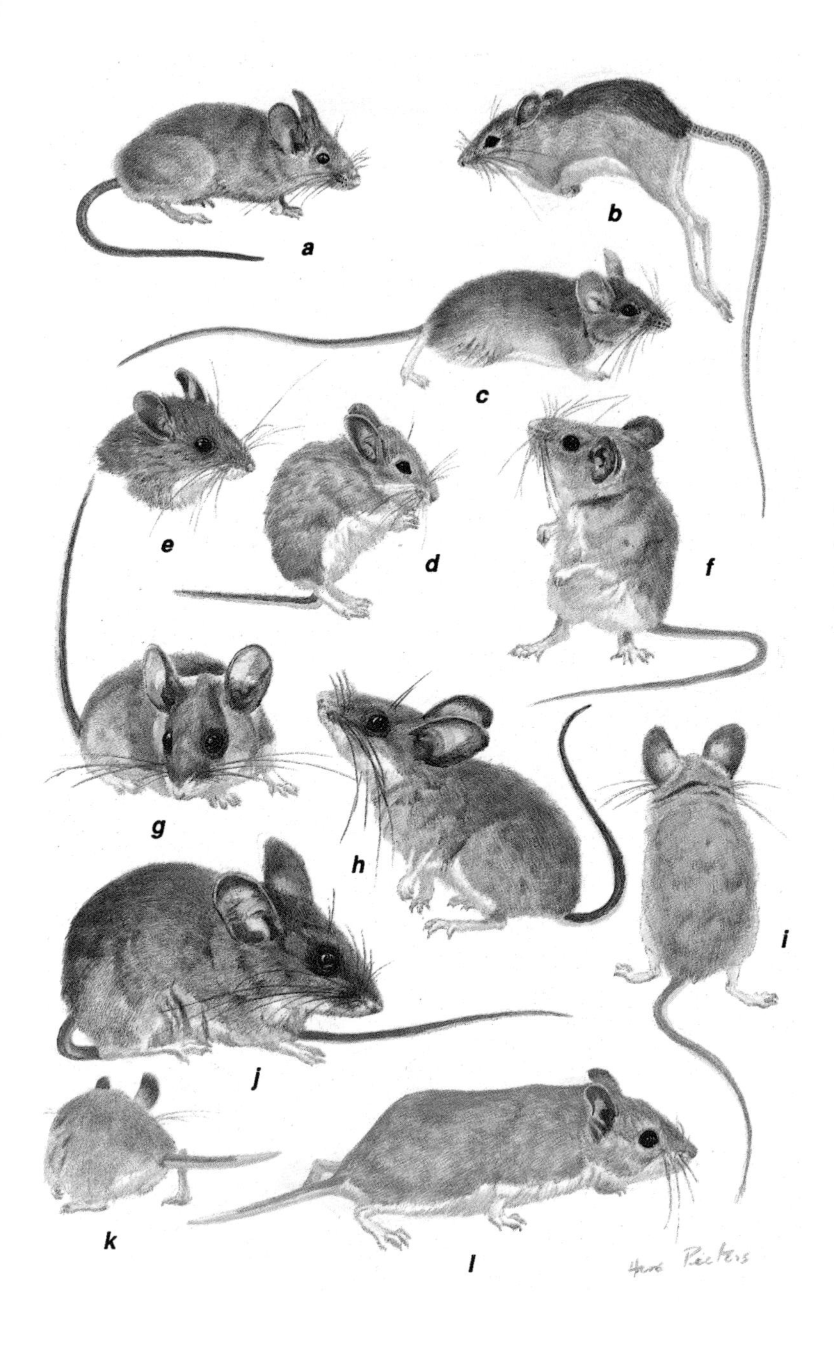
a
b
c
e
d
f
g
h
i
j
k
l

PLATE 16 Pika and Rabbits

	Map	Text
a. Pika (*Ochotona princeps*)	383	332
b. Snowshoe Rabbit (*Lepus americanus*). This species turns white in winter, with dark-tipped ears.	383	335
c. Brush Rabbit (*Sylvilagus bachmani*)	385	340
d. Audubon's Cottontail (*Sylvilagus audubonii*)	384	339
e. Black-tailed Jackrabbit (*Lepus californicus*)		337
f. Pigmy Rabbit (*Sylvilagus idahoensis*)	384	340
g. White-tailed Jackrabbit (*Lepus townsendii*). This species also has a white winter coat. Tail color is a good field characteristic in summer.	384	338

a
b
c
d
e
f
g

Distribution: Breeds on coastal islands from Baja California north to the Farallon Islands. North to Alaska during nonbreeding season. One other species in the Southern Hemisphere.

Food: Apparently bottom fish, such as ratfish and cusk eels as well as squid. Known to dive at least 200 m.

Reproduction: Unlike most other seals, a male will dominate a group of breeding females. Adults gather on coastal islands in December, and the single pup is born within days, usually early January. Nursing is completed in about a month, during which period the pup triples its natal weight. Such rapid growth results from the fat-rich milk, which is characteristic of many pinnipeds. Until the pup is weaned, the mother does not feed; by the time lactation ceases, therefore, she has lost a great deal of weight. After her pup is weaned, she mates; gestation lasts eleven months, until she returns once again to the rookery. Mating takes place on land, in contrast to the aquatic sexual pursuit characteristic of most seals. This difference may reflect the constant territorial defense by the bulls, preventing their entering the water.

The Northern Elephant Seal has had a very interesting history, for like the Guadalupe Fur Seal it was once very scarce. Its large size made it a source of a great amount of oil, and commercial exploitation of this species almost caused its extinction. A small population remained on Guadalupe Island and prospered, and breeding populations now occur as far north as the Farallon Islands. With continued protection, one can expect this northward movement to continue.

Harbor Seal (*Phoca vitulina*) [pl. 5b]

Description: Extremely variable in color from nearly white to almost black; usually some shade of brown or gray and spotted. Shape chunky, appearing neckless. Males up to almost 3 m and more than 100 kg; females slightly smaller.

Distribution: Common along the coast of California, frequently close to shore, and sometimes on isolated beaches; occurs in San Francisco Bay. They may forage close to shore and commonly haul out on rocky islets close to beaches. Common in the North Pacific and North Atlantic. Mostly coastal and known to enter rivers for 100 mi (160 km) or more.

Food: A wide variety of medium-sized fish are taken. In addition to bivalves and crabs, it eats squid, octopus, herring, flounders, and cod. Salmon comprise a very small amount of its natural diet, but the Harbor Seal may damage gill nets in trying to remove captured fish.

Reproduction: A single pup is born in the spring. The period of nursing is rather brief, as it is in most seals; mating occurs after the pup is weaned. This pattern is in contrast to sea lions, in which mating follows birth of the young by one or two weeks and long before it is weaned. Mating takes place in the water. A delay in implantation regulates development so that the next pup is born just about one year later. Breeding assemblages may contain a smaller number of both males and females.

This seal is well distributed about the northern seas of both hemispheres, although there are local variations. Probably the Atlantic and Pacific populations were in contact until the emergence of the Isthmus of Panama some 3 million years ago. This little seal seems to have remained common throughout encroachment of civilization. It tends to be timid and enters the water as people approach. Its small size is not a great inducement to commercial exploitation.

Cetacea

These large mammals are almost totally adapted for aquatic life, except for the retention of lungs and the breathing of air. Body streamlined, with tail broadened horizontally into two lobes, or flukes. Forearms are always present as paired fins, or flippers, but external hind limbs are always absent. Some species have a dorsal fin, which may be small or large.

Aspects of streamlining include the absence of hair, eyelids, and an external ear. Nostrils have moved, over evolutionary time, to the front or top of the head. The skin is thin but covers a thick insulating layer of fat, the "blubber." Mammary glands are modified also: the nipples are normally retracted into long grooves and protrude only during periods of suckling. In a female with a nursing young, there is scarcely any external evidence of mammary swelling.

The cetaceans are divided into two distinct groups, sometimes called suborders. The Odontoceti or toothed whales have well-developed teeth in one or both jaws; these teeth number from at least two to many. The nostrils are joined to form a single blowhole, which is placed dorsally but sometimes near the front of the head. There are six families of Odontoceti, four of which have been reported from California coastal areas. The Mysticeti include the whalebone or baleen whales, which lack teeth but possess many long thin baleen plates through which they strain food from water. By swimming with the mouth open, a whalebone whale engulfs small crustaceans and some small fish (up to the size of a herring), and the closely spaced baleen plates retain the food while water passes out. The Mysticeti are further characterized by paired nostrils or blowholes on top of the head. There are three families of Mysticeti, and all have representatives in California.

References

Coffee, D. J. 1977. *Dolphins, whales and porpoises: An encyclopedia of sea mammals.* New York: Macmillan.

Gaskin, D. E. 1982. *The ecology of whales and dolphins.* Exeter, N.H.: Heinemann Educational Books, Inc.

Jones, M. L.; Swartz, S. L.; Leatherwood, S.; and Folkens, P. A. 1984. *The gray whale.* New York: Academic Press.

Leatherwood, S., and Reeves, R. R. 1983. *The Sierra Club handbook of whales and dolphins.* San Francisco: Sierra Club Books.

Leatherwood, S.; Reeves, R. R.; Perrin, W. F.; and Evans, W. E. 1982. *Whales, dolphins and porpoises of the eastern North Pacific and adjacent arctic waters: A guide to their identification.* Washington, D.C.: U.S. Department of Commerce.

Matthews, L. H. 1975. *The whale.* New York: Crescent Books.

Matthews, L. H. 1978. *The natural history of the whale.* London: Weidenfeld & Nicolson.

Mörzer Bruyns, W.F.J. 1972. *Fieldguide of whales and dolphins.* Amsterdam: Tor.

Rice, D. W. 1977. *A list of marine mammals of the world.* 3d ed. National Oceanic and Atmospheric Administration Technical Report NMFS SSRF-711. Washington, D.C.: U.S. Government Printing Office.

Slijper, E. J. 1976. *Whales.* London: Hutchinson & Company.

Slijper, E. J. 1976. *Whales and dolphins.* Ann Arbor: University of Michigan Press.

Watson, I. 1981. *Sea guide to whales of the world.* New York: E. P. Dutton.

Winn, L. K., and Winn, H. E. 1985. *Wings in the sea: The humpback whale.* Hanover, N. H.: University Press of New England.

SPERM WHALES, DOLPHINS, AND PORPOISES (ODONTOCETI)

The Odontoceti or toothed whales are mostly small species, but the huge Sperm Whale has long been of commercial significance. The dolphins and porpoises are numerous in both species and number; they are extremely varied in distribution and, presumably, also their biology. There are seven families of Odontoceti, and four of these have been recorded from our coast.

Dolphins and Allies (Delphinidae)

A large, loosely knit family of variable form and size. Usually with many similar conical teeth. Some thirty-two species of which eleven have been observed in our waters. Although the family constitutes a rather heterogeneous assemblage, these species can be placed into three groups on the shape of the head: blunt-nosed, gradually pointed snout, or a distinct slender beak clearly set apart from the "brow" area. There are four species in which the head is rather large and the snout bluntly rounded: Pilot Whale, False Killer Whale, Killer Whale, and Risso's Dolphin.

Common Dolphin (*Delphinus delphis*) [fig. 55]

Description: A small (2½ m) dolphin with a distinct beak. Black or brownish dorsally with a narrow dark line from corner of mouth to base of flipper. Dorsal fin falcate (sickle-shaped) and pigmented, broader and shorter than flipper. Many similar conical teeth in both jaws.

Distribution: Found thoughout the seas of the world, avoiding polar regions. Sometimes in large schools. More common in the southern part of the state than farther north.

Food: Eats a variety of fish and squid which it pursues at some depths (perhaps 10 to 25 m) at night.

Reproduction: Gestation from ten to thirteen months; long period of lactation. Young born in summer, and pregnancies recur every two to three years.

This is one of the most familiar small dolphins and one very likely to be seen swimming at the prow of ships. It is usually not found in shallow waters but may occur, sometimes together with other kinds of dolphins, in herds in the hundreds or even thousands.

FIG. 55. Common Dolphin (*Delphinus delphis*).

Pilot Whale (*Globicephala macrorhynchus*) [fig. 56]

Description: A medium-sized dolphin, maximum length about 6 m. Dorsal fin falcate, placed in the anterior third of body. A lightly pigmented saddle just behind dorsal fin. Flippers slender, pointed, and slightly curved, about three times length of mouth. Seven to twelve pairs of teeth in each jaw. Similar in some aspects to the False Killer Whale, which has dorsal fin just slightly anterior of midpoint of body and relatively short flippers, about same length as mouth. Risso's Dolphin has dorsal fin higher than basal length.

Distribution: Found most commonly along the southern waters of the state. Worldwide. The same or similar species occur in the western Pacific and south of the equator.

Food: Mostly squid and cuttlefish; some fish.

Reproduction: Mating seems to be concentrated in the summer season; gestation occupies fifteen to sixteen months. The single young is not weaned until nearly two years later but begins to capture its own food before that time.

The Pilot Whale presumably gets its name from the frequency with which entire schools, or pods, swim ashore as a unit. Because of their great weight, they cannot breathe out of water and quickly die. Without the support of water, their mass

FIG. 56. Pilot Whale (*Globicephala macrorhynchus*).

collapses their lungs. The tendency to beach themselves has been exploited by some cultures that herd them ashore. Because of its size, it provides a substantial amount of red meat. This whale is also known as the Blackfish or Pilot Blackfish.

Risso's Dolphin (*Grampus griseus*) [fig. 57]

Description: Dark or gray (but not black) medium-sized dolphin with a blunt head. Maximum size about 4 m. Falcate dorsal fin higher than basal length just anterior to middle of body. Skin commonly scarred, presumably from fighting. Lower jaw with two to seven pairs of teeth; upper jaw without teeth. Flippers moderate and falcate, longer than mouth.

Distribution: Occurs in warm and temperate regions of the world but usually pelagic, not coastal. Found along the west coast of the United States but not common north of southern California.

Food: Feeds largely on squid.

Reproduction: Little is known of the reproductive pattern of this whale. The young is usually darker than the adult, apparently because the accumulation of scars results in a loss of pigmentation which does not return.

FIG. 57. Risso's Dolphin (*Grampus griseus*).

White-sided Dolphin (*Lagenorhynchus obliquidens*) [fig. 58]

Description: Graceful lines with a falcate fin placed midway on the back. About 2 m long. Upper and lower jaw with eighteen to twenty conical teeth. Dorsum dark or black and venter mostly white, with a narrow dorsal stripe dividing the black back into an upper and a lower part. Dorsal fin usually largely white except on anterior margin. Beak short.

Distribution: Common along California coast; may occur in large numbers. Migratory with a movement in winter from northerly populations and a departure in the spring to higher latitudes. Throughout the northern Pacific Ocean.

Food: Squid and small fish.

Reproduction: The calf is born in summer, but little is known of the details of their breeding cycle.

FIG. 58. White-sided Dolphin (*Lagenorhynchus obliquidens*).

Northern Right Whale Dolphin [fig. 59]
(*Lissodelphis borealis*)

Description: A medium-sized (3 m) dolphin with a black dorsum and a white venter. Distinctive in lacking a dorsal fin and in having the lower jaw protruding beyond the snout. Body slender and flukes small. Many small conical teeth in upper and lower jaws.

Distribution: Not common close to shore; more pelagic in occurrence. Found across the Pacific, south at least to Baja California.

Food: Feeds on squid and small fish.

Reproduction: Little is known of the breeding of this species. The young is duller than the adult and without the distinctive contrasting black-and-white pattern.

The Northern Right Whale Dolphin moves to offshore waters in the North Pacific in the summer and may occur in schools of hundreds, sometimes together with other species.

FIG. 59. Northern Right Whale Dolphin (*Lissodelphis borealis*).

Killer Whale (*Orcinus orca*) [fig. 60]

Description: Perhaps the most familiar of whales, easily recognized by its conspicuous black-and-white pattern with a large white blotch over the eye. The triangular dorsal fin stands erect and is about 2 m in height; in females it is lower and falcate. Behind the dorsal fin is a lightly pigmented saddle. Males may reach 9 m and females 7 m. Both jaws with a series of large similar teeth, ten to twelve to a side.

Distribution: Found in all seas and commonly ranging far into polar waters. Not uncommon along the California coast.

Food: Unusual in its occasional capture of birds and mammals. Often observed killing and eating seals and small dolphins, but squids and many fish are also important food items.

Reproduction: Apparently nonseasonal in breeding, but its reproductive biology is little known.

False Killer Whale (*Pseudorca crassidens*) [fig. 61]

Description: Almost entirely black. Some 6 m in length. Moderate-sized dorsal fin, slightly in front of midpoint of body. Flippers short, not longer than mouth, with a swelling on anterior margin. Eight to eleven conical teeth in each jaw.

Distribution: Mostly tropical in occurrence, the False Killer Whale reaches the southern shores of California. Whether migratory or nomadic, it sporadically appears in some numbers. Worldwide in warm oceans of the world.

FIG. 60. Killer Whale (*Orcinus orca*).

FIG. 61. False Killer Whale (*Pseudorca crassidens*).

Food: Mostly cuttlefish and squid but also bonito, snappers, and jacks.

Reproduction: Apparently mating is nonseasonal. Very little is known of the breeding pattern of this whale.

FIG. 62. Pacific Spotted Dolphin (*Stenella attenuata*).

Pacific Spotted Dolphin (*Stenella attenuata*) [fig. 62]

Description: A smallish dolphin of more or less uniform gray or dull color with poorly defined clusters of spots about the eye and on the sides behind the dorsal fin. Moderate-sized dorsal fins and flippers. A large number of small conical teeth, up to 160.

Distribution: Known from southern California. A subtropical and tropical dolphin occurring commonly as far north as Baja California.

Food: Mostly small fish of many species; also squid and cuttlefish.

Reproduction: Calving is nonseasonal; gestation is estimated to be approximately one year.

Blue and White Dolphin (*Stenella coeruleoalba*) [fig. 63]

Description: A strikingly marked, black (or dark) and white dolphin; distinctive stripes run from the eye to base of the flipper and from the eye to the belly. Beak distinct but short;

FIG. 63. Blue and White Dolphin (*Stenella coeruleoalba*).

falcate fin midway on the body. Flippers short and slightly curved. About 3 m long.

Distribution: Characteristic of warm waters and apparently prefers pelagic waters. Known from strandings along the coast of Washington, Oregon, and California but more common from Baja California southward. The same or a similar species occurs in the western Pacific.

Food: Small fish, squid, and pelagic crustaceans.

Reproduction: Mating is nonseasonal and followed by a gestation of about one year. The young takes milk for more than one year, and reproduction may occur about every three years.

The Blue and White Dolphin, also called the Striped Dolphin, is sometimes seen in large numbers, from hundreds to one or two thousand.

Rough-toothed Dolphin (*Steno bredanensis*) [fig. 64]

Description: A small (2½ m) dolphin. Snout long and slender with twenty-four to thirty-two teeth per side in both upper and lower jaws. Teeth with rough crowns and fine vertical ridges. Dorsal fin conspicuous, midway on back. Dorsum dark gray or black; venter pink and spotted.

Distribution: Widely occurring in warm seas of the world; apparently a rare species in California. It has been found on the coast of Marin County.

FIG. 64. Rough-toothed Dolphin (*Steno bredanensis*).

Food: Small fish, squid, and pelagic octopuses. They have been seen to swim slowly near the surface, perhaps a feeding maneuver.

Reproduction: Unknown.

This is a poorly known cetacean and appears nowhere to be common. It has been observed in small groups, sometimes with other species.

Bottlenosed Dolphin (*Tursiops truncatus*) [fig. 65]

Description: A dull-colored dolphin without a sharp distinction between dorsal and ventral pigmentation. About 4 m long. Beak distinct and rather stout. Falcate fin midway or slightly anterior to midway on back. About twenty conical teeth in both upper and lower jaws.

Distribution: A coastal species occurring in small groups all along our coast and along most of the coastal waters of the Pacific Ocean. Virtually worldwide and known under several names, the relationship of which is not well understood. Characteristic of warm and temperate waters.

Food: A broad spectrum of marine fish and invertebrates, especially squid and shellfish.

Reproduction: One young is born after a gestation lasting approximately a year. Mating may occur every second or third year. Weaning is a gradual process, usually complete in about a year.

FIG. 65. Bottlenosed Dolphin (*Tursiops truncatus*).

Porpoises (Phocoenidae)

Small cetaceans with blunt, rounded heads. Teeth laterally compressed and spatulate, or spadelike. Some species have a well-developed dorsal fin; a dorsal fin is present in the two species known from California.

Harbor Porpoise (*Phocoena phocoena*) [fig. 66]

Description: A small (1.5–1.8 m) cetacean with a rather small dorsal fin placed in the middle of the back. Dorsum black and venter white or whitish with a slender dark stripe above upper jaw. Weight: 45–55 kg.

Distribution: Worldwide in temperate or cold coastal waters. Frequently abundant and sometimes assembles in schools of 50 to 100 or more. Migrates to southern part of its range in autumn. Common along the California coast, south to Monterey, and sometimes enters San Francisco Bay.

Food: Takes a broad spectrum of small fish such as herring, sardines, small cod, and whiting. Also captures squid and crustaceans. Presumably its distinctive spatulate teeth have a dietary significance, but their special function remains a mystery.

Reproduction: Births may occur every other year; mating

FIG. 66. Harbor Porpoise (*Phocoena phocoena*).

sometimes follows immediately after birth of the young, so that lactation and embryonic development are concurrent. A single young.

This little cetacean is shy and not easily observed. Throughout the world it seems to be declining but for unknown reasons. It is seldom pursued by humans but may sometimes become entrapped in gill nets. It is known to be preyed upon by the Great White Shark and the Killer Whale.

Dall's Porpoise (*Phocoenoides dalli*) [fig. 67]

Description: A conspicuous black-and-white porpoise about 2 m in length. Triangular dorsal fin sometimes whitish at the tip. Middle third of sides with a large clear-cut white patch. Other features as described for the family. Weight: 200 kg or more.

Distribution: From Baja California northward along our coast to Alaska and east to Japan. Frequently encountered in interisland waters between Washington and Alaska but occurs also in deeper seas to east coast of Eurasia, where it is captured for food.

Food: Mostly cuttlefish and squid; also small fish.

FIG. 67. Dall's Porpoise (*Phocoenoides dalli*).

Reproduction: A single young. Mating is apparently not seasonal.

This is one of the small cetaceans that swims next to moving ships. It is also one of the fastest species, said to attain speeds of up to 30 knots.

Sperm Whales (Physeteridae)

Lower jaw with eight to twenty-five teeth per side, depending on the species; upper jaw without teeth. The head is blunt and the jaw ventral. Only the left blowhole functions in breathing; the right breathing passage produces sound. This family includes the Sperm Whale and the much smaller Pigmy Sperm Whale.

Pigmy Sperm Whale (*Kogia breviceps*)

Description: In many respects a miniature edition of the Sperm Whale. About 3 m long. Head not as large, relatively, as in *Physeter* but mouth entirely ventral. With a small dorsal fin. Color dark above, lighter ventrally. About fifteen curved teeth in lower jaw.

Distribution: Worldwide but not in polar waters. Little is known of their occurrence except when found dead on beaches. Apparently not scarce, though seldom seen.

Food: Various marine invertebrates, especially crabs, squid, and octopuses.

Reproduction: Little is known of the breeding of this uncommon species. Females have been known to be both pregnant and nursing at the same time, a phenomenon not uncommon in many terrestrial mammals.

The Dwarf Sperm Whale (*Kogia simus*) was long confused with the Pigmy Sperm Whale, and distinctive aspects of its biology are not known. The Dwarf Sperm Whale has been recorded from the coastal waters of California.

Sperm Whale (*Physeter macrocephalus*) [fig. 68]

Description: Distinctive in the large bulbous head, which has a vertical front margin. Head with an oil-filled "spermaceti organ," the function of which is not clearly established. Length to 18 m. Lower jaw with a row of about twenty teeth. Males much larger than females.

Distribution: Worldwide but favoring warmer seas. Females seem to avoid polar regions; males prone to wander widely. This species sometimes assembles in large numbers, especially females with young.

Food: The Sperm Whale is one of the most specialized cetaceans in its selection of food. It has a penchant for squid of various sizes, including the Giant Squid (*Architeuthis* spp.) but also takes octopuses and a variety of medium-sized fishes. It frequently hunts in water 500 m and deeper; in the region of perpetual darkness, it locates its prey by sound, including echolocation. One proposed function of the spermaceti organ is a role in echolocation. The rear of this cavity is bounded by a vertical bony crest, which is curved in a way that directs reflected sound to a single point. Feeding is continuous throughout the year.

Reproduction: Mating usually occurs in summer; gestation is prolonged, lasting some fifteen to sixteen months. The calf is

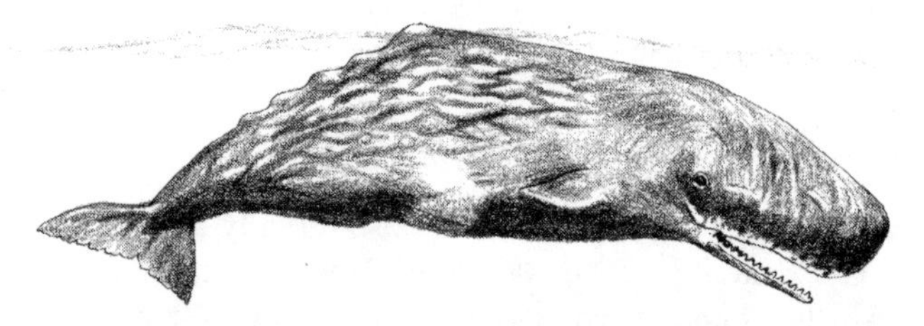

FIG. 68. Sperm Whale (*Physeter macrocephalus*).

some 4 m at birth, and births are four to five years apart for a given female. Because the Sperm Whale does not perform regular annual migrations between feeding areas and calving grounds, there is no need to restrict gestation to a period of slightly less than a year. Lactation is prolonged in the Sperm Whale, and the calf may nurse for up to two years.

As a result of the Sperm Whale's consumption of large numbers of squid, the horny beaks accumulate in its stomach and are eventually regurgitated as a floating mass called ambergris. This material is used as a base to carry fragrances and has long been of great value in the perfume industry. Despite its replacement by other more readily available materials, ambergris still has value.

Beaked Whales (Ziphiidae)

Medium-sized whales, 4 to 12 m in length, with a long, slender snout. Teeth variable in number, but usually only one or two teeth on each side of the lower jaw and none in the upper jaw. Throat with two longitudinal grooves which diverge posteriorly. Dorsal fin small and well behind midpoint of dorsum. Lower jaw projecting beyond tip of snout in some species. Teeth frequently not erupting through the gum in females.

This distinctive family is poorly known. They seem to be more pelagic than coastal, and most species are known from stranded carcasses. Some species are rather large among the Odontoceti and have been taken commercially for a long time.

A recently reported beaked whale (*Mesoplodon hectori*) stranded on beaches of southern California was previously known only from the Southern Hemisphere and probably does not regularly occur in our waters.

Baird's Beaked Whale (*Berardius bairdii*)

Description: Color dark or blackish dorsally and somewhat lighter ventrally, frequently disrupted by long scars. Snout narrow and round in cross section, hence the alternate name "bottlenosed whale." Length up to 12 m; females somewhat larger than males. Males with a pair of large teeth in each lower jaw; front teeth placed in front of the snout. Dorsal fin low, much longer at the base than high, and placed well behind midpoint of body.

Distribution: Apparently mostly pelagic. Occasionally stranded on the coast. Generally North Pacific in occurrence. In temperate and cold waters.

Food: Largely squid but also other deep-water invertebrates and fish.

Reproduction: Gestation occupies seventeen months, perhaps the longest of any whale. Mating occurs in autumn and parturition the spring of the second following year.

Because of its great size, Baird's Beaked Whale has occasionally been the object of commercial pursuit. In Japan it is one of several whales sought for their flesh.

Stejneger's Beaked Whale (*Mesoplodon stejnegeri*)

Description: A blackish whale with a low dorsal fin placed slightly behind midpoint of body. Beak may be white. In males, the single pair of stout teeth protrude outside the snout (not in front as in some other beaked whales). Up to 5 m in length.

Distribution: Known from several strandings on our coast but apparently more frequent in colder waters. Occurs across the North Pacific west to the Sea of Japan.

Food: Unknown.

Reproduction: Unknown.

FIG. 69. Cuvier's Beaked Whale (*Ziphius cavirostris*).

Cuvier's Beaked Whale (*Ziphius cavirostris*) [fig. 69]

Description: Color darkish or black; face white in adults. Dorsal fin triangular, in the rear third of the back. Frequently with numerous scars. Lower jaw protrudes beyond tip of snout, and the sole pair of teeth emerges in front of the snout. Mouth small, reaching about halfway to eye. Length to 7 m.

Distribution: The most common member of the beaked whales along the California coast. Worldwide in warm and temperate waters.

Food: Squid and small fish.

Reproduction: Virtually unknown.

WHALEBONE WHALES OR BALEEN WHALES (MYSTICETI)

This group includes the largest creatures the world has ever seen. The baleen whales have supported whale fisheries in many nations, but dwindling stocks of most species have prompted protective measures. Today most countries support the recommendation of the International Whaling Commission for a total ban on the commercial pursuit of the large baleen whales.

Gray Whale (Eschrichtidae)

Large, rather slender whales with dorsum irregular but no dorsal fin. Mouth short and curved or arched. Throat with only two or three grooves allowing for expansion of the skin in that area. Flippers and flukes broad. Color dark throughout. Only one species.

Gray Whale (*Eschrichtius robustus*) [fig. 70]

Description: Separable from rorquals by the absence of a dorsal fin. Gray Whale has a rough, knobby dorsal ridge. Color grayish both dorsally and ventrally. Only two to four throat grooves. Up to 15 m long.

Distribution: North Pacific. Occurred in North Atlantic at least until 1500 years ago and perhaps as recently as the early 1700s. Summers in far north from Bering Sea north to Arctic Ocean. Migrates south to Baja California and Korea to winter calving areas. Enters shallow waters in Gulf of California where a single calf is born.

Food: Unlike other whalebone whales, the Gray Whale forages on or near the bottom in shallow waters. It sucks in bottom-dwelling amphipods and strains them from the water with the upper-side baleen plates. Takes in food through the right side of the mouth; thus the right side tends to become worn and the baleen plates on the right side become shorter than those on the left.

FIG. 70. Gray Whale (*Eschrichtius robustus*).

Reproduction: The single calf is some 5 m long and weighs some 450 kg. It is weaned the following summer while in the productive seas of the arctic summer. Pregnancies probably occur every other year. Gestation lasts between twelve and thirteen months.

Rorquals (Balaenopteridae)

Large, rather slender whales with distinctive longitudinal grooves on the throat and a small curved dorsal fin. These ventral grooves allow for expansion of the mouth cavity when filled with food. Modern whaling methods are successful in the pursuit and capture of these whales; consequently their populations are now greatly reduced. Possibly as a result of the decline in the numbers of the large species of finback whales (*Balaenoptera*), they have more rapid growth rates and a marked decline in the age of sexual maturity. Prior to 1930 these large whales first bred at about ten years of age; by 1960, however, the age of sexual maturity had dropped to five or six years.

Five of the six known species of rorquals occur in the coastal waters of California.

Minke Whale (*Balaenoptera acutorostrata*) [fig. 71]

Description: A rather small rorqual about 10 m long. Dorsum blue or blue-gray; venter whitish to near base of tail. Flipper white in middle, dark at base and tip (or flipper with a white band). Dorsal fin relatively far forward. Snout sharp and pointed, rather like a broad knife blade. Baleen yellowish; black or dark in other rorquals.

Distribution: In the Pacific Ocean, ranging north into the Bering Sea in summer and south of California in winter. Worldwide.

Food: Primarily a planktonic crustacean called krill; but more prone than other rorquals to take anchovies and other small fish.

FIG. 71. Minke Whale (*Balaenoptera acutorostrata*).

Reproduction: Generally like the Blue Whale but differs in that the Minke Whale may produce young every year, mating shortly after parturition.

Sei Whale (*Balaenoptera borealis*) [fig. 72]

Description: Maximum size around 20 m but usually much less. Dorsal fin about one-third the distance from the tail. Color blue-black dorsally. Venter lighter except under flippers. Dorsal fin, because of its position, more visible than in the Blue Whale. Head with rostrum arched.

Distribution: Worldwide but favors warm waters. More common on California coast than Blue Whale but less common than Fin Whale.

Food: Dives less deeply than some other rorquals and feeds nearer the surface. Feeds on various planktonic crustaceans such as copepods, amphipods, and krill. Baleen is finer than in other rorquals and food is smaller.

Reproduction: Pattern similar to Blue Whale. About 1 percent of pregnancies result in twins. Young about 5 m long at birth.

Blue Whale (*Balaenoptera musculus*) [fig. 73]

Description: The largest animal ever known to have existed. Maximum length around 31 m. Dorsal fin small and far back on the body. Dorsum bluish and venter sometimes pale yellow.

FIG. 72. Sei Whale (*Balaenoptera borealis*).

FIG. 73. Blue Whale (*Balaenoptera musculus*).

When it breaks the water, the dorsal fin remains beneath the surface until just before the flukes are raised for the next dive.

Distribution: Worldwide and highly migratory. Not generally common in coastal waters when in our latitude. Summers in North Pacific.

Food: Their food consists primarily of the euphausiid shrimp called krill.

Reproduction: Gestation adjusted to migratory pattern: mating occurs in low latitudes in winter, after which the whales move to the plankton-rich arctic waters, and birth occurs some ten or eleven months later in warm waters of the low latitudes. Fetal development rapid, for the young is only 400 cm long at birth. Weaning occurs in the high latitudes some six or seven months later. Timing results in rapid embryonic growth and rapid growth of the nursing young, both occurring in the highly productive waters of the north. Pregnancies occur ever other year.

FIG. 74. Fin Whale (*Balaenoptera physalus*).

Fin Whale (*Balaenoptera physalus*) [fig. 74]

Description: A large rorqual up to 27 m but usually less. Dorsal fin more anteriorly placed than in Blue Whale. Dorsum black or blackish with a large, dark chevron mark dorsally just behind the head. Left side of lower jaw dark, right side light-colored. Venter light, sometimes yellowish (due to growth of diatoms). Rostrum flat.

Distribution: The most common rorqual off our coast. Migrates far to the north into the Bering Sea in summer. Winters south to the Gulf of California. Worldwide.

Food: Apparently feeds beneath the surface in areas of greatest concentration of krill. Also takes small fish (herring) and copepods.

Reproduction: Probably not greatly unlike Blue Whale. One young: twinning in about 1 percent of pregnancies. Young 6–7 m and 1500 kg at birth.

Humpback Whale (*Megaptera novaeangliae*) [fig. 75]

Description: At once recognized by its extremely long and rather slender flippers, which it often raises well out of the

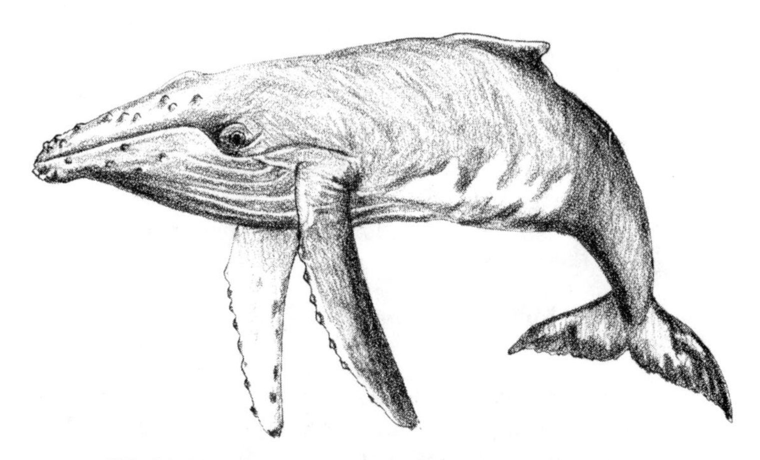

FIG. 75. Humpback Whale (*Megaptera novaeangliae*).

water. Head and flippers with rough surface. Dorsum dark and venter light. Up to 16 m long. Body stout and flukes broad. Also recognized by its habit of breaching, or jumping out of the water. A middorsal fin.

Distribution: Worldwide with clear-cut migration routes to low latitudes in winter and polar seas in summer. North to the Bering Sea and Gulf of Alaska in summer; south to California and Hawaii in winter. Recently of regular occurrence near the Farallons.

Food: Largely krill but not infrequently eats fish from the size of anchovies to cod or salmon. Seen to employ a unique method of driving fish into dense schools by a "bubble net": one or two individuals circle below a school of fish and release a constant stream of bubbles which rise as a coil, apparently frightening the fish into a dense aggregation. The whale then rises quickly to the surface with its mouth open, engulfing some of the fish so driven into the coil of bubbles.

Reproduction: Mating occurs in winter with rather conspicuous display of courtship. Conception is biennial as in other large rorquals. The single calf weighs some 900 kg and is 5 m long. The young nurses for about eleven months.

This species has received much publicity because of recordings of its "songs." These extended vocalizations may last twenty or thirty minutes and are individually distinctive. The significance of their squeaky music is not clearly established.

Right Whales (Balaenidae)

Large whales with head about one-third total length and mouth strongly arched. Flippers rather short and rounded and no dorsal fin. Throat without longitudinal grooves. One species known from our waters.

Northern Right Whale (*Balaena glacialis*) [fig. 76]

Description: A large black whale some 15–18 m long in large specimens. Sometimes white patches on throat. Other characteristics as for the family.

Distribution: Scarce in modern times; consequently distribution is poorly known. Summers in Gulf of Alaska and Aleutians to area of Bering Strait. Winter range known only from fragmentary observations but observed as far south as Baja California and Hawaii Islands.

Food: Small surface-dwelling copepods, krill, and some small fish.

Reproduction: One young perhaps every third or fourth year.

FIG. 76. Northern Right Whale (*Balaena glacialis*).

After a gestation of about one year, a calf some 6 m in length is born.

This whale is very rare today but was once very common and the most eagerly sought after species. Because it floats after being killed (in contrast to the fin whales, which sink at death), the right whale was most easily dealt with in the days of open-boat whaling and hand-held harpoons. In recent years this whale has been observed in the Santa Barbara Channel.

PERISSODACTYLA

Commonly known as odd-toed ungulates; living forms with either one or three toes. With well-developed limbs. Dentition modified for crushing plant material. Without bony horns or antlers.

Only one family, the Equidae (Horses), occurs in California. Horses have had a long history and flourished long before the development of the various families of the Artiodactyla. As the artiodactyls diversified through evolutionary time, the horses declined; today modern horses and other perissodactyls represent a small remnant of a once major group.

Reference

Berger, J. 1986. *Wild horses of the Great Basin: Social competition and population size.* Chicago: University of Chicago Press.

Horses (Equidae)

Hoofed mammals in which the middle toe is greatly developed and constitutes the hoof. Neck with a ridge of stiff, heavy hairs, the mane. Tail with long hairs. Body hair short. Legs and neck long. Dentition: 3/3, 0–1/0–1, 3–4/3, 3/3.

It is ironic that there are no longer native horses in the Western Hemisphere, where the family flourished for many millions of years. There are today two feral species in our borders: feral or "wild" horses and the Burro. In limited numbers, each species has desirable features. Each is also capable of doing harm to the environment and, indirectly at least, depressing populations of native birds and mammals.

The problems of wild horses and Burros were recognized by the "Wild Horse Annie Act" of 1959, in which Congress halted the pursuit of these animals by motorized vehicles on public lands, a small protective step. In 1971 the protection was extended by the "Wild Free-roaming Horse and Burro Act," which delegated to the Department of Interior and the Department of Agriculture the authority to control the method and extent of capture of these feral mammals. They are now protected on public lands and cannot be taken without permission.

Burro (*Equus asinus*)

Description: A small horse with large ears. Mane short and erect. A dark brown slender dorsal strip with lateral branches at the shoulder.

Distribution: Arid lands from Inyo to Imperial counties up to 3500 m or more. Native to northeast Africa.

Food: Basically a plant feeder but very adaptable. Begs food from motorists and will eat almost anything offered.

Reproduction: A single young. Breeding not known to be seasonal.

The Burro is a quaint addition to our fauna. Present populations have descended from stock discarded by miners. Fortunately these little horses are not common, for they do eat some of the native plants in Death Valley. They seem to be very well established and have few enemies.

Because they feed heavily on some of the most desirable perennial grasses and forbs, they compete with the native and less aggressive Bighorn Sheep. The burro, moreover, drives the sheep from waterholes, and there is no question that the Burro is at least partly responsible for the low densities and actual disappearance of the sheep from some of their original range.

Feral Horse (*Equus caballus*)

Description: One mammal familiar to everyone. Wild horses vary widely in color and size. Many are spotted and tend to be rather small or runty, less than 375 kg. Some are fine specimens of 500 kg or more.

Distribution: Complete distributional pattern unknown, but small herds are established in many scattered localities. Substantial populations have persisted in Modoc and Lassen counties. There is also a large population in Inyo and Kern counties (Naval Weapons Center).

Food: Plant food, both grasses and forbs; also browses leaves of many species of shrubs. Depends upon the presence of standing water.

Reproduction: A fairly rigid social organization of a dominant male with several (up to ten) females and occasionally immature males. Gestation 330–345 days. Most births in the spring in April and May. Approximately 3 percent of births are twins.

Herds of wild horses are few and widely scattered, and most are rather small. Large herds of feral horses exist in some areas of the northeastern part of the state and may compete with domestic stock and damage wildlife habitat. Similarly, horses on the Naval Weapons Center are undesirably numerous so that periodically the Bureau of Land Management removes excess horses and makes them available for adoption.

ARTIODACTYLA

Herbivores with two toes (the third and fourth) well developed. Side toes (second and fifth) rather small and not bearing weight. Upper incisors frequently lacking (see fig. 78). Many forms have a stomach of several chambers with a population of bacteria and protozoa for digestion of cellulose. Frequently with horns or antlers, usually only in the male.

This order includes deer, pigs, cattle, and sheep in the New World and antelopes and their allies elsewhere.

References

Barrett, R. H. 1978. The feral hog on the Dye Creek Ranch, California. *Hilgardia* 46:283–355.

Caton, J. D. 1977. *The deer and antelope of America.* Boston: Hurd & Houghton.

Chapman, D., and Chapman, N. 1975. *Fallow deer: Their history, distribution and biology.* Lavenham, Suffolk: Terence Dalton, Ltd.

Geist, V. 1971. *Mountain sheep: A study in behavior and evolution.* Chicago: University of Chicago Press.

Goss, R. J. 1983. *Deer antlers: Regeneration, function and evolution.* New York: Academic Press.

Jones, F. L. 1950. A survey of the Sierra Nevada Bighorn. *Sierra Club Bulletin* (June): 29–76.

McQuivey, R. P. 1978. *The desert Bighorn Sheep of Nevada.* Biological Bulletin No. 6. Reno: Nevada Department of Wildlife.

Monson, G., and Sumner, L. (Eds.) 1980. *The desert Bighorn.* Tucson: University of Arizona Press.

Murie, O. J. 1951. *The elk of North America.* Harrisburg: Stackpole Books; Washington, D.C.: Wildlife Management Institute.

Schaller, G. R. 1967. *The deer and the tiger: A study of wildlife in India*. Chicago: University of Chicago Press.
Taylor, W. P. (Ed.) 1965. *The deer of North America*. Harrisburg: Stackpole Books; Washington, D.C.: Wildlife Management Institute.
Thomas, J. W., and Toweill, D. E. (Eds.) 1982. *Elk of North America: Ecology and management*. Harrisburg: Stackpole Books.
Wallmo, O. C. (Ed.) 1981. *Mule and Black-tail Deer of North America*. Lincoln: University of Nebraska Press.
Yoakum, J. 1967. *Literature of the American Pronghorn Antelope*. Reno: U.S. Department of the Interior, Bureau of Land Management.

Pigs (Suidae)

Feet with four toes, the median pair functional. Head with a truncate, oval snout. Stomach simple, nonruminant, but with partial divisions. With well-developed incisors and canines. Dentition: 3/3, 1/1, 4/4, 3/3.

This family is not native of the New World but has been widely introduced and is adaptable, so that feral populations exist in many regions.

Feral Hog and Wild Boar (*Sus scrofa*) [fig. 77]

Description: Two-toed ungulate with a barrel-shaped (or laterally compressed) body, short legs, and a truncate snout. Legs in feral animals tend to be longer than those of domestic strains. Up to 1 m in height and weighing approximately 265 kg. Variously colored, for feral hogs and Wild Boar readily interbreed. Canine teeth large and present in both sexes.

Distribution: Open woodlands and wildlands throughout much of the state, except at higher elevations in the mountains. Populations widely scattered in North and South Coast Ranges, in foothills on western slopes of Sierra Nevada, and south to Riverside County. Channel Islands.

Food: Largely plant materials. Near irrigated land they favor green perennial grasses and forbs, especially clover; in foothills they graze on grasses in the spring and eat such native foods as acorns and manzanita berries and browse on native bushes. When acorns are scarce, they may root in soil for bulbs. Little animal food is taken.

FIG. 77. Wild Boar (Feral Hog) (*Sus scrofa*).

Reproduction: Breeding is not clearly seasonal, but most mating is in autumn. Usually five or six young in a litter, many fewer than in domestic hogs, perhaps reflecting a lower nutritional level in the feral animals. Young partly altricial and at first confined to a nest. A sow experiences estrus shortly after giving birth; if she does not conceive at that time, mating occurs when the young are weaned. Among adult boars there is vigorous competition for a sow in estrus. This rivalry does not necessarily result in hostility: when the dominant boar has completed mating, the sow mates with the next subordinate boar and so on until the dominant boar has recovered his strength and repeats his performance. Because estrus recurs repeatedly until the sow conceives, most adult sows are either pregnant or lactating. Feral hogs in California average two litters a year.

In California hogs exist both as introduced Wild Boars and as domestic stock that has become wild; the two interbreed. Feral hogs have existed in California since the early 1800s. The Wild Boar, already with some domestic blood, was first brought to California in 1924 by George Gordon Moore. A dozen animals were obtained from North Carolina and released on the San Francisquito Ranch near Carmel. Some eight years later two dozen more were released in the nearby Los Padres National Forest. Consequently the Wild Boar in our state all contain some genetic contribution from domestic stock, but not all feral hogs contain infusions of Wild Boar stock.

Today the Wild Boar and feral hog are important big game

mammals, and the total population may exceed 70,000. The annual take is an estimated 28,000 to 36,000, second only to the Black-tailed Deer and Mule Deer.

Deer (Cervidae)

Small to large two-toed ungulates, usually with bony antlers in the male, rarely in both sexes. Antlers are shed annually. Ears long and rather mobile. Legs long and tail relatively short. With several dermal glands (suborbital, tarsal, metatarsal, and interdigital), some of which produce scent. Hairs hollow in many species. With two functional toes and an outer and an inner dewclaw that do not touch the ground. Upper canines normally absent in North American species. Dentition (for species in California): 0/3, 0–1/1, 3/3, 3/3 (fig. 78).

The family of deer includes such dissimilar species as Moose, Wapiti (or "Elk"), Caribou, and the familiar Black-tailed Deer and Mule Deer. Most regions of California have at least one species, except in most parts of the Central Valley, and the Black-tailed Deer and the Mule Deer are the major large game species of our state.

Axis Deer or Chital (*Cervus axis*) [fig. 79]

Description: A medium-sized deer with permanent white spots and a middorsal dark stripe. Reddish brown at all sea-

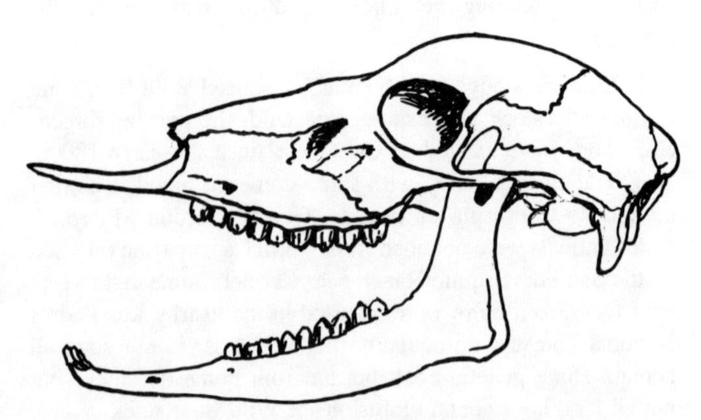

FIG. 78. Skull of the Mule Deer (*Odocoileus hemionus*).

FIG. 79. Axis Deer (*Cervus axis*).

sons. Tail and venter white. Antlers long (up to 1 m) and directed posteriorly and upward; with a small brow tine and a terminal fork. Height at shoulders about 1 m in grown bucks. Weight: 45–85 kg.

Distribution: Mostly pastures and open grassy areas. Point Reyes Peninsula. Native to India, Nepal, and Sri Lanka.

Food: Largely grasses during the rainy season; some forbs and browse at other times.

Reproduction: Apparently breeds throughout the year in California, but most fawns are born in spring. Does may conceive at less than one year of age. A single fawn is the rule.

The Axis Deer or Chital was introduced from stock from the San Francisco Zoo in the 1940s and has increased continuously since then. The population is increasing and expanding its area.

Fallow Deer (*Cervus dama*) [fig. 80]

Description: A medium-sized deer of highly variable color from nearly white to almost black; individual coat color darkens in winter but is spotted on back and sides at all seasons. Spots tend to be less clear in winter. A distinct diagonal white line on flank; rump white with a black border. Legs and belly white. Antlers with distinctive palmate or expanded distal half. Dewclaws present above hooves. Scent glands between the hooves, behind the hock, and beneath the eye. Weight: 68–92 kg (bucks), 43–72 kg (does).

Distribution: Generally open lands, Point Reyes Peninsula, central Mendocino County, Tehama County, and San Mateo County; possibly small populations occur elsewhere. Native to Europe.

Food: Grasses, forbs, and browse.

Reproduction: Rut occurs in late summer and autumn. At this time bucks are attracted to the scent of urine from does in estrus. Most mating occurs in October in California. Usually a single young born in June.

The Fallow Deer has been widely introduced throughout the world. In the Old World it is a popular game mammal. The beauty of its large, palmate antlers and spotted coat have made it a favorite in many public and private parks.

Wapiti or Elk (*Cervus elaphus*) [fig. 81]

Description: The largest of California deer, occurring in three subspecies in our state. A large, brown deer with a conspicuous mane. Bulls with large, posteriorly projecting antlers with five to seven (usually six) tines, including well-developed basal (brow and bez) tines. Head and neck darker brown than back; rump light, almost white from a distance. Unique among North American deer in having an upper canine tooth, which

FIG. 80. Fallow Deer (*Cervus dama*).

FIG. 81. Tule Elk (*Cervus elaphus nannoides*): note "bugling" posture of bull during rut.

has traditionally provided the Elks Club with a pendant. The Tule Elk (*Cervus elaphus nannoides*) is the smallest of the three forms: bulls range in weight from 195 to 250 kg. Bulls of the Roosevelt Elk (*Cervus elaphus roosevelti*) vary from 450 to 550 kg but are sometimes smaller. The Rocky Mountain Elk (*Cervus elaphus nelsoni*) is somewhat intermediate between the first two subspecies in size and is paler than the Roosevelt Elk. Weights vary with age of the individual and condition of the range.

Distribution: Originally widespread in California. The Tule Elk formerly ranged from an area north of Red Bluff, in Shasta County, southward throughout much of the Central Valley and west to the coast. The Roosevelt Elk was found from the northwest corner of the state south to about Fort Ross. The Rocky Mountain Elk is native to northeastern California and regions to the north and east. Today the Tule Elk occurs in numbers in the Owens Valley from Bishop southward; others herds have been established from Mendocino and Lake counties south to Marin County and on Grizzly Island. South of San Francisco the Tule Elk is found in scattered localities from Alameda and Contra Costa counties south to San Luis Obispo and Kern counties. Some of these herds are small, but they seem to be increasing annually. The Rocky Mountain Elk occurs at Shasta Lake, in Shasta County, part of its original range. It has been released in San Luis Obispo and Kern counties, where it is slowly increasing. (See p. 385.)

Food: Grasses and forbs, moving to new growth with the change in seasons. Browse becomes more important as low annual and perennial herbaceous plants dry out in the summer. Browses on terminal growth of many broad-leaved and some coniferous trees. Also mast, such as acorns.

Reproduction: Sexually mature in the second year, sometimes as yearlings. Bulls mature at approximately the same age as cows but are much less likely to participate in mating activity until older. A large bull dominates a small group of cows and expels other bulls. The rut and mating occur in late summer; a single calf is born the following spring after a gestation of about 250 days. Twinning is rare.

The Elk or Wapiti is the North American representative of the Red Deer of Europe. Populations of this stately deer are strictly managed by the California Department of Fish and Game. There have been supervised hunts of the Tule Elk in the Owens Valley, and the herds there number about 470 animals. The other Tule Elk herds have not been open to public hunting. Both the Rocky Mountain and the Roosevelt Elk have been subjected to small quota-type hunts in the past.

Black-tailed Deer and Mule Deer [fig. 82; pl. 7]
(*Odocoileus hemionus*)

Description: Different geographic populations of this species are generally distinguished in the field. The form in the Coast Ranges is known as the Black-tailed Deer; the Sierran form, the Mule Deer, is larger. A medium-sized deer (small in

FIG. 82. Mule Deer (*Odocoileus hemionus*).

the Coast Ranges), dorsally reddish in summer and gray-brown in winter. Ears dark; much gray about face. Tail black-tipped and sometimes with dorsal surface black; length about 130 mm. Antlers dichotomously branched, often with a small spike medially, near the base. Size varies with sex, age, and locality. Moreover, variation reflects morphological differences of the six subspecies that are known from California.

Distribution: Forests, brushfields, and meadows throughout most of the state. Absent from most of the San Joaquin Valley and from some desert areas of the southeastern part of California. Much of western North America from northern Mexico (including Baja California) north throughout the Rocky Mountains and the coniferous forests of western Canada. The Mule Deer is migratory, moving to lower elevations in the fall. (See p. 386.)

Food: A broad variety of grasses, forbs, and leaves of shrubs and small trees. Also acorns.

Reproduction: In late summer or early autumn a swelling in the neck of the mature buck marks the start of the rut, the sexual pursuit of the doe. The rut extends into January in some herds but ends in November in others. Does breed first when about a year and a half old; bucks normally become sexually mature a year later. One or two fawns are born in May or June and continue to nurse until late summer.

This species is California's most important big game mammal and a familiar sight to most hikers. They are frequently so abundant as to overbrowse their habitat and reduce the availability of their food. Deterioration of their range lowers survival of the young as well as the strength and size of those that do survive. The winter range of the Mule Deer, in the Sierra foothills, is becoming increasingly urbanized. The increase of permanent homes, with dogs, is an encroachment on the breeding range of these deer.

White-tailed Deer (*Odocoileus virginianus*) [fig. 83]

Description: A medium-sized deer of rich chestnut-red color in summer and gray in winter. Antlers with unbranched tines

FIG. 83. White-tailed Deer (*Odocoileus virginianus*).

coming dorsally from the beam. Tail snow-white ventrally and commonly held erect when in retreat; about 250 mm long. The Black-tailed Deer, by contrast, has antlers that branch dichotomously. Whether the tail of the Black-tailed Deer is held down or erect, a considerable amount of the black is clearly visible, and the tail is relatively short (about 130 mm).

Distribution: Reported in a narrow area in extreme northeastern California. From the southernmost part of Central America north through most of the United States (except the Great Basin) and including much of the Canadian coniferous forest.

Food: Probably grasses, forbs, and leaves or shrubs.

Reproduction: Little is known in our area.

There is a tantalizing history of reports of the White-tailed Deer from a narrow area where northeastern California joins Nevada. Although antlers and occasional reports of sightings indicate that populations may have existed in the past, there is no evidence that the White-tailed Deer exists in our state today.

The nearest populations of this species are near Roseburg,

in southwestern Oregon, and also near the Snake River in the northeastern region of that state.

Sheep, Goats, and Cattle (Bovidae)

Two-toed ungulates. Usually with a pair of permanent, hollow horns. These horns are unbranched and may be long and coiled or twisted. In most species both sexes are horned. Canine small or absent in skull, but may be present in lower jaw.

In North America the Bovidae include such large ungulates as Bison, Pronghorn, Bighorn Sheep, and Mountain Goat. There is also a large number of species in the Old World, especially Africa. The diverse antelopes and gazelles of the Old World all belong to the family Bovidae.

Pronghorn (*Antilocapra americana*) [fig. 84]

Description: Medium-sized antelopelike creature with prominent shiny black horns in both sexes. Horns often with two prongs in male and usually unbranched in female. General body color buff to russet on back and whitish ventrally; with blackish facial and throat markings. Rump with conspicuous patches of white, which can be "flashed" by erection of hairs. Weight: 20–70 kg.

Distribution: High-elevation sagebrush of northeastern California and also sagebrush plains about Big Pine. From central Mexico and Baja California north through Great Basin to southern Canada. (See p. 386.)

Food: An extensive fare of leaves of shrubs, including sagebrush and bitterbrush. Forbs predominate in the summer diet.

Reproduction: Mating is from late summer to early autumn. Dominant bucks are territorial, and many immature males do not mate. A strong buck may defend a territory containing some eight to fifteen does during the breeding season. One or two young born in May or June after a gestation of about 250 days. The young are extremely precocial and stand to suckle even before the embryonic fluids are dry. Sexual maturity is normally at about sixteen months of age.

The Pronghorn is a very popular game animal, and some 40,000 are taken annually in the western United States. De-

FIG. 84. Pronghorn (*Antilocapra americana*).

spite this hunting pressure, the population continues to increase. Early in this century the Pronghorn became extirpated in the region about Big Pine and Bodie; the population that exists there today is the result of an introduction in 1949 by the Department of Fish and Game. Originally the Pronghorn was common in the Central Valley from at least the Sutter Buttes southward to the desert. The Pronghorn survived in Los Angeles County (Antelope Valley) until late 1933.

Bighorn Sheep (*Ovis canadensis*) [frontispiece]

Description: A large sheep with relatively short pelage. Dorsum brown with rump buff or white; belly white. Lambs born nearly white and darken with age. Both sexes with large horns, curved in ewes and coiled in rams. Tail short. With nonfunctional lateral toes. With both tarsal and facial glands. TL 1380–1600, T 70–115, HF 390–420, E 105–120. Weight: 70–190 kg.

Distribution: High elevations (3800–4500 m) of southern

Sierra Nevada to Owens Valley (2200 m). Also desert mountains of southeastern California, San Gorgonio Mountains (near Los Angeles), and Warner Mountains. Northern Mexico (including Baja California) north through the Rocky Mountains to southern British Columbia. (See p. 386.)

Food: Grasses, sedges, and forbs.

Reproduction: Mostly from mid-November to mid-December but may vary locally. If the Mountain Sheep follows the same environmental responses that govern reproductive activity of domestic sheep, gonadal growth and breeding are stimulated by short days. Following a gestation of approximately 180 days, a single lamb is born. Older rams (three years or more) dominate the mating, but in their absence young rams are sexually active. Ewes mate in their second autumn and drop their first lamb when just two years of age.

Bighorns are among the most prized of all North American game mammals. Not only are the horns of an old ram most impressive, but their home is frequently difficult to reach. Today no hunting of Bighorn Sheep is allowed in California, but populations in general have not responded to protection. There is very little loss to poachers.

Previously the Bighorn was more common and widespread in our state. John Muir found a large number of heads and horns of Bighorn Sheep in a cave near Sheep Rock, on the north slope of Mt. Shasta. This was an area frequented by the Bighorn in winter, and it is likely that the remains found by Muir represented the accumulation of sheep killed by Indians.

RODENTIA

Mice and squirrels and their allies are diverse but united by certain readily apparent features. All have a single pair of both upper and lower incisors, growing continuously, and frequently with orange or yellow enamel on the anterior surface. The canine teeth are absent, and a large toothless space separates the incisors from the cheek teeth (premolars and molars).

References

Hall, E. R. 1981. *The mammals of North America.* 2d ed. 2 vols. New York: John Wiley & Sons.

Larson, E. A. 1981. *Merriam's Chipmunk on Palo Escrito.* Big Pine, Calif. Wacoba Press.

Linsdale, J. M. 1946. *The California Ground Squirrel: A record of observations made on the Hastings Natural History Reservation.* Berkeley and Los Angeles: University of California Press.

Linsdale, J. M., and Tevis, L. P., Jr. 1951. *The Dusky-Footed Wood Rat: A record of observations made on the Hastings Natural History Reservation.* Berkeley and Los Angeles: University of California Press.

MacClintock, D. 1970. *Squirrels of North America.* New York: Van Nostrand Reinhold Co.

McCabe, T. T., and Blanchard, B. D. 1950. *Three species of Peromyscus.* Santa Barbara: Rood Associates.

Prakash, I., and Ghosh, P. K. (Eds.) 1975. *Rodents in desert environments.* The Hague: Dr. W. Junk b.v. Publishers.

Rue, L. L. III. 1964. *The world of the Beaver.* Philadelphia and New York: J. B. Lippincott Co.

Warren, E. R. 1927. *The Beaver.* Monographs of the American Society of Mammalogists, No. 2. Baltimore: Williams & Wilkins Co.

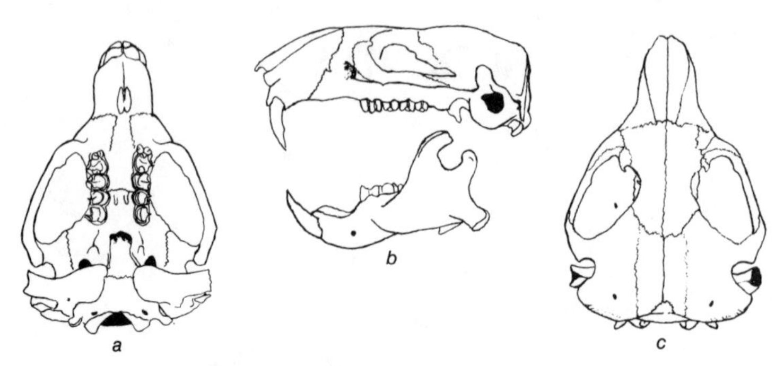

FIG. 85. Skull of the Mountain Beaver (*Aplodontia rufa*): *a,* ventral view; *b,* lateral view; *c,* dorsal view.

Mountain Beaver (Aplodontidae)

Medium-sized, stocky rodents with small eyes, reduced external ears, and a minute tail. Skull wide and shallow at the rear, producing a triangular aspect when viewed dorsally (fig. 85C). No postorbital process. Cheek teeth growing continuously. Dentition: 1/1, 0/0, 2/1, 3/3. There is only one living species.

Mountain Beaver (*Aplodontia rufa*) [fig. 86]

Description: A rabbit-sized rodent with the family characters given above. Dark grizzled brown, both dorsally and ventrally. With a white spot by each ear. Feet with soles naked. TL 300–465, T 20–35, HF 32–60, E 15–20. Weight: 800–1000 g. (See fig. 85.)

Distribution: Creekside thickets along north coast; also near mountain creeks south to Mono Lake region. Sporadically common on brush-covered hillsides, usually close to water. Their presence often shown by their burrows in moist earth. North through western Oregon and western Washington to southern British Columbia. (See p. 359.)

Food: Leaves of a broad spectrum of forbs and terminal twigs of such shrubs and small trees as deerberry, maples, dogwoods, and alders. Also berries of wild species.

Reproduction: One litter annually, usually from two to three young. Nearly naked and blind at birth.

FIG. 86. Mountain Beaver (*Aplodontia rufa*).

The Mountain Beaver is the sole survivor of a long lineage of very primitive rodents; it is not related to the true beavers (Castoridae). This family has been known as fossils from the past 20 million years. Their ancestors were once much more widely distributed and lived in Eurasia and to the east of California. It lives today on the coast and at elevations up to 2300 m in the Sierra Nevada and is active throughout the year.

Squirrels (Sciuridae)

Small to medium-sized rodents with well-developed postorbital processes. Ears erect, sometimes tufted. Tail usually rather long and bushy.

Squirrels are among the most familiar of native mammals, for most members of this family are diurnal. They include not only the tree squirrels, which have long, bushy tails, but also chipmunks and ground squirrels. Only the Flying Squirrel is nocturnal.

Key to Genera of Squirrels (Sciuridae) in California

1. Forelimbs and hind limbs not joined to each other and to sides of body by a flap or fold of skin 2

   Forelimbs and hind limbs joined to each other and to sides of body by a fold of skin; pelage very soft, light brownish gray; tail flattened, with long hairs on side and short hairs

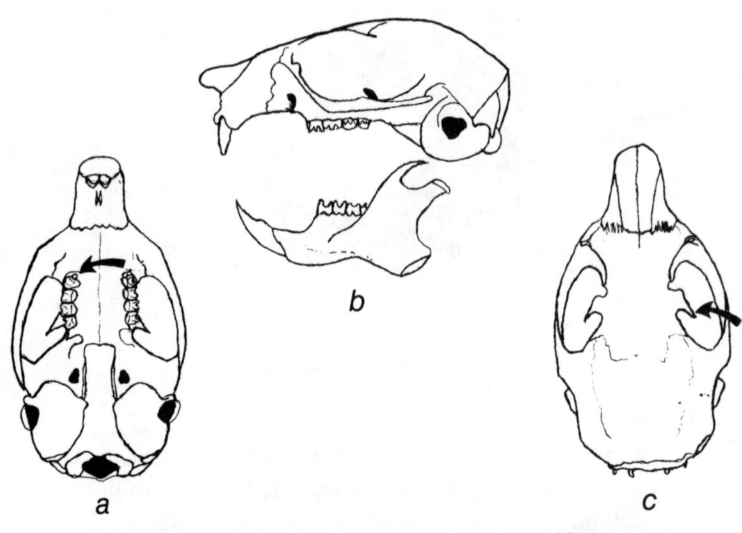

FIG. 87. Skull of the Northern Flying Squirrel (*Glaucomys sabrinus*): *a*, ventral view (arrow to peglike premolar); *b*, lateral view; *c*, dorsal view (arrow to small postorbital process).

on top and bottom; with peglike upper premolar (fig. 87b) . . . . *Glaucomys*

2. Head without distinctly contrasting light and dark stripes 3

   Head with contrasting light and dark stripes; skull with laterally directed postorbital process (fig. 88c) . . . *Tamias*

3. Tail bushy, broader than thigh; postorbital process directed distinctly toward rear rather than to the side (figs. 89c and 90c) 4

   Tail pelage not bushy but narrower than thigh; postorbital process usually directed laterally, rear margin approximately at right angles to longitudinal axis of skull (if directed toward rear, a pale lateral stripe is present) (figs. 91c and 92c) 5

4. Body without a discrete dark lateral stripe; with or without a small peglike first premolar (see fig. 89a) . . . *Sciurus*

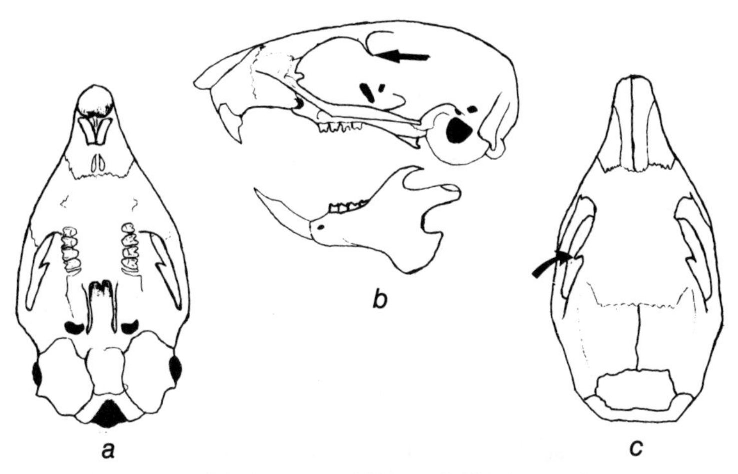

FIG. 88. Skull of the Long-eared Chipmunk (*Tamias quadrimaculatus*): *a*, ventral view; *b*, lateral view; *c*, dorsal view (arrows on *b* and *c* to postorbital process).

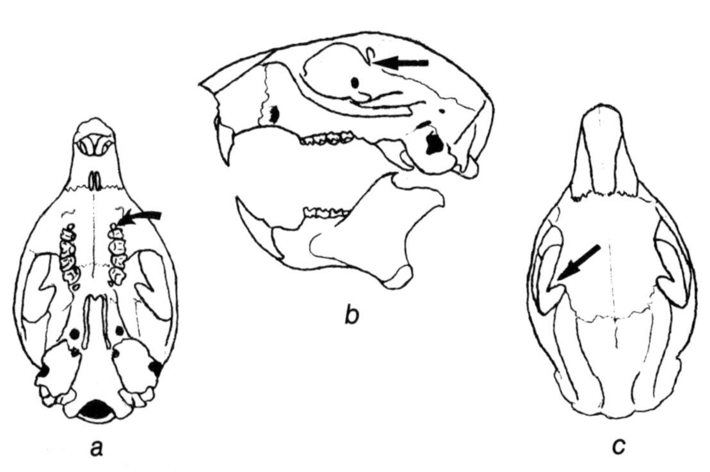

FIG. 89. Skull of the Western Gray Squirrel (*Sciurus griseus*): *a*, ventral view (arrow to peglike premolar); *b*, lateral view; *c*, dorsal view (arrows on *b* and *c* to postorbital process).

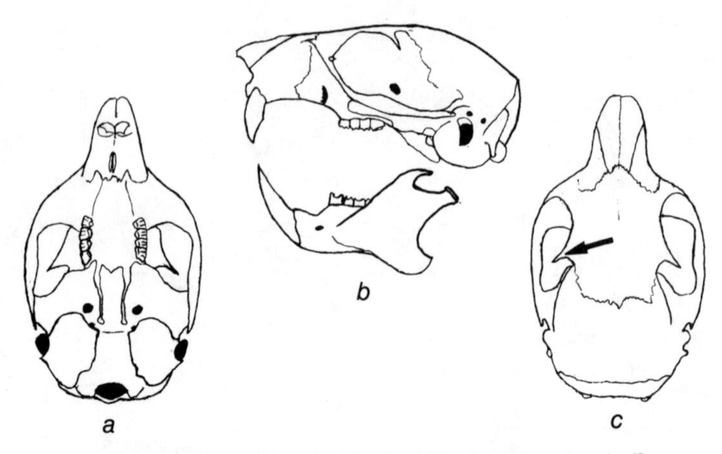

FIG. 90. Skull of Douglas' Squirrel (*Tamiasciurus douglasii*): *a,* ventral view; *b,* lateral view; *c,* dorsal view (arrow to postorbital process).

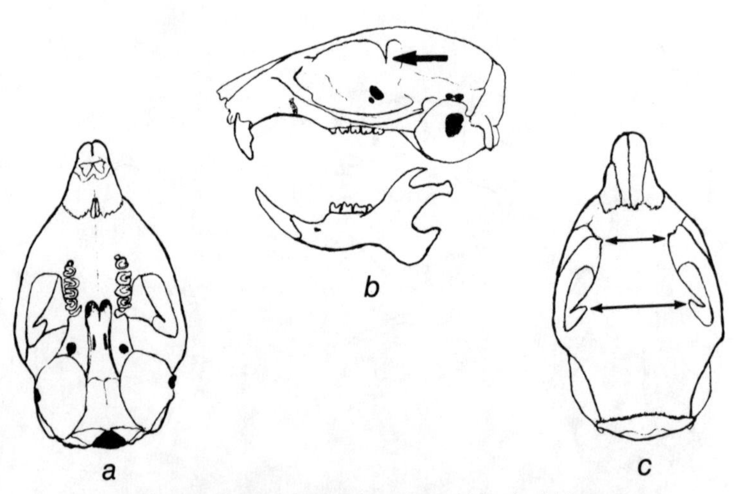

FIG. 91. Skull of the Antelope Ground Squirrel (*Ammospermophilus leucurus*): *a,* ventral view; *b,* lateral view (arrow to postorbital process); *c,* dorsal view (arrows to interorbital and postorbital constrictions).

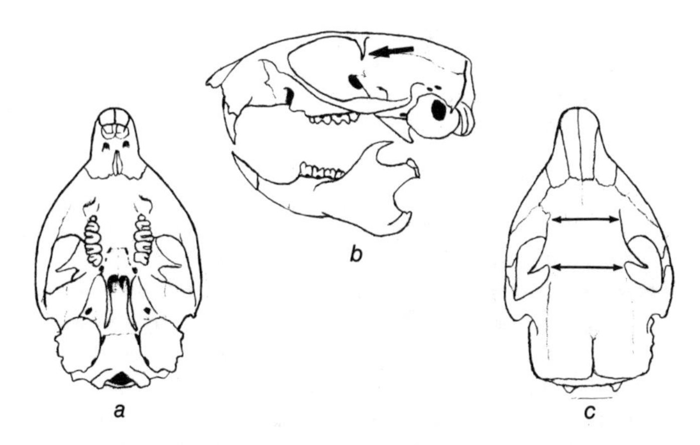

FIG. 92. Skull of the California Ground Squirrel (*Spermophilus beecheyi*): *a*, ventral view; *b*, lateral view (arrow to postorbital process); *c*, dorsal view (arrows to interorbital space and postorbital process).

Body with a distinct black lateral stripe; without a small peglike premolar (see fig. 90a) . . . . . . *Tamiasciurus*

5. Skull dome-shaped, markedly convex dorsally; postorbital process below dorsal margin of skull when viewed from the side 6

   Skull flat; postorbital process above dorsal margin of skull when viewed from the side . . . . *Marmota*

6. Interorbital constriction narrower than postorbital constriction (see fig. 91c) . . . . *Ammospermophilus*

   Interorbital and postorbital constrictions about equal (see fig. 92c) . . . . *Spermophilus*

Ground Squirrels (*Ammospermophilus* and *Spermophilus*)

Small or large-sized squirrels of various color patterns: even, flecked, spotted, or striped, but no stripes on head. Tail moderately or scantily furred, seldom bushy. Pelage coarse. Dentition: 1/1, 0/0, 1–2/1, 3/3 (see figs. 91a, b and 92a, b).

The many species of ground squirrels have long been known

under the generic name of *Citellus,* and this name is still applied to these squirrels by mammalogists in Europe and Asia. Much of the important literature on the research of these rodents lists them as species of *Citellus.*

Most of these squirrels live in open spaces and do not usually climb trees. Typically they live in the open, often near large rocks, stumps, or hillocks from which they can view the surrounding countryside. Many species are abundant and may be pests in forests or on ranches. In forests they may feed on conifer seeds and remove them when set out by foresters in their attempt to reseed logged or burned areas. Some kinds of ground squirrels are carriers of plague. In wild rodents plague is usually referred to as "sylvatic plague," but the causative organism is that which produces bubonic plague and pneumonic plague, the latter being the form known as the black death. Fleas transmit the plague organism (a bacterium) from one squirrel to another and, near human settlements, from squirrel to rat or vice versa. Thus, it is prudent to avoid close contact with wild free-living ground squirrels. If ill or dead squirrels are seen, county or state health authorities should be notified.

A ground squirrel's weight fluctuates markedly from season to season. Therefore, weights given in the following species accounts usually indicate a broad range and are not very critical in the identification.

Antelope Ground Squirrel [pl. 8a]
(*Ammospermophilus leucurus*)

Description: A small ash-colored ground squirrel with a conspicuous white lateral stripe. Tail light-colored with a dark margin. Dorsum may be a cinnamon brown, especially in summer. TL 211–233, T 63–71, HF 35–40, E 8–11. Weight: 74–103 g. Interorbital breadth less than postorbital breadth. (See fig. 91c.)

Distribution: Arid regions of sagebrush, greasewood, and other Great Basin shrubs, far from trees. East of Sierra Nevada in Great Basin and Mojave and Colorado deserts. From Oregon and Idaho east to Colorado and south to Arizona, New Mexico, and Baja California. (See p. 359.)

Food: Much green plant material when available and seeds of many desert plants, including fruits of prickly pear, buffalo berry, mesquite, and saltbrush. Also insects and small vertebrates such as mice and lizards.

Reproduction: One, or sometimes two, litters of five to eight young. Born in May but may remain in the nest until August.

The Antelope Ground Squirrel is a conspicuous member of the rodent fauna of the western edge of the Great Basin. Although it does not hibernate, its activity does seem to be reduced by inclement weather. This squirrel appears to be somewhat intolerant of others of its own kind, and it occurs widely scattered and not clustered in colonies. Nevertheless, they sometimes huddle together at night, a behavior that reduces their energy expenditure by an estimated 40 percent.

Nelson's Antelope Ground Squirrel [pl. 8b]
(*Ammospermophilus nelsoni*)

Description: A small fawn-brown ground squirrel with a white lateral stripe. Dorsum buffy, sometimes yellowish. Very similar to *A. leucurus,* which is grayish dorsally and slightly smaller than *nelsoni*. The two species do not occur together. TL 218–240, T 63–79, HF 40–43, E 8–9.

Distribution: Open grassy areas, sometimes with scattered brush. Valley and coast range foothills from southwest Merced County through western Kern County. (See p. 359.)

Food: Green leaves in winter and spring; insects and seeds throughout the year.

Reproduction: Breeds from the middle of January; young are born in mid-March after a gestation of almost four weeks. Litters range from six to twelve.

California Ground Squirrel [pl. 8d]
(*Spermophilus beecheyi*)

Description: A large ground squirrel of a general gray-brown color mottled by light flecks. Commonly a large dark mantle, which may vary from black to a mere slightly darkened area.

Rarely with faint light lateral stripes next to the dark mantle. The size, dark mantle, and usual absence of stripes distinguish this species from all other ground squirrels in California. TL 375–500, T 135–195, HF 50–64, E 17–22. Weight: 300–650 g. Interorbital and postorbital widths about equal. (See fig. 92c.)

Distribution: Common in fields of stubble, along roadsides, and on well-grazed pastures; avoids ungrazed grasslands where the cover is sufficiently tall to obstruct the view. Most of the state except the Great Basin and the southeastern desert regions. From sea level to 2200 m or perhaps above. From Washington south through Baja California. (See p. 360.)

Food: A broad spectrum of seeds, berries, and leaves of grasses, forbs, and woody plants. Corms and tubers. Often eats road-killed carrion.

Reproduction: From three to ten young in a single litter annually. Time of birth varies with locality.

This is almost strictly a ground-dwelling species that apparently needs an open area to provide a clear view of the surrounding cover. Along roadsides where the grass has become tall this squirrel is frequently seen on fenceposts or awkwardly grappling wire fencing.

Hibernates from late summer until early spring. Males enter hibernation in the summer and adult females later. Young of the year remain active until autumn.

This ground squirrel is sometimes known as the Beechey Ground Squirrel.

Belding's Ground Squirrel [pl. 8g]
(*Spermophilus beldingi*)

Description: A stout ground squirrel of a gray-brown color with a darker brown dorsal saddle. Venter and legs pinkish. Tail reddish ventrally; scantily furred. Body stocky and tail rather short. Like Townsend's Ground Squirrel, but Belding's Ground Squirrel has a pink venter and a larger hind foot (hind foot less than 41 mm in *townsendii*). TL 268–296, T 60–75, HF 42–47, E 9–10. Weight: 126–305 g.

Distribution: A common resident of high mountain meadows in Sierra Nevada up to 3000 m and sagebrush flats of Great Basin in northeastern California. From western Oregon and southern Idaho to north central Nevada. (See p. 360.)

Food: Leafy parts of growing plants, most grasses and forbs.

Reproduction: An annual litter of four to twelve young.

Belding's Ground Squirrel enters its annual dormant cycle in summer and emerges in late winter or spring when green grasses and forbs flourish. Typically it sits quite erect, either on its haunches or stretching up to the length of its hind legs, in a manner quite unlike most ground squirrels. This is its method of keeping a constant lookout, for it rarely climbs trees.

Golden-mantled Ground Squirrel [pl. 8c]
(*Spermophilus lateralis*)

Description: A medium-sized, brightly colored ground squirrel distinguished by a bright white lateral stripe bounded by black stripes and by a yellow or orange mantle over the head and shoulders. These features separate it from all other ground squirrels. Vaguely resembling chipmunks, which have striped heads, the Golden-mantled Ground Squirrel has the head unstriped. TL 235–295, T 83–102, HF 37–41, E 13–16. Weight: 136–245 g.

Distribution: In coniferous forests from Klamath and Warner ranges south through the Sierra Nevada from 1800 to 3000 m. Also San Bernardino Mountains. Extending into sagebrush flats of the Great Basin east of Sierra Nevada. From central British Columbia south through the Rocky Mountain region into Arizona and New Mexico. (See p. 361.)

Food: Many leaves and seeds of forbs and grasses. Occasionally makes heavy use of nuts. Also roots, bulbs, and other underground plant parts, including underground fungi. Some insects.

Reproduction: A single litter of four to eight young born about thirty days after the end of hibernation. Young emerge from the nest when about six weeks old.

The Golden-mantled Ground Squirrel is a very common and easily observed forest dweller. It is active through the summer holiday season, and its beautiful and unique colors make it one of the best known of California squirrels. It is a profound hibernator, and much research on hibernation has used this species as a model.

Mojave Ground Squirrel [pl. 8f]
(*Spermophilus mohavensis*)

Description: A pale, evenly pink-brown ground squirrel of rather small size. Without distinctive spots or stripes. Tail moderately furred but flattened. May occur together with the Antelope Ground Squirrel (*Ammospermophilus leucurus*), which has distinctive white stripes on the sides. TL 210–230, T 57–72, HF 32–38. Weight: 85–130 g.

Distribution: Open areas in eastern and northern parts of Mojave Desert. Reaches foothills of southern Sierra Nevada to Harper Dry Lake and Searles Dry Lake. (See p. 361.)

Food: Many seeds and vegetative parts of desert plants. Fruits of the Joshua Tree are a favorite food.

Reproduction: Little known. Breeds from early March; litters contain four to six young.

The Mojave Ground Squirrel is a relatively shy and secretive animal and seems not to be abundant anywhere. It is a hibernator and thus is dormant when most people visit its desert habitat. It emerges in March in the southern part of the Mojave Desert but may remain dormant until May farther north.

Round-tailed Ground Squirrel [pl. 8e]
(*Spermophilus tereticaudus*)

Description: A small, evenly colored, pale cinnamon-brown ground squrrel. Venter white. Tail with hair evenly distributed and very sparse, so that the appearance of the tail is more ratlike than squirrellike. Its sparsely furred tail separates it from the other ground squirrels in the state. TL 204–266, T 60–107, HF 32–40, E 5–6. Weight: 116–133 g.

Description: On sandy and coarse soils in Mojave and Colorado deserts. Generally in open areas but known to climb low shrubs. From southern Nevada through southern Arizona to northern Mexico. (See p. 360.)

Food: Green leaves and seeds of desert plants.

Reproduction: Litter of six to twelve born in June.

Townsend's Ground Squirrel (*Spermophilus townsendii*)

Description: A small, uniformly colored, or faintly flecked, ground squirrel. Gray or buff-cinnamon dorsally (two distinct color phases) and white ventrally. Ventral surface of tail reddish; tail scantily furred. Similar to *S. mohavensis,* with which it does not occur; underside of tail white in *mohavensis.* May occur with *S. beldingi,* which is larger and has a hind foot longer than 41 mm. TL 167–270, T 32–72, HF 29–38, E 8–9. Weight: 185–315 g.

Distribution: Open sagebrush flats of Great Basin. From southern Washington east through southern Idaho and south through western Utah and most of Nevada. (See p. 361.)

Food: Fresh leafy vegetation, sprouting seeds, bulbs, and insects.

Reproduction: A single litter annually, from five to twelve young, born in April. Females with fifteen embryos have been reported.

This little squirrel emerges from its "winter sleep" in winter and may be seen as early as February, when precipitation has stimulated plant growth. By early summer it has begun to enter estivation, and adult males may have gone underground by June, by which time herbaceous plant growth has dried. This pattern is biologically reasonable for a species that seems to depend so heavily on herbaceous plants for its sustenance, but it illustrates the arbitrary distinction between estivation and hibernation.

Rock Squirrel (*Spermophilus variegatus*) [pl. 8h]

Description: A large, mottled or variegated, gray-brown ground squirrel lacking the black mantle normally present on *Spermophilus beecheyi*. With neither stripes nor patches. Tail well furred. Occurs with the Antelope Ground Squirrel (*Ammospermophilus leucurus*), which is small and conspicuously striped, and the Round-tailed Ground Squirrel (*S. tereticaudus*), which is also small and has a scantily furred tail. TL 434–510, T 198–235, HF 53–60, E 15–19. Weight: 580–795 g.

Distribution: Usually in the proximity of large rocks or rocky outcrops but sometimes in burrows in open country. Providence Mountains in Mojave Desert. From Nevada to extreme western Oklahoma and south into central Mexico. (See p. 361.)

Food: Largely leafy material in spring and seeds and berries in autumn. Some insects.

Reproduction: Possibly two litters a year; from five to seven young.

This distinctive species is normally an inhabitant of rocky hillsides, and in California it is to be found only in the Providence Mountains in Mojave Desert. From Nevada to extreme western Oklahoma and south into central Mexico. (See p. 361.) known.

Flying Squirrels (*Glaucomys*)

Rather small, dull-colored squirrels modified for gliding: there is a broad membrane which forms a plane when the four legs are stretched out. The tail is conspicuously flattened, dorsoventrally, and probably guides the animal during its flight. The fur is very soft and the eyes large.

In many regions of the world one or more species of flying squirrels inhabit mature forests. These species occur from the tropics to the Arctic and are always nocturnal. They do not hibernate but are active at all seasons.

Northern Flying Squirrel (*Glaucomys sabrinus*) [pl. 9a]

Description: A medium-sized brownish-gray squirrel modified for gliding and nocturnal activity. Forelimbs and hind limbs joined by a flap of skin forming a "gliding sail" when the limbs are outstretched. Tail fur conspicuously flattened with long hairs on sides and very short hairs on top and bottom. Postorbital processes small, acute, and posterolaterally directed (see fig. 87). TL 250–310, T 115–150, HF 34–40, E 20–29. Dentition: 1/1, 0/0, 2/1, 3/3; with a peglike upper premolar (fig. 87a).

Distribution: Coniferous forests, especially with mature stands of fir (*Abies* spp.) with open understory. North Coast Ranges, Klamath, Cascade, and Sierra Nevada ranges south to Sequoia National Park; also San Jacinto and San Bernardino mountains. (See p. 362.)

Food: Various nuts and berries; some leaves, birds' eggs, and insects. In northern California the winter food may consist largely of the lichens growing on large conifers: they may feed heavily on "hair moss" (*Alectoria fremontii*) and less commonly on a staghorn lichen (*Usnea ceratina*). Hypogeous fungi are known to constitute an important summer food.

Reproduction: Nests in hollow trees. Annually one litter of two to five young born between May and June. Autumnal breeding has been recorded.

This is the only nocturnal squirrel in California. Its fondness for birds was well known to early trappers who baited Marten traps with feathers. While seldom observed in the daytime unless disturbed, it is occasionally seen by campers after dark and by entomologists pursuing moths. Like other tree squirrels, the flying squirrel does not hibernate.

It does not actually fly but glides from one tree to another. By sailing from tree to tree, losing a little height each time, the flying squirrel may quickly go well over 100 ft (about 30 m). This requires tall trees with unobstructed spaces between them, and in such a habitat these squirrels may abound.

Yellow-bellied Marmot (*Marmota flaviventris*) [pl. 8i]

Description: A large, heavy-bodied squirrel with russet grizzled pelage on the dorsum; venter yellow. Tail moderately bushy and rather short. Eyes small. Not likely to be confused with any other California mammal. TL 470–700, T 130–220, HF 70–90, E 18–21. Weight: 1.5–4.0 kg.

Distribution: Common about rocky outcrops and along rocky embankments of mountain roads, 2000 m and above. Cascades and Sierra Nevada in California. Occurring widely from an area south of southern British Columbia east to Montana and south to New Mexico. (See p. 362.)

Food: Many green plants, grasses, and forbs. They forage very close to their dens beneath rocks and boulders.

Reproduction: Mates after emergence from hibernation in April, or later, depending on the elevation and snow cover. After thirty-day gestation, three to eight young born in the underground den. Young emerge from nest at four or five weeks of age. Some females mate after the first hibernation, when nearly one year old.

Marmots are not infrequently seen basking in the early morning sun along roadside rock exposures which afford den sites. When fed by tourists, marmots become very tame. They hibernate from late summer or early autumn until the following spring.

Tree Squirrels (*Sciurus* and *Tamiasciurus*)

Slender, bushy-tailed squirrels with small but erect ears. Skull dome-shaped; snout short. Pelage without series of alternating stripes. Postorbital processes short and directed latero-caudad.

Tree squirrels are genuine forest-dwellers from sea level to the tree line. Conspicuous and loquacious, they are familiar to everyone who walks through wooded areas. They are active on the ground, while foraging for fallen seeds and underground fungi, and high in the trees, cutting cones of pines and firs. Tree squirrels do not hibernate.

Eastern Gray Squirrel (*Sciurus carolinensis*) [pl. 9d]

Description: A large, grayish squirrel with a suffusion of yellow or reddish brown dorsally; venter tawny or whitish. Tail long and bushy. Similar to Western Gray Squirrel but distinguished by the yellow or rusty color in the dorsum. Tail relatively shorter than in Western Gray Squirrel. Resembling also the Fox Squirrel, which has an essentially reddish or russet dorsum and a very long tail. TL 445–500, T 184–231, HF 61–70, E 28–35. Weight: 500–625 g.

Distribution: Urban parks and wooded streets in some California cities and adjacent woodlands. Introduced from eastern United States. Mississippi Valley and along East Coast from southern Canada to Gulf of Mexico.

Food: Leaves and nuts of woody vegetation.

Reproduction: Two to six young born in early spring after a gestation of forty to forty-five days. Two broods a year in times of abundant food.

This species is found in a number of cities, including Chico, Sacramento, and Stockton, sometimes together with the Fox Squirrel. Both of these tree squirrels are native to eastern United States. Our native Western Gray Squirrel is less likely to be observed in urban areas.

Western Gray Squirrel (*Sciurus griseus*) [pl. 9b]

Description: A large, silver-gray squirrel with a grizzled "salt and pepper" aspect to the dorsal fur. A very large, bushy tail. Occasionally with a yellowish hue. Underparts white. With a small, peglike upper premolar (see fig. 89a). Somewhat similar to the Eastern Gray Squirrel, which has been introduced in many urban areas of California, but the eastern species has consistently some yellow or brown on both the back and venter and is much smaller with a smaller tail, although it also has the peglike upper premolar. Also resembles the Fox Squirrel, another introduced species that occurs in California cities; the Fox Squirrel lacks the peglike upper premolar, however, and is extremely variable in color. The Fox Squirrel is usually (in California) very russet and frequently has the head irregularly colored, sometimes with light or dark patches. TL 500–575, T 240–280, HF 72–80, E 28–36. Weight: 735–900 g.

Distribution: An abundant tree squirrel in woodlands from sea level to approximately 1500 m in the central Sierra Nevada. At low elevations it is common in groves of native walnuts (*Juglans* spp.) and at higher elevations it is associated with Black Oak. Throughout most of the state except extremely high mountains and the deserts of the southeast. From Washington south to Baja California. (See p. 363.)

Food: A broad variety of fruits and green foliage. Forages both in trees and on the ground. Seeds of pines, oaks, and California Bay. Feeds heavily on hypogeous fungi (truffles) throughout the year; these fungi seem to constitute their main fare in most of the state. In contrast to mushrooms, which form a minor part of their diet, hypogeous fungi are heavy, solid, and highly nutritious.

Reproduction: One or probably two litters a year, depending on the food supply. Litters vary from two to six young.

These squirrels, among the most beautiful of California mammals, are diurnal and active all year. Unfortunately, they can become a serious pest in orchards of almonds, walnuts, and filberts.

Fox Squirrel (*Sciurus niger*) [pl. 9e]

Description: A rather large, russet-colored tree squirrel with a long bushy tail. Color patterns extremely variable, but most California individuals lack the extensive black areas which occur commonly in specimens in eastern United States. For comparisons between Eastern Gray Squirrel and Western Gray Squirrel see accounts of those species. Lacking a peglike upper premolar, typically present in both species of gray squirrels. TL 475–580, T 220–280, HF 51–80, E 26–33. Weight: 590–700 g.

Distribution: Urban parks and woodlands of the Central Valley and Coast Ranges. From northern Mississippi Valley south to Gulf of Mexico to Atlantic Ocean; Atlantic states from extreme western New York and Pennsylvania southward.

Food: Largely nuts of oaks and walnuts; leaves and buds of large trees.

Reproduction: Probably two broods annually. A litter of one to six, but usually two to four, is born in a nest of leaves and sticks built high in the branches of a tree. The first litter may be born in March and a second breeding may occur in late summer.

This squirrel is present in a number of California cities and is likely to be the species one sees running about lawns collecting peanuts and other tidbits from admirers. Fox Squirrels can be destructive in walnut and almond orchards, however, and in cities they may gnaw their way into the attics of homes.

Douglas' Squirrel (*Tamiasciurus douglasii*) [pl. 9c]

Description: A medium-sized tree squirrel with a chestnut-brown or olive dorsum, white venter, and a black stripe between dorsal and ventral pelage in summer. With seasonal and geographic variation in color. Ears tufted in winter. Skull lacking the peglike premolar (see fig. 90a), which is characteristic of the two species of gray squirrels. TL 330–370, T 110–140, HF 48–55, E 20–28. Weight: 200–300 g.

Distribution: Coniferous forests from North Coast Ranges east to Klamath, Cascade, and Sierra Nevada ranges south to Kern County. From southwestern British Columbia (not including Vancouver Island) south through Washington and Oregon. (See p. 363.)

Food: Various conifer seeds, including those of hemlock, pines, firs, and Douglas Fir. Also berries and mushrooms. In early spring, buds and leaves of broad-leaved trees. Birds' eggs in season.

Reproduction: One or two litters of four to seven young; first litter born in June. Nest usually in a hollow tree but sometimes a tightly built structure of leaves and twigs wedged high in the branches of a tall tree.

Chipmunks (*Tamias*)

Small ground-dwelling squirrels. Skull delicate with small postorbital processes directed posterolaterally (fig. 88). Tail long, more than 40 percent of total length. Conspicuously striped: five darkly pigmented stripes alternate with four rather lightly pigmented stripes, and these nine stripes run from the snout to the rump and nearly to the rump, sometimes interrupted behind the head. Dentition: 1/1, 0/0, 1/1, 3/3.

These beautiful little squirrels are familiar to campers and picnickers. For the most part they dwell in the mountains, but coastal species live at low elevations, and the Least Chipmunk (*Tamias minimus*) occurs among the scattered and open sagebrush of the Great Basin. The thirteen species of chipmunks in California are characterized mostly by slight differences in color, size, and color proportions—features that cannot easily be contrasted in a conventional key. For this reason we have provided comparative comments for the species occurring in different regions of the state (Table 2); more extensive descriptions are given in the accounts of species.

Although chipmunks eat many sorts of nuts, conifer seeds, and berries, they also dig up large numbers of highly nutritious underground fungi (truffles). They may also feed heavily on insects: many years ago the naturalist Vernon Bailey noted that the Least Chipmunk fed on caterpillars of butterflies and moths on sage and other chaparral species. They are frequently a nuisance because of their seed-eating tendencies. When foresters reseed coniferous trees on burned areas, chipmunks are prone to dig up the seeds and carry them to their nests.

Chipmunks are carriers of both plague and relapsing fever and should be kept out of mountain cabins. Plague is transmitted by bites of fleas, which may live in the nests of small rodents. Relapsing fever is carried by a tick that inhabits chipmunk nests. This night-feeding tick sometimes enters human bedding while the occupants are asleep and returns to the chip-

TABLE 2. Distribution and Diagnostic Features of California Chipmunks (Refer to Map 1 for Localities)

Locality	Species	Features
Regions north of a line from San Francisco Bay to Lake Tahoe		
Extreme northwestern part of state	*siskiyou*	Light stripes with many gray and brown hairs; lateral and median light stripes similar
	amoenus	Outer light stripes nearly white and much lighter than inner pair
Cascade, Klamath, and North Coast ranges	*amoenus*	Median light stripes broader than outer light stripes
	sonomae	Dorsal stripes all of about equal width
Between Klamath and Van Duzen rivers	*senex*	Light dorsal stripes very dark, about equally pigmented
Between Van Duzen River and Bodega	*ochrogenys*	Not separable from *senex* in the field
Northern Sierra Nevada (north of Lake Tahoe)		
Mostly above 2200 m	*amoenus*	Median light stripes broader than outer light stripes, outer stripes nearly white; sides russet; ear 13–17 mm; dark head stripes brown
Mostly above 2200 m	*speciosus*	Medium-sized, larger than *alphinus;* dark head stripes black; outer dorsal stripes white; sides of body reddish; ear medium (18–20 mm) with distinct light patch behind ear

TABLE 2. (*Continued*)

Locality	Species	Features
Mostly above 2200 m	*quadrimaculatus*	Colors bright and ears long (18–24 mm); conspicuous white stripe from snout and below eye to rear of ear, showing as bright white patch behind ear; lateral light stripes nearly white
Mostly above 2500 m	*umbrinus*	Medium-sized with grayish tones to pelage; lateral light stripes white, outer dark stripes faint; ear not conspicuously long (16–19 mm)
From 1500 to 2500 m (including Great Basin)	*panamintinus*	Reddish except for top of head, which is gray; lateral light stripes not white; ears moderate (16–18 mm)
Mostly above 1500 m	*senex*	Rather large and stocky, darkly colored; lateral and median light stripes with many brown and gray hairs; gray behind ear
From 250 to 1500 m	*merriami*	Large, dull-colored; all light stripes grayish; gray behind ear
Great Basin (east of southern Sierra Nevada)	*minimus*	Very small; grayish with mantle of gray over shoulders; lateral light stripes conspicuous
Mostly above 2000 m	*senex*	Rather large and stocky, dark colors; lateral and median light stripes with many brown and gray hairs; gray behind ear
Mostly above 2000 m	*quadrimaculatus*	Colors bright and ears long (18–24 mm); conspicuous white stripe from snout and below eye to rear of ear, showing as bright white patch behind ear; lateral light stripes nearly white

Locality	Species	Features
Great Basin (east of northern Sierra Nevada)	*minimus*	Very small; grayish with mantle of gray over shoulders; lateral light stripes conspicuous
Regions south of a line from San Francisco Bay to Lake Tahoe		
South Coast Ranges	*merriami*	Large, dull-colored; all light stripes grayish and not white; grayish spot behind ear
Peninsular Ranges	*merriami*	Characters as for *merriami* above
	obscurus	Much like *merriami;* color paler; dark dorsal stripes reddish; may not always be separable from *merriami* in the field
Transverse Ranges	*merriami*	Characters as for *merriami* above
	speciosus	Similar to *merriami* but sides russet and lateral light stripes white
Southern Sierra Nevada (south of Lake Tahoe)		
Mostly above 3000 m	*alpinus*	Size very small; dark head stripes brown; dorsal stripes indistinct, light stripes not white
Mostly above 2200 m	*amoenus*	Median light stripes broader than outer light stripes, outer light stripes nearly white; sides russet; ear 13–17 mm

munk nests before morning. Relapsing fever is not usually a serious problem, but in 1973 there were sixty-two human cases acquired by visitors staying in vacation cabins on the north rim of the Grand Canyon.

The taxonomic relationships of some California chipmunks have been drastically altered by recent studies. These studies involve arrangement and number of chromosomes and also details of the baculum or os penis, a bone that lies within the penis and is characteristic of certain species. Since these structures cannot be employed in field identification, they are omitted from our discussions. Fortunately, however, species which differ only in features of the baculum and chromosomal arrangement usually do not occur together.

Alpine Chipmunk (*Tamias alpinus*) [pl. 10j]

Description: A small, drab-colored chipmunk of yellowish hue. Light and dark body stripes not strongly contrasting. Stripes on head brown. Lateral light stripes grayish, not white. Ears short. Smaller and darker than other species with which it occurs. In *T. speciosus,* lateral light stripes are white and sides of body reddish or russet. TL 166–195, T 70–85, HF 28–31, E 12–14. Weight: 30–42 g.

Distribution: In relatively open coniferous forests at very high elevations, generally above 3000 m. Southern Sierra Nevada. (See p. 364.)

Food: Seeds of many kinds of forbs and grasses. Perhaps because this species dwells at elevations where conifer growth is thin, seeds of pines do not constitute a major source of food. Known also to gather fungi.

Reproduction: From four to five young born in June.

This species ranges higher than any other chipmunk in California and can be seen well above timberline. It probably has a greatly restricted period of activity, from late spring to early autumn. Very little is known of the biology of the Alpine Chipmunk, however.

Yellow-pine Chipmunk (*Tamias amoenus*) [pl. 10a]

Description: Size rather small but larger than *minimus;* the tail of *amoenus* is shorter than in *minimus*. Lateral brown sides distinctly more reddish than in *minimus* or *alpinus*. TL 188–202, T 73–85, HF 29–31, E 10–12. Weight: 36–50 g.

Distribution: Yellow-pine and Jeffrey Pine forests from the Yosemite area northward in the Sierra; in the Klamath, Shasta, and Warner mountains in the north. Ranges east to Colorado and Wyoming and north to British Columbia. (See p. 364.)

Food: Seeds of many forbs and shrubs, especially manzanita, ceanothus, and conifer seeds. Also eats substantial amounts of insects, especially in the spring, and fungi in the fall.

Reproduction: A single litter of four to six young is born in early June.

This species accumulates fat in autumn as do other California chipmunks, but apparently *T. amoenus* is a nonhibernator. In captivity it remains active in the winter under conditions that allow torpidity in *T. speciosus, quadrimaculatus,* and *senex*.

Merriam's Chipmunk (*Tamias merriami*) [pl. 10b]

Description: One of the larger species of California chipmunks; colors rather dull and drab and with light stripes grayish (not white). Head grayish (not brownish); light patch behind ear grayish (not white). Tail relatively long. TL 233–277, T 110–124, HF 30–39, E 16–18. Weight: 60–82 g.

Distribution: Lower western slopes of southern Sierra, Coast Ranges, and in wooded areas (generally above chaparral) south to Baja California. (See p. 364.)

Food: Seeds of various forbs and shrubs; insects.

Reproduction: Breeds rather early. Probably a single litter annually of three to five young born in late April.

Perhaps because this chipmunk occurs at low elevations, it is active for most of the year and breeding is early. Like the Lodgepole Chipmunk, Merriam's Chipmunk may climb trees and is at home on rail fences.

Least Chipmunk (*Tamias minimus*) [pl. 10f]

Description: The smallest chipmunk in California. Dorsal stripes less distinct than in most species. Underside of tail rather yellow; in *T. amoenus,* with which *minimus* could be confused, underside of tail is reddish. Dark dorsal stripes wider than adjacent light stripes. TL 184–203, T 71–90, HF 28–32, E 12–14. Weight: 27–38 g.

Distribution: Sagebrush areas, generally east of the Sierra Nevada. From Arizona and New Mexico north through the Rocky Mountains, east to northern Minnesota and Wisconsin, and in Canada from Yukon east along the southern margin of Hudson Bay to western Quebec. (See p. 364.)

Food: Seeds of desert grasses and forbs. Like other chipmunks, they sometimes feed on insects, especially caterpillars.

Reproduction: A litter of four to six young is born early in the year (April in California).

Chaparral Chipmunk (*Tamias obscurus*)

Description: A large, dull-colored chipmunk, probably not separable from *T. merriami* in the field. Light body stripes grayish, not white; patch behind ear gray, not white. This species is distinguished by bony parts of the genitalia of both sexes. TL 208–240, T 75–120, HF 30–37, E 13–40. Weight: 70–90 g.

Distribution: Peninsular Ranges, extreme southern California. Also northern part of Baja California.

Food: Unknown.

Reproduction: Breeding begins early in the year (February); probably more than two litters of three or four young characterize this species.

This chipmunk is active throughout the year.

Redwood Chipmunk (*Tamias ochrogenys*)

Description: A large, dark chipmunk with body stripes rather indistinct. General tone olive in winter and tawny in summer; colors dull, never bright. Light stripes of about equal pigmentation. TL 252–277, T 107–126, HF 37–39, E 15–18. Weight: 65–92 g.

Distribution: Coastal conifer region from Van Duzen River south to Bodega within approximately 25 mi (40 km) of ocean. (See p. 365.)

Food: A broad variety of small seeds, berries, and flowers of shrubs.

Reproduction: A single litter of three to five young.

Panamint Chipmunk (*Tamias panamintinus*) [pl. 10h]

Description: A medium-sized chipmunk of bright coloration. Rump gray. Larger than *T. minimus,* which has gray shoulders and adjacent part of back. Ventral surface of tail reddish; yellowish in *minimus*. Smaller than *T. quadrimaculatus* and without the long ears and white postauricular ear patches of that species. TL 190–214, T 80–95, HF 28–31, E 17–18. Weight: 55–70 g.

Distribution: Open piñon–juniper forest of eastern slope of Sierra Nevada; also Kingston Range and northern edge of Mojave Desert. (See p. 365.)

Food: Seeds of piñon pine, spring forbs, and also insects.

Reproduction: A single litter of three to six young.

Long-eared Chipmunk (*Tamias quadrimaculatus*) [pl. 10d]

Description: A medium-sized, rather brightly colored species. Ears conspicuously long and slender with a bright white patch behind ears. Underside of tail russet. Lateral light stripes white or nearly so, neither grayish or brownish. The colors, long slender ears, and white postauricular patches distinguish this species from all others with which it occurs. TL 200–250, T 85–118, HF 34–37, E 18–24. Weight: 52–100 g. Postorbital processes directed laterally. (See fig. 88c.)

Distribution: Open mature coniferous forests and logged areas in the Cascades and Sierra Nevada from Susanville south to the Yosemite area. Also immediately east of Lake Tahoe in Nevada. (See p. 365.)

Food: Seeds and leaves of many plants. Fresh leaves and sprouting seeds taken in the spring, along with insects and fungi. Feeds heavily on seeds of pines and Douglas Fir when available. When conifer seeds are not available, it may subsist on nutritious hypogeous fungi.

Reproduction: A single litter of four to six young is born in June.

Shadow Chipmunk (*Tamias senex*) [pl. 10g]

Description: A large, rather dull-colored chipmunk similar to both *T. siskiyou* and *T. ochrogenys*, with which it does not occur. Light stripes grayish or brownish, not white. Larger than *T. speciosus,* which has reddish or russet sides and white lateral light stripes. Ears relatively short and without a bright white postauricular patch. TL 229–258, T 95–112, HF 35–38, E 18–24. Weight: 70–98 g.

Distribution: Most commonly in dense growth and streamside thickets in coniferous forests. From the Klamath Mountains east to the Warner Range and south through the Cascades and Sierra Nevada to the Yosemite area. North into central Oregon. (See p. 365.)

Food: Seeds of forbs and various woody plants, including manzanita, ceanothus, and conifers. Also fresh leaves in the springtime. At times, especially in autumn, may feed heavily on fungi.

Reproduction: Probably a single litter annually. From two to five young are born in May or June.

This species was previously known as *T. townsendii,* which is now known not to occur in California. The common name Shadow Chipmunk refers not only to the dark coloration but also to its tendency to make its home and forage in forest thickets and other heavily shaded areas. It is a profound hibernator and in the northern Sierra Nevada emerges in early March, when there is still much snow on the ground.

Siskiyou Chipmunk (*Tamias siskiyou*)

Description: A large, dull-colored chipmunk similar to both *senex* and *ochrogenys,* with which it does not occur. Outer lateral light stripes grayish or brownish, not white, but lighter than inner light stripes. May occur with *T. amoenus,* which is much smaller, is reddish on the sides of the body, and has lateral light stripes white. TL 250–268, T 98–117, HF 35–38, E 16–19. Weight: 65–85 g.

Distribution: Usually in heavy growth of coniferous forests in extreme northwestern California up to about 1200 m. North into central Oregon. (See p. 365.)

Food: Wild berries and nuts, including pine nuts and acorns.

Reproduction: One litter of four to six young born in June.

Sonoma Chipmunk (*Tamias sonomae*) [pl. 10e]

Description: A rather large, light-hued chipmunk. Lateral light stripes grayish, not white. Head stripes brownish or reddish, not black. Underside of tail reddish. May occur together with *T. ochrogenys,* which is a much darker species. TL 220–227, T 93–103, HF 33–39, E 15–22. Weight: 42–65 g.

Distribution: Dry, open chaparral at low elevations up to about 800 m. North Coast Ranges and southern part of Klamath Mountains. (See p. 364.)

Food: Probably seeds and leaves of chaparral plants, but little is actually known.

Reproduction: Breeds from February to end of August, with

at least two broods a year. From three to five (usually four) young in a litter.

Lodgepole Chipmunk (*Tamias speciosus*) [pl. 10c]

Description: A medium-sized chipmunk with outer dorsal stripes white. Lateral brown patches russet. Ear shorter than in *quadrimaculatus;* light area behind ear not white as in *quadrimaculatus.* TL 197–218, T 80–95, HF 30–36, E 18–20. Weight: 30–64 g.

Distribution: Sierra Nevada south to Yosemite area, ranging from forests of Yellow and Jeffrey pines to Lodgepole Pine. Also occurs in Nevada in the Lake Tahoe area. (See p. 366.)

Food: Seeds of many grasses, forbs, and trees, especially pines. Among insects it favors caterpillars.

Reproduction: A litter of three to six young is born in early July.

Among the species with which this chipmunk may occur, it most nearly resembles *T. quadrimaculatus,* which has considerably longer ears set off by a bright white patch behind each ear, both of which characters are apparent in the field. *T. speciosus* is also distinctive in spending a large amount of time in trees, and it will sometimes climb when frightened.

Uinta Chipmunk (*Tamias umbrinus*) [pl. 10i]

Description: A medium-sized chipmunk of rather dark, mostly grayish hue. Top of head gray. Lateral light stripe white; lateral dark stripes obscure. TL 210–225, T 86–103, HF 30–34, E 16–19. Weight: 52–71 g.

Distribution: Eastern slopes of southern Sierra Nevada up to 3000 m. High-elevation open forests. From northern Arizona to Wyoming, Montana, and Colorado. (See p. 366.)

Food: Seeds of various montane trees and shrubs. Piñon nuts are a favorite food.

Reproduction: From four to six young are born in late June or early July; young may continue to nurse until mid-August.

Pocket Gophers (Geomyidae)

Medium-sized, rather stocky rodents greatly modified for burrowing. Eyes very small and ear pinna (earflap) reduced. Tail nearly nude. Forefeet enlarged for burrowing. Resembling Heteromyidae in possession of fur-lined cheek pouches. Skull broad, flat, and heavy. Auditory bullae (ear capsules) small. Dentition: 1/1, 0/0, 1/1, 3/3. Closely related to Heteromyidae.

Pocket gophers are widely distributed in North America from central Mexico north to the edges of the continental coniferous forests; absent from northeastern North America. Their diggings are apparent on many light soils. Although pocket gophers and moles may occur on the same soils, moles favor moist ground and pocket gophers are more common in rather arid regions where the ground allows burrowing.

They build elaborate tunnels and, as they dig new branches, habitually place some of the dirt into old or abandoned tunnels. In winter they come to the surface and burrow through the snow; later they fill these tunnels with dirt from tunnels through the ground. When the snow melts, these "earth cores" remain in testimony to the gophers' activity during the winter.

Pocket gophers are virtually always solitary, socializing only during the breeding season. One individual occupies a system of burrows and alone may account for the destruction of many plants and vegetables in one's garden. The fortunate aspect to this unpleasant situation is that the offenders can be removed. Trapping by gopher traps is best for this purpose. Poison can be effective, especially when applied on fresh baits, such as carrots, and placed within a tunnel. Commercially prepared baits such as various pellets are much less attractive to pocket gophers.

These little rodents are notoriously variable. Innumerable populations with minor morphological distinctions, and sometimes rather major differences, have been named by taxonomists. In the past many of these geographic variants were described as separate species, but careful study reveals that the vast majority of these "species" intergrade with each other. Very few distinct species are recognized today.

Generally, different species of pocket gophers are mutually exclusive and do not occupy the same habitat in an area. In certain regions, such as northern California, there may be several species, each dwelling in the habitat to which it is best suited. Although it favors light soils, *Thomomys bottae* may occur on clay soils.

Key to Species of Pocket Gophers (*Thomomys*) in California

(Modified from Hall, 1981)

1. Sphenoidal fissure absent (figs. 93b, 94b, 95b) 2
   Sphenoidal fissure present (figs. 96b and 97b) 4
2. Ear relatively long, more than 6.9 mm from notch . . . 3
   Ear relatively short, less than 6.9 mm from notch . . . . *talpoides*
3. Reddish brown to black; ear from notch 7–8.5 mm . . . . *mazama*
   Dark brown; ear from notch 8–9 mm . . . . . *monticola*
4. Hind foot 30–40 mm; northeastern part of state . . . . *townsendii*
   Hind foot 22–35 mm; widely distributed but absent from northeastern part of state . . . . *bottae*

Botta's Pocket Gopher (*Thomomys bottae*) [pl. 11g]

Description: A highly variable species both in size and color. Usually a dull brown but sometimes yellow, buff, or black. Skull with a distinct and deep sphenoidal fissure (see fig. 97b). Incisive foramina mostly behind rear margin of infraorbital canal (see fig. 97a). TL 190–300, HF 26–38, E 5–8. Weight: 102–209 g.

Distribution: Usually on light soils but occasionally even on clay. Virtually statewide; absent from higher elevations in Sierra–Cascade ranges. May occur in some areas with *T. monticola;* in northwestern California one may also find it with *T. mazama*. North to Oregon, southern Idaho, and Colorado and south to northern Mexico, including Baja California. (See p. 366.)

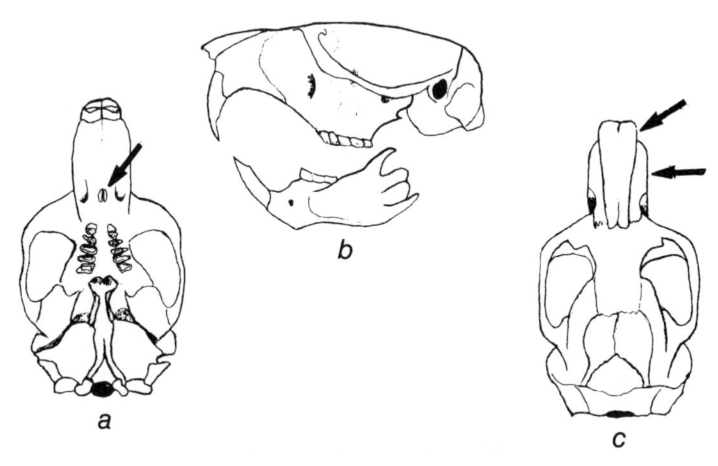

FIG. 93. Skull of the Northern Pocket Gopher (*Thomomys talpoides*): *a,* ventral view (arrow to incisive foramina); *b,* lateral view; *c,* dorsal view (top arrow to nasal bones, bottom arrow to premaxillary bones).

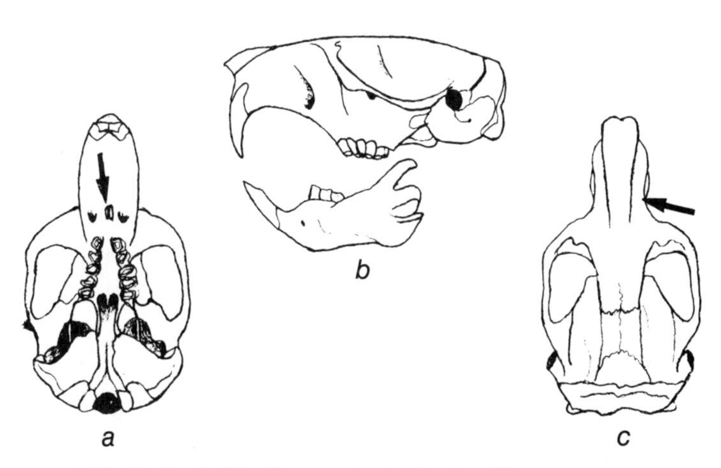

FIG. 94. Skull of the Mazama Pocket Gopher (*Thomomys mazama*): *a,* ventral view (arrow to incisive foramina); *b,* lateral view; *c,* dorsal view (arrow to premaxillary bone).

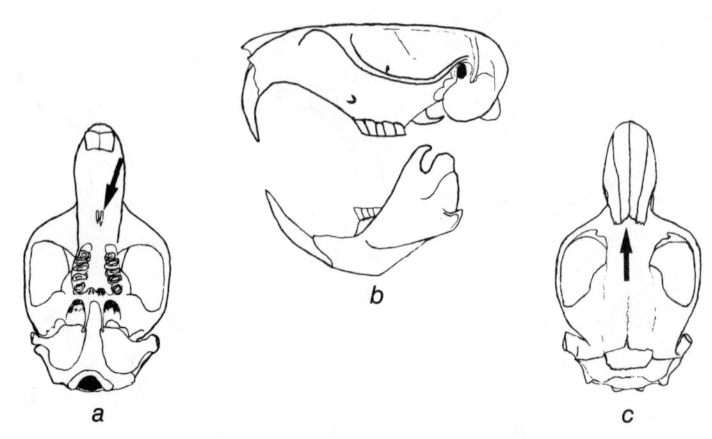

FIG. 95. Skull of the Mountain Pocket Gopher (*Thomomys monticola*): *a,* ventral view (arrow to incisive foramina); *b,* lateral view; *c,* dorsal view (arrow to nasal bones).

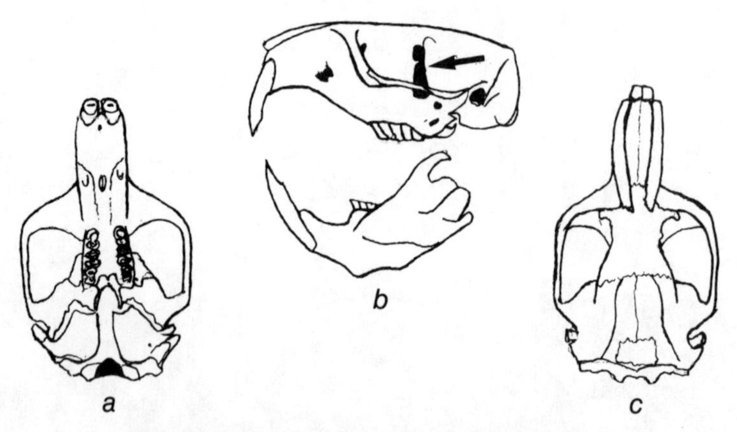

FIG. 96. Skull of Townsend's Pocket Gopher (*Thomomys townsendii*): *a,* ventral view; *b,* lateral view (arrow to sphenoidal fissure); *c,* dorsal view.

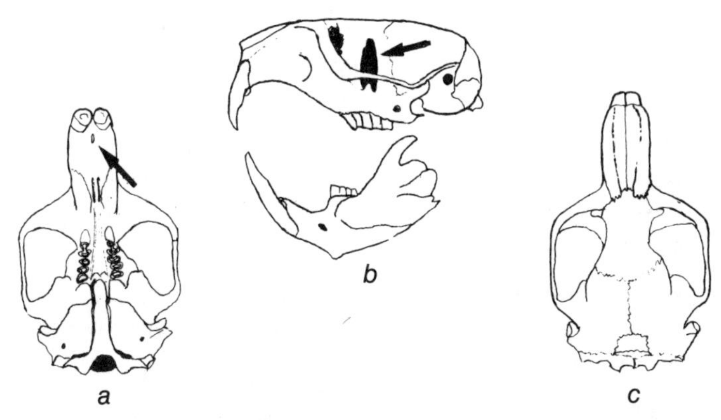

FIG. 97. Skull of Botta's Pocket Gopher (*Thomomys bottae*); *a*, ventral view (arrow to incisive foramina); *b*, lateral view (arrow to sphenoidal fissure); *c*, dorsal view.

Food: A pest on agricultural lands and in home gardens and lawns. Attracted to a continuously growing root system; eats roots, bulbs, and tender bases of growing plants. Especially fond of alfalfa; can also be very destructive to potatoes, sugar beets, and carrots.

Reproduction: Breeds from late winter to summer; but reproductive season is prolonged where land is irrigated. There may be from one to four litters of two to twelve young.

Although this species is a dietary generalist, on serpentine soils, where there is a very restricted flora, it feeds almost exclusively on the corms of the small lily *Brodiaea* spp.; in such a habitat, it may indeed depend upon this food.

Mazama Pocket Gopher (*Thomomys mazama*) [pl. 11h]

Description: A small, reddish brown pocket gopher. Dorsum especially reddish on head area. Black patches about eyes. Feet gray, scarcely lighter than on forearms. Ear shorter than in *T. monticola*. Premaxillary bones extending far beyond nasal bones (see fig. 94c). Incisive foramina anterior to infraorbital canal (see fig. 94a). No sphenoidal fissure. TL 175–262, T 45–79, HF 21–32, E 7–8.5. Weight: 90–120 g.

Distribution: On light soils and cultivated areas in northern parts of California. In Cascades west to mountains of northwestern California and east to west side of Mt. Shasta. (See p. 369.)

Food: Underground parts of plants and some leaves. Known to feed on potatoes, clover, alfalfa, and carrots, both leaves and roots or tubers.

Reproduction: Breeds through spring and summer. Litter size varies from five to seven, and females probably produce at least two litters a year.

Mountain Pocket Gopher (*Thomomys monticola*) [pl. 11i]

Description: A dark brown pocket gopher of medium size. Fur on back of forefeet white; fur about ears black. Tail with more hairs than in other species. Ear relatively long (8–9 mm) and pointed in contrast to *T. mazama* (7–8.5 mm). Posterior margin of nasal bones forming a V; premaxillary bones extending approximately as far back as nasals (see fig. 95c). Skull lacking sphenoidal fissure; incisive foramina at least partly anterior to rear margin of infraorbital foramina (see fig. 95a). TL 190–220, T 55–80, HF 26–30, E 8–9. Weight: 50–105 g.

Distribution: On light sandy or occasionally on gravelly soils in Cascade–Sierra Nevada ranges. From west side of Mt. Shasta to central Modoc County. A separate population south of Pit River in central Shasta County to Fresno County from 1000 to 2700 m elevation. Also extreme western Nevada. (See p. 367.)

Food: Leaves and underground parts of plants. Feeds on leaves as it burrows aboveground under a protective cover of snow in the winter.

Reproduction: Breeds in spring and early summer; probably one litter a year. Three to four young in a litter. Young become sexually mature at one year of age.

Northern Pocket Gopher (*Thomomys talpoides*)

Description: A rather small, brown or yellow-brown pocket gopher with gray feet. With black ear patches. Fur grayer in winter in contrast to brownish summer coat. Sphenoidal fissure absent; incisive foramina anterior to rear margin of infraorbital canal (see fig. 93a). Posterior ends of nasal bones square-ended, not V-shaped as in *monticola,* and premaxillary bones extending slightly behind nasals. (see fig. 93c). TL 170–225, T 47–70, HF 24–30, E 5–6. Weight: 74–110 g.

Distribution: Generally on sandy or loose soils in Modoc Plateau, south to Honey Lake. Also Big Valley (Bieber) in Lassen County. A widely distributed species from southern Canada to northern Arizona and New Mexico east to the Dakotas. (See p. 366.)

Food: Largely underground parts of plants of both forbs and woody species. Known to eat such species as dandelion, yarrow, sagebrush, and penstemon.

Reproduction: Breeds in June and July; probably one litter a year. From four to six young in a litter.

Townsend's Pocket Gopher (*Thomomys townsendii*)

Description: The largest pocket gopher in our state. Occurs in two color phases: gray phase has black ears; black phase has small white patches on chin and/or feet. Sphenoidal fissure present (see fig. 96b). Ears relatively small. TL 240–305, T 72–100, HF 35–40. E 5–8. Weight: 220–280 g.

Distribution: Northeastern California, usually on deep soils and frequently abundant on irrigated land. Does not occur with any other species of pocket gopher. East to Nevada and north to Oregon. (See p. 367.)

Food: Roots and aboveground parts of many forbs. Feeds heavily on grasses. Sometimes destructive on irrigated land, where it may damage such crops as potatoes and alfalfa.

Reproduction: Breeds in late winter or early spring in undisturbed environments but probably brings forth young throughout the year on irrigated land. Usually from six to eight young; sometimes simultaneously pregnant and lactating.

This very large pocket gopher is notable for the size of its burrows. The mounds of dirt it brings to the surface may measure up to 3 ft (1 m) across. Because it seems to favor moist soils, it is sometimes destructive to irrigation levees.

Kangaroo Rats, Kangaroo Mice, and Pocket Mice (Heteromyidae)

Small rodents variously modified—morphologically, physiologically, and behaviorally—for a nonfossorial life in a semiarid or arid environment. Forelimbs reduced and hind limbs enlarged. Tail elongated. Fur-lined cheek pouches. Auditory bullae (ear capsules) greatly enlarged. Dentition: 1/1, 0/0, 1/1, 3/3.

These little rodents are typical of rather sparsely vegetated, seasonally arid or desert lands and frequently constitute a substantial part of the total small rodent population. Like many animals of open areas, kangaroo rats and kangaroo mice are adapted for a saltatory, or jumping, locomotion. They use their well-developed external, fur-lined cheek pouches to carry seeds, which they take to their underground nests or hide in surface caches.

Kangaroo rats (*Dipodomys* spp.) occur commonly on rather loose, sandy soils. They are conspicuously marked with dark spots behind the eyes and ears and a distinctive white stripe on the thigh. The tail commonly has both a dorsal and a ventral stripe and a conspicuous tuft of hairs at the tip. Their appearance is in marked contrast to pocket mice (*Perognathus* spp.) and kangaroo mice (*Microdipodops* spp.), which are dull in color. In many areas pocket mice and kangaroo rats occur together; where kangaroo mice are found, all three may live on the same ground. Kangaroo rats are additionally distinctive in having a dorsal skin gland; this is lacking in the other two genera. Kangaroo mice possess a constriction at the base of the tail, so that their tail is slightly swollen in the middle.

Key to Genera of Heteromyidae in California

1. Tail without a conspicuous lateral white stripe; pelage rather evenly colored dorsally (sometimes grizzled but not spotted or striped); no dermal gland on back 2

   Tail with a conspicuous lateral white stripe; pelage of contrasting light and dark spots or stripes above eyes and behind ears, with a white stripe on thigh; a clearly marked dermal gland on neck between shoulder blades . . . . *Dipodomys*

2. Tail of even width throughout; sole of hind foot sometimes hairy; auditory bullae only slightly enlarged . . . . *Perognathus*

   Tail wider at middle than at base; sole of hind foot well furred; auditory bullae greatly enlarged . . . . *Microdipodops*

Kangaroo Rats (*Dipodomys*)

Molars flat-crowned with enamel forming a circle or oval not interrupted by a groove (in adults); rootless. Dorsal pelage with conspicuous spots and striped. Tail with a dorsal (and usually a ventral) pigmented stripe. Tail usually tufted. Dorsum with a dermal gland. Ear capsules (auditory bullae) very large, creating a triangular skull (or nearly heart-shaped when viewed from above). Fifth toe on hind foot vestigial or lacking. The earflap or pinna is rather small.

Kangaroo rats are well adapted for a desert life and are physiologically suited to aridity. They survive water scarcity by releasing metabolic wastes in a very concentrated urine. This ability to secrete high concentrations of salts is so well developed that the kangaroo rat can drink seawater and still maintain proper water balance in its tissues. In nature some species can obtain water from leaves, which are very salty. In addition to its concentrated urine, the kangaroo rat releases very dry feces. It can thereby subsist on water bound chemically and released by digestion of its dry food; such "metabolic water" maintains the body fluids of this rodent.

Behavioral adaptations provide an escape from the desert heat. Their burrows are deep enough to allow the occupants to escape midday temperatures. When the air reaches 45° C just above the surface of the ground and the soil surface an intolerable 70° C, the kangaroo rat's burrow may remain a comfortable 30–34° C. This burrow temperature, though warm, is not dangerous, for in the laboratory these rodents can adjust their metabolism to an air temperature of 37° C. Thus they adapt to desert heat mainly by avoiding it.

The kangaroo rats possess a conspicuous dorsal gland, which seems to have a role in scent marking. This may replace urine marking, which would be metabolically expensive for a mammal that must conserve water. Sand bathing is a characteristic activity of kangaroo rats; among captives, odor in sand seems to attract them to habitual sand-bathing sites.

The reproductive season of at least some kinds of kangaroo rats seems to be correlated with rainfall and the subsequent growth of forbs and grasses. Recently ecologists have pointed out the increased intake of fresh leaves by breeding kangaroo rats, a phenomenon noted by Joseph Grinnell in 1932.

In California deserts these animals are the most commonly seen mammals crossing the road at night, especially along roads that are not bordered by ditches. To the uninitiated they present a startling sight, appearing like small ghosts hopping about the pavement in kangaroo fashion. In campgrounds, kangaroo rats quickly learn to approach humans and accept proffered tidbits, not in the least disturbed by flashlights.

In areas of creosote bush (*Larrea*) or other large desert shrubs, the burrows of these rodents are easily located. Each rat owns a castlelike mound of soil or sand surrounding the base of a bush, and there are usually several entrances, the most frequently used made obvious by a profusion of radiating tracks.

Key to Species of *Dipodomys* in California
(Modified from Hall, 1981)

1. Hind foot with four toes 2

   Hind foot with five toes (fifth toe represented by a claw above inside of foot) 5

2. Tail length 160 mm or less 3

   Tail length 161 mm or more 4

3. North of Tehachapi Mountains; San Joaquin Valley . . . . *nitratoides*

   South and east of Tehachapi Mountains . . . *merriami*

4. Dorsal color dark brown . . . . *californicus*

   Dorsal color pale buff or whitish . . . . *deserti*

5. Head and body usually less than 130 mm 6

   Head and body usually more than 130 mm; Merced and Kern counties south to Santa Barbara County . . . *ingens*

6. Occurring east of Sierra Nevada–Transverse Ranges (Tehachapi)–South Coast Ranges 7

   Occurring west of Sierra Nevada–Transverse Ranges (Tehachapi)–South Coast Ranges 9

7. Lower incisors rounded on anterior surface 8

   Lower incisors flat on anterior surface (fig. 98c) . . . . *microps*

8. Hind foot usually less than 44 mm; total length less than 280 mm . . . . *ordii*

   Hind foot usually more than 44 mm; total length more than 280 mm . . . . *panamintinus*

9. South of Coast Ranges, Point Conception southward 10

   Northward from Point Conception; Coast Ranges and Central Valley 11

10. Ear pinna completely yellow-brown; dorsal and ventral tail stripes with many white hairs . . . . *stephensi*

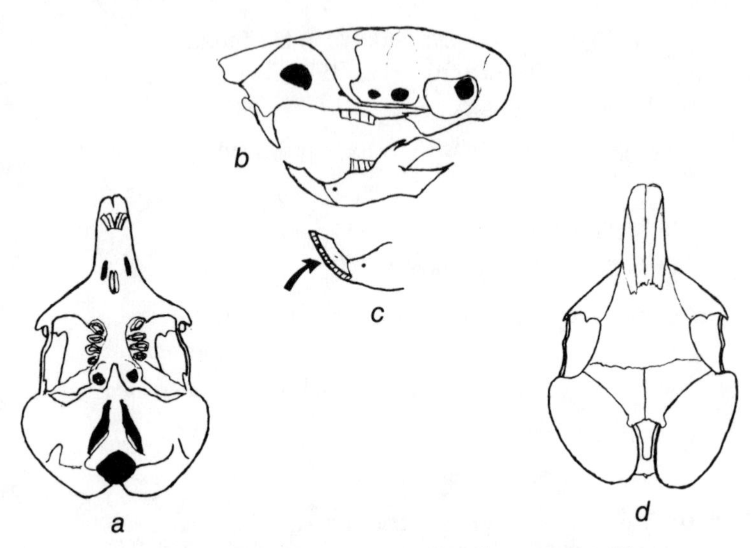

FIG. 98. Skull of the Chisel-toothed Kangaroo Rat (*Dipodomys microps*): *a*, ventral view; *b*, lateral view; *c*, lower incisor enlarged (arrow to anterior surface); *d*, dorsal view.

Ear pinna with white (or light) spot at base; dorsal and ventral tail stripes with very few white hairs . . . *agilis*

11. Zygomatic arch with posterior margin rounded, shallowly concave (fig. 99a, c); ear 15–18 mm 12

    Zygomatic arch with posterior margin straight, distinctly set off from a ventral concavity (fig. 100a, c); ear 12–15 mm . . . . *heermanni*

12. Ear mostly brown, 16–18 mm from crown; ventral dark tail stripe narrower than lateral white stripes at midpoint of tail . . . . *elephantinus*

    Ear mostly black, 15–16 mm from crown; ventral dark tail stripe wider than lateral white stripe at midpoint of tail . . . . *venustus*

Pacific Kangaroo Rat (*Dipodomys agilis*)

Description: A medium to large-sized kangaroo rat. Hind foot with five toes. Ear pinna with white (or light) spot at base.

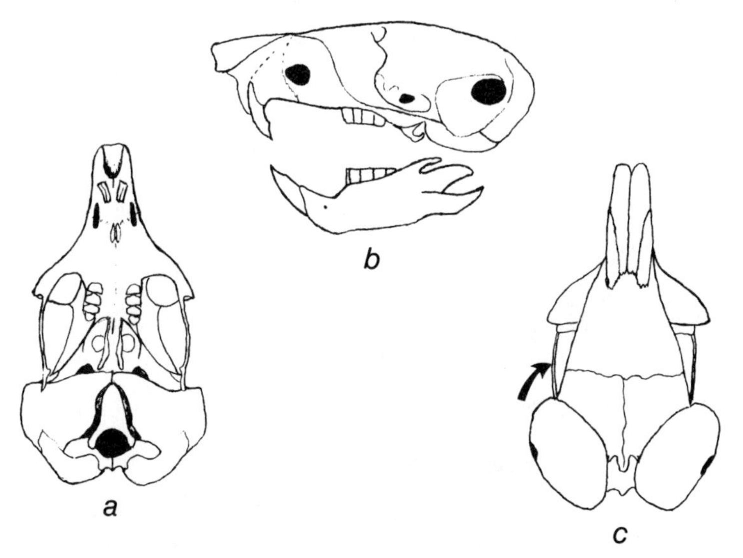

FIG. 99. Skull of the Narrow-faced Kangaroo Rat (*Dipodomys venustus*): *a,* ventral view; *b,* lateral view; *c,* dorsal view (arrow to zygomatic arch).

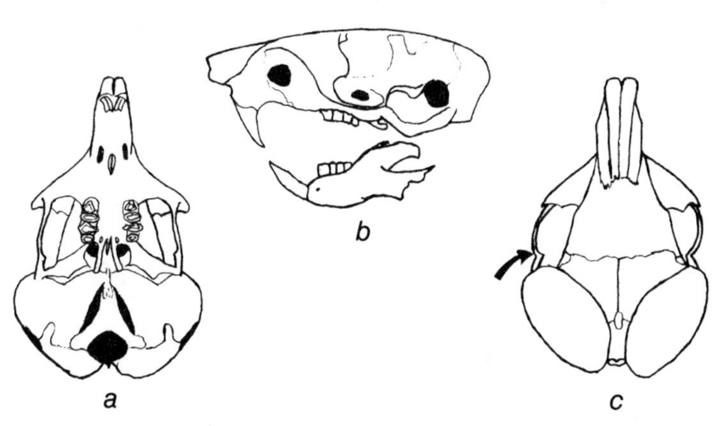

FIG. 100. Skull of Heermann's Kangaroo Rat (*Dipodomys heermanni*): *a,* ventral view; *b,* lateral view; *c,* dorsal view (arrow to zygomatic arch).

Similar to *D. stephensi* in which the ear lacks such a light basal spot. Tail with a brown or black tuft; dorsal and ventral dark tail stripes with very few white hairs. TL 265–319, T 155–197, HF 40–46. Weight: 67–76 g.

Distribution: Sagebrush and chaparral on sandy soils of South Coast Ranges. Up to 2200 m in San Gabriel Mountains. South into Baja California. (See p. 367.)

Food: Seeds of forbs, grasses, and shrubs. Known to store seeds of Laurel-sumac (*Rhus laurina*) and Chamise (*Adenostoma fasiculatum*). Also eats some insects.

Reproduction: Two to four young born in June or early July.

California Kangaroo Rat (*Dipodomys californicus*)

Description: A rather large, dark-colored kangaroo rat with four toes on the hind foot. Tip of tail white. TL 270–340, T 174–194, HF 39–47. Weight: 57–78 g.

Distribution: Brushy areas and chaparral from North Coast Ranges (in the west) and up to 1300 m in El Dorado County northward into extreme south-central Oregon. (See p. 367.)

Food: Berries and seeds of grasses, forbs, and shrubs such as manzanita (*Arctostaphylos* spp.), buckthorn (*Ceanothus* spp.), and rabbitbrush (*Chrysothamnus* spp.). Also green leaves.

Reproduction: Breeds from late winter to summer. From two to four young in a litter.

Desert Kangaroo Rat (*Dipodomys deserti*) [pl. 12e]

Description: A very large, buff-colored, very pale kangaroo rat. Tail without a black tip; ventral dark stripe usually absent. Hind foot with four toes. TL 305–377, T 180–215, HF 50–58. Weight: 83–138 g.

Distribution: Areas of extremely loose and dry sand in deserts of southeastern California, southern Arizona, and northern Mexico. (See p. 368.)

Food: The Desert Kangaroo Rat feeds on seeds and leaves of desert forbs; green vegetation may be consumed in quantity in winter and spring.

Reproduction: Breeding occurs from midwinter to summer; from three to four young in a litter. Breeds throughout the year in some areas.

The tracks of this kangaroo rat are frequently noticed in the loose sand before the animal itself is seen. Its burrow entrances are often found at the bases of the Creosote Bush (*Larrea tridentata*).

Big-eared Kangaroo Rat (*Dipodomys elephantinus*)

Description: A large, dark-colored kangaroo rat with five toes on hind foot. Ear large. Separable from *D. venustus* by the dark ventral stripe which is narrower than the lateral white stripe in the terminal half of the tail (wider in *venustus*). TL 305–336, T 183–210, HF 44–50, E 16–18 mm (from crown). Weight: 81–97 g.

Distribution: Sagebrush and chaparral of eastern slope of Coast Ranges in San Benito County. (See p. 368.)

Food: Seeds and leaves of forbs and shrubs.

Reproduction: One or two litters of two to three young born in late winter and spring.

Heermann's Kangaroo Rat [pl. 12b]
(*Dipodomys heermanni*)

Description: A medium large, dark-colored kangaroo rat. Hind foot with five toes. Zygomatic arch with upper rear margin straight and sharply set off from a distinctly lower concavity (see fig. 100a, c). TL 250–340, T 160–200, HF 38–47. Weight: 56–74 g.

Distribution: Brushy and grassy slopes and flats; chaparral-covered hillsides in Southern Coast Ranges and most of Central Valley north to San Francisco Bay area and foothills of El Dorado County. (See p. 369.)

Food: Known to gather seeds from a broad spectrum of forbs and grasses including Red Brome (*Bromus rubens*), lupines (*Lupinus* spp.), Red-stem Filaree (*Erodium cicutarium*), as well as green leaves and occasionally insects. One student noted that it gleaned seeds from the entrance to dens of the Black Harvester Ant (*Veromessor pergandei*).

Reproduction: Litters of two to five (usually four) young are born from February to August. Young born early in the year breed in the same year.

Giant Kangaroo Rat (*Dipodomys ingens*)

Description: A very large, brownish kangaroo rat with a light brown tip to the tail. Hind foot with five toes. Tail relatively short (usually less than 130 percent of head and body). TL 311–348, T 157–198, HF 46–55. Weight: 131–180 g.

Distribution: Found in open areas on fine soils of southwest side of San Joaquin Valley. Merced and Kern counties south to Santa Barbara County. (See p. 368.)

Food: Seeds and sometimes leaves of such plants as Peppergrass (*Lepidium nitidum*), Red-stem Filaree (*Erodium cicutarium*), saltbush (*Atriplex*), and cudweed (*Gnaphalium*). Known to store seeds.

Reproduction: Breeds from February to June or later. Commonly has litters of four to five or even six young, a rather large number for a kangaroo rat.

This species has declined with the increase of cultivation of its range. Also, cattle probably destroy the burrow system close to the surface.

Merriam's Kangaroo Rat (*Dipodomys merriami*) [pl. 12f]

Description: A rather small, buff-colored kangaroo rat with four toes on hind foot. Rather similar to *D. ordii,* from which it can be distinguished by the absence of a minute fifth toe on the hind foot. TL 220–260, T 123–160, HF 34–39. Weight: 39–52 g.

Distribution: On light sandy soils across much of the southern half of the state except in Coast Ranges. Mojave and Colorado deserts and a small area in extreme northeastern California. North to northern Nevada, east to Texas, and south into Mexico, including Baja California. (See p. 369.)

Food: Seeds of grasses (genera *Oryzopsis, Bouteloua,* and *Hilaria*) and forbs, probably exploiting what species are available. Known also to feed on leaves of *Franseria* and other forbs.

Reproduction: May breed from January to June or later. From one to five young in a litter. Reproduction seems to depend on the growth of winter annuals, the leaves of which are eaten by the kangaroo rats. The growth of winter annuals is much greater following heavy rain in the previous autumn. Thus an extended breeding season follows heavy fall rains, and reproduction may not occur in dry years.

When found in sand dunes together with *Dipodomys deserti,* Merriam's Kangaroo Rat tends to occur on rocky patches, even though the burrows may be separated from those of the Desert Kangaroo Rat by only a few feet (about 1 m).

Chisel-toothed Kangaroo Rat (*Dipodomys microps*)

Description: A medium-sized kangaroo rat of dark color. Hind foot with five toes. Anterior face of lower incisors flat (see fig. 98c). Similar to *D. ordii,* which has lower incisors with rounded or awl-shaped anterior surfaces. Also may be found with *D. panamintinus,* which has a larger hind foot (usually more than 44 mm) and rounded lower incisors. TL 244–290, T 140–173, HF 38–44. Weight: 55–75 g.

Distribution: Piñon–juniper associations on rather light soils in southeastern California, mostly Mojave Desert. Also sagebrush flats. Ranges through Nevada, north to southeastern Oregon, east to Salt Lake, and south to northern Arizona. (See p. 369.)

Food: Seeds and leaves of desert grasses and forbs. Known to gather seeds of *Atriplex, Lepidium,* and *Trifolium* and leaves of sagebrush (*Artemisia* spp.). Fresh leaves constitute the bulk of this species' diet, and it is known to climb bushes to gather food.

Reproduction: Breeds from late winter through the spring and, in wet years, into summer. From one to four (usually two) young to a litter.

The flattened lower incisors are used to shave off the outer, saline layers of the leaves of saltbush (*Atriplex* spp.), exposing the inner succulent tissue which is eaten.

San Joaquin Kangaroo Rat [pl. 12d]
(*Dipodomys nitratoides*)

Description: A somewhat small-sized kangaroo rat with four toes on hind foot. Color rather dark. Distinguished also by its relatively short tail (less than 160 mm), which is buff-tipped. Most resembling *D. merriami,* which does not occur in the San Joaquin Valley. TL 211–253, T 120–152, HF 33–37. Weight: 39–47 g.

Distribution: Occurs in alkali sink communities of western Fresno and Tulare counties south to Kern County. (See p. 367.)

Food: Apparently not known. Presumably feeds on seeds of such annuals as filaree (*Erodium*), Shepherd's Purse (*Capsella bursa-pastoris*), and saltbush (*Atriplex*), which are abundant in its habitat.

Reproduction: Known litters have consisted of two young. Breeds throughout the year.

Ord's Kangaroo Rat (*Dipodomys ordii*)

Description: A medium-sized brownish kangaroo rat. Anterior edge of lower incisors rounded. Hind foot shorter than 44 mm (in California); hind foot with five toes. TL 208–281, T 100–163, HF 33–41. Weight: 50–61 g.

Distribution: Sagebrush deserts of Great Basin. A wide-ranging species occurring north to Washington and southern Canada, east to central Kansas, and south to central Mexico. (See p. 367.)

Food: Eats seeds of grasses (*Andropogon*), forbs (*Helianthus* and *Ambrosia*), and mesquite (*Prosopis*). Also tender growing leaves of forbs and some insects.

Reproduction: From one to six young born from late winter to early summer. Perhaps more than one brood a year.

Panamint Kangaroo Rat (*Dipodomys panamintinus*) [pl. 12c]

Description: A rather large kangaroo rat with five toes on each foot. Color varies from ashy gray to dark brown or cinnamon brown. Lower incisors rounded on anterior surface. Distinguished from the Desert Kangaroo Rat, with which it may occur, by the rather dark color (*D. deserti* is very pale) and by the presence of five toes on the hind foot (*D. deserti* has four). TL 285–334, T 156–202, HF 42–48. Weight: 64–81 g.

Distribution: Joshua Tree and juniper–piñon associations in the southern part of the state. Also east of the Sierra Nevada to the area south of Lake Tahoe. Also west-central Nevada. (See p. 368.)

Food: Known to feed on green leaves of forbs and also juniper berries.

Reproduction: Breeds from March through July; more than a single brood a season is the rule. A second mating occurs shortly after the first birth of the year, for not infrequently females are nursing the first litter while the second is developing. Litters vary from three to five young (usually four).

Stephens' Kangaroo Rat (*Dipodomys stephensi*)

Description: A medium-sized kangaroo rat with rather dark color. Five toes on hind foot. Tail conspicuously striped and with a crest; the black stripes with many white hairs. Similar to *D. agilis,* with which it may occur, but *agilis* has very few white hairs in black tail stripes. Ear pinna of an even yellow-brown color; in *D. agilis* ear pinna has a light spot at base. TL 277–300, T 164–180, HF 39–43. Weight: 60–74 g.

Distribution: In areas of sagebrush and grassy patches on

sandy or gravelly soils. Known from San Jacinto Valley of San Diego, Riverside, and San Bernardino counties. (See p. 369.)

Food: Species of *Artemisia, Eriogonum, Marrubium, Rumex,* and *Haplopappus* are common in its habitat; presumably their seeds are the staple diet of this kangaroo rat.

Reproduction: A litter of two or three young is born in late spring.

Narrow-faced Kangaroo Rat [pl. 12a]
(*Dipodomys venustus*)

Description: A large, dark kangaroo rat. Hind foot with five toes. Ears larger (15–16 mm from crown) than in most members of the genus except *D. elephantinus* (16–18 mm from crown). This species may be a geographic race of *D. agilis,* but the two species can be separated on the basis of their distributions. TL 293–332, T 175–203, HF 44–47. Weight: 66–74 g. (See fig. 99a, c.) Zygomatic arch with posterior margin rounded.

Distribution: South Coast Ranges from San Francisco Bay to Point Conception; in open sandy areas or dense chaparral. (See p. 369.)

Food: Seeds of grasses and shrubs (such as Bur Clover).

Reproduction: One or two litters of two to four young.
This form may be conspecific with *D. elephantinus.*

Kangaroo Mice (*Microdipodops*)

Large-headed, long-tailed rodents with small forelegs and much larger hind legs. Tail constricted at base and tapered toward the tip; not tufted. Hind feet with fur on soles. With fur-lined cheek pouches. Dorsum without a skin gland (present in *Dipodomys* but absent in *Perognathus*). Auditory bullae (ear capsules) extremely large (see fig. 101a).

The two species of kangaroo mice are unique to the arid sagebrush deserts of the Great Basin. They are similar to the kangaroo rats (*Dipodomys*) in being bipedal, jumping rodents, but they may also scurry about on all fours like a pocket mouse (*Perognathus*). Species of the three genera frequently occur together in the same habitat.

Like other heteromyid rodents of arid lands, kangaroo mice can subsist on water in such "dry" foods as seeds, which may contain from 5 to 8 percent free water, for they release a highly concentrated urine. In nature kangaroo mice undoubtedly obtain a great deal of water from the insects they eat. Kangaroo mice hibernate.

Key to Species of Kangaroo Mice (*Microdipodops*) in California

1. Dorsum blackish brown or dark gray with hairs dark at base (next to the skin); incisive foramina wider posteriorly (fig. 101d) . . . . *megacephalus*

   Dorsum pink-cinnamon with hairs white at base (next to the skin); incisive foramina with outer margins parallel (fig. 101e) . . . . *pallidus*

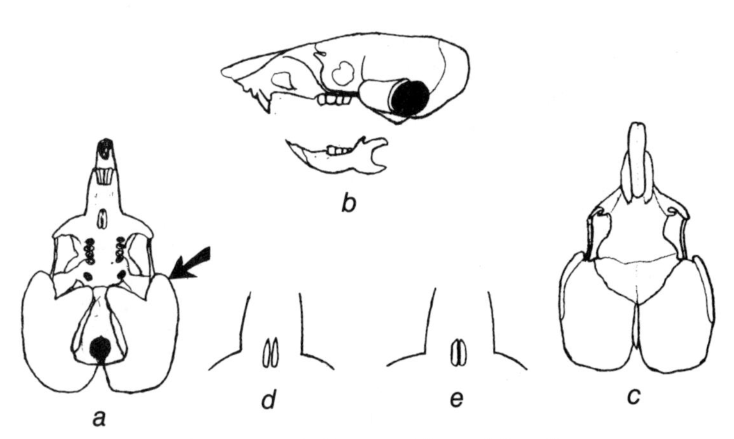

FIG. 101. Skull of the Dark Kangaroo Mouse (*Microdipodops megacephalus*): *a*, ventral view (arrow to auditory bullae); *b*, lateral view; *c*, dorsal view; *d*, incisive foramina of *Microdipodops megacephalus; e*, incisive foramina of *Microdipodops pallidus*.

Dark Kangaroo Mouse (*Microdipodops megacephalus*) [pl. 12g]

Description: Dorsum brownish or blackish or gray; venter paler but base of hairs dark or smoky in color. Incisive foramina pointed and converging anteriorly, wider posteriorly (see fig. 101d). Upper part of tail clearly darker than body. Distinguished from the Pale Kangaroo Mouse by color and by a smaller hind foot. TL 140–177, T 68–103, HF 23–25. Weight: 10–17 g.

Distribution: Sagebrush deserts east of Sierra Nevada, characteristically on somewhat coarse gravelly soils, where it may occasionally be very common. From central Oregon through Nevada east to Utah. (See p. 370.)

Food: This mouse takes a variety of seeds of desert forbs and grasses as well as insects. Insects are found in the stomach of this rodent although apparently they are not carried in cheek pouches.

Reproduction: The Dark Kangaroo Mouse breeds from April to September; litters consist of two to seven young. Young may breed in the year of their birth.

Pale Kangaroo Mouse (*Microdipodops pallidus*) [pl. 12h]

Description: Dorsum rather pale pink, venter white, with hairs white at the base (next to the skin). Tail about the same color as body, not blackish or dark at the tip. Hind foot 25 mm or longer (longer than in the Dark Kangaroo Mouse). Weight: 10–17 g. Incisive foramina with outer margins parallel (see fig. 101e).

Distribution: In Mono and Inyo counties, generally in regions of very fine sand; at times becomes extremely common. Extreme eastern margin of state, east of Yosemite National Park. Also adjacent western Nevada. (See p. 370.)

Food: The diet seems similar to that of the Dark Kangaroo Mouse: a mixture of seeds and insects.

Reproduction: Breeds from late winter until June or July. From two to six young in a litter; more than one litter a season.

Pocket Mice (*Perognathus*)

Small, long-faced mice with nearly equal forelegs and hind legs; tail moderately long to long. With external (fur-lined) cheek pouches. Pelage rather stiff and "spiny" in some species and soft or "silky" in others. Skull with small to moderately sized auditory bullae (ear capsules). Snout relatively longer than in either *Dipodomys* or *Microdipodops*. External ear rather short. Molars tuberculate and rooted, with a transverse groove. Hind foot with five toes.

Pocket mice are small relatives of kangaroo rats, and both genera occur together in many parts of arid and seasonally arid western North America. Both groups have behavioral and physiological modifications for living in hot, dry habitats. Some species have been carefully studied in the laboratory. Like most desert rodents, pocket mice are nocturnal.

Seeds and insects are both important in the diets of these mice and provide the water they need. The free water in fresh seeds can be substantial, and there is also water, chemically bound, which is released from seeds during digestion. As these mice forage, they place seeds in their cheek pouches for later storage near or in their homes (fig. 102). Insects are also eaten, and the stomachs of pocket mice may be crammed with grasshoppers or cutworms. They seem not to place insects in their cheek pouches.

Some species of pocket mice are known to become dormant and spend long periods of the winter underground. *Perognathus longimembris* is a profound hibernator and tends to become torpid with a decline in temperature at any season. Torpidity in several species results from deprivation of food. *Perognathus flavus,* for example, may be active in subzero temperatures, but torpidity is induced by withholding food (in captivity). While torpid, the pocket mouse has reduced breathing and heartbeat. Such an energy-saving suspension of bodily activity is valuable in the dry, harsh deserts.

Observers have noted that pocket mice tend to forage under

FIG. 102. Pocket mouse burrow.

the cover of shrubs; kangaroo rats in contrast, readily move much greater distances when gathering food. At least some kinds of pocket mice (*Perognathus penicillatus*) have been known to climb up at least 25 cm in bushes, possibly in search of insects.

Key to Species of Pocket Mice (*Perognathus*) in California

1. Sole of hind foot hairy; pelage soft; auditory bullae converging and usually meeting anteriorly; interparietal bone about 1½ times wide as long, narrower than interorbital width of skull (fig. 103c) 2

   Sole of hind foot hairless; pelage harsh, bristly or spiny; auditory bullae entirely separate; interparietal bone nearly twice as wide as long, as wide or wider than interorbital width of skull (fig. 104c) 5

2. Hind foot longer than 20 mm; skull longer than 24 mm 3

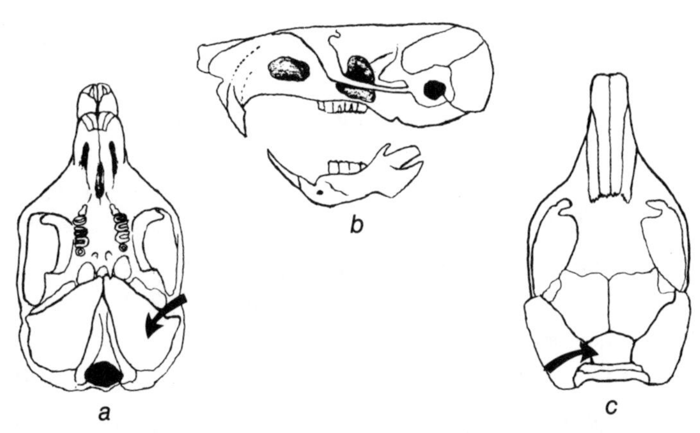

FIG. 103. Skull of the San Joaquin Pocket Mouse (*Perognathus inornatus*): *a,* ventral view (arrow to auditory bulla); *b,* lateral view; *c,* dorsal view (arrow to interparietal bone).

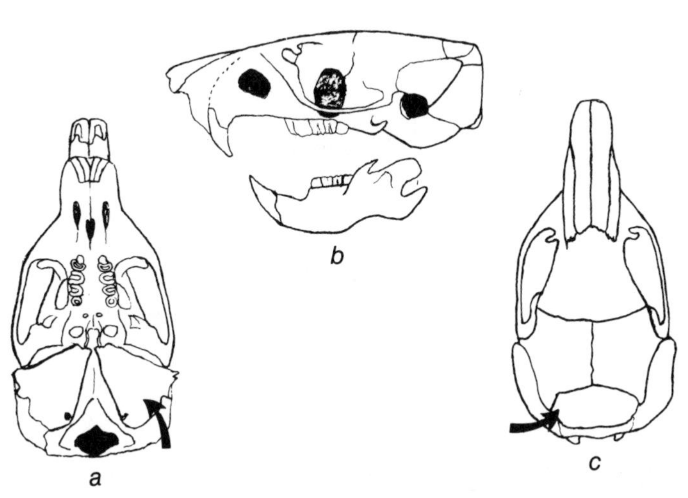

FIG. 104. Skull of the California Pocket Mouse (*Perognathus californicus*): *a,* ventral view (arrow to auditory bulla); *b,* lateral view; *c,* dorsal view (arrow to interparietal bone).

Hind foot shorter than 20 mm; skull less than 24 mm 4

3. Tail with long dorsal hairs near the tip; pelage coarse . . . . *formosus*

   Tail scantily haired except for an apical crest of long hairs projecting beyond the tip; pelage soft . . . . *parvus*

4. Auditory bullae large, in contact anteriorly (see fig. 103a) . . . . *inornatus*

   Auditory bullae smaller, not in contact anteriorly (fig. 105a) . . . . *longimembris*

5. Pelage on rump conspicuously spiny or bristly 6

   Pelage on rump without spines or bristles 8

6. With a well-marked lateral line; spines only on rump 7

   With a poorly marked lateral line or none; spines from

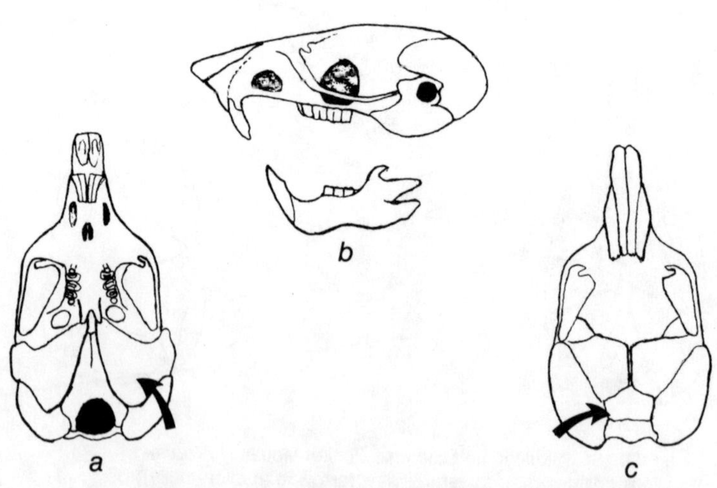

FIG. 105. Skull of the Little Pocket Mouse (*Perognathus longimembris*): *a*, ventral view (arrow to auditory bulla); *b*, lateral view; *c*, dorsal view (arrow to interparietal bone, which is only slightly broader than long).

shoulders to rump; southeastern California west to southeast San Diego County . . . . *spinatus*

7. Ear short and rounded, less than 9 mm; southwestern California . . . . *fallax*

   Ear longer, with sides partly parallel, more than 9 mm; southern half of California . . . . *californicus*

8. Dorsum grayish; total length usually more than 210 mm; extreme southern California . . . . *baileyi*

   Dorsum yellow-brown; total length usually less than 210 mm; desert areas of southern and eastern California . . . . *penicillatus*

Bailey's Pocket Mouse (*Perognathus baileyi*) [pl. 13h]

Description: A rather large, gray-buff pocket mouse with black spines on rump. Tail bicolored with a distinct tuft. Sole of hind foot naked. Auditory bullae not touching anteriorly. TL 201–230, T 110–125, HF 26–28. Weight: 15–20 g.

Distribution: Rocky deserts of extreme southern California; often associated with mesquite. South throughout Baja California, east to New Mexico, and along eastern coast of Gulf of California. (See p. 370.)

Food: Known to eat seeds from a broad spectrum of desert grasses, forbs, and woody plants. Also feeds on insects. A genuine dietary generalist.

Reproduction: One or more litters of two to five young; breeds from June to October.

California Pocket Mouse (*Perognathus californicus*) [pl. 13e]

Description: A rather large pocket mouse with dorsal pelage a mixture of yellow and black hairs; strong spiny hairs on sides and rump. Ears unusually long (9–14 mm). Sole of hind food naked. Auditory bullae separated anteriorly (see fig. 104a). Interparietal bone nearly twice as wide as long (see fig. 104c). TL 190–235, T 103–143, HF 24–29. Weight: 16–21 g.

Distribution: This species occurs in the San Joaquin Valley and coastal areas south of San Francisco Bay. It has been found in a variety of habitats from open scrub oak to sagebrush and may occur up to 2800 m in the southern part of the state. Also in northern Baja California. (See p. 370.)

Food: Seeds of grasses and shrubs, such as *Salvia* spp.

Reproduction: From two to five young in a litter; breeds from April through June.

This species is well known to spend brief periods in torpor with lowered breathing, heartbeat, and body temperature. It does not appear aboveground in the winter.

San Diego Pocket Mouse (*Perognathus fallax*) [pl. 13f]

Description: A medium-sized, brownish pocket mouse. Dorsum and sides with conspicuous white, spiny hairs; black spines on rump. Tail bicolored and crested. A buffy stripe or "lateral line" between venter and dorsum. Hind foot with sole naked. Auditory bullae well separated anteriorly. TL 176–200, T 88–118, HF 21–26. Weight: 15–18 g.

Distribution: Restricted to the southwestern region of the state; common on compact soils in open desert areas. Frequently associated with species of yucca. Up to 2000 m. (See p. 371.)

Food: Seeds of plants with which it occurs. Known to collect and store seeds of yucca, sage (*Salvia* sp.), ryegrass (*Lolium* sp.), and other grasses.

Reproduction: Two to four young; known to breed in the autumn.

Although this species may not enter prolonged periods of dormancy, it does become inactive in cold weather.

FIG. 106. Long-tailed Pocket Mouse (*Perognathus formosus*) sand bathing and perhaps simultaneously scent marking.

Long-tailed Pocket Mouse [fig. 106; pl. 13g]
(*Perognathus formosus*)

Description: A rather large, dark pocket mouse with a bicolored tail conspicuously tufted at tip. Sole of hind foot with hair. TL 172–211, T 86–118, HF 22–26. Weight: 19–25 g.

Distribution: This pocket mouse occurs on mixed sandy and rocky soils in eastern California. East of the Sierra Nevada in relatively open arid areas. Typically occurring where there are many small rocks which resemble this mouse in size and color. East to Utah and northern Arizona and south to Baja California. (See p. 371.)

Food: Seeds of grasses and forbs; insects; green leaves in springtime.

Reproduction: There are commonly five to six young in a litter. Breeding follows the emergence of grasses and forbs in the spring.

This species is active in midwinter and presumably does not hibernate. Some students, however, report that this species becomes torpid in the cold months. Apparently winter activity varies with the locality.

San Joaquin Pocket Mouse [pl. 13a]
(*Perognathus inornatus*)

Description: A small, buff-orange pocket mouse with sprinkling of dark guard hairs (but no spiny hairs) on dorsum. Body with an indistinct lateral line. Auditory bullae rather large and in contact anteriorly. Interparietal bone about 1½ times as wide as long (see fig. 103c). Hind foot with hair on sole. Similar to Little Pocket Mouse, but adults average larger; auditory bullae not in contact in Little Pocket Mouse. TL 128–163, T 63–78, HF 18–21. Weight: 15–18 g.

Distribution: Flat ground and low hills in Central Valley north to Marysville Buttes and south to Carrizo Plain; also Salinas Valley. Not known outside California. (See p. 371.)

Food: Seeds of grasses, forbs, and shrubs (such as *Artemisia* and *Atriplex*). Also soft-bodied insects such as cutworms.

Reproduction: Breeds from March to July. At least two litters a season; from four to six young in a litter.

Little Pocket Mouse (*Perognathus longimembris*) [pl. 13c]

Description: A small, pink-buff pocket mouse. Tail either bicolored or evenly pale; more heavily furred on distal third and tufted on distal 3–7 mm. Auditory bullae (ear capsules) not in contact anteriorly (see fig. 105a). Interparietal bone about 1½ times as wide as long (see fig. 105c). Hind foot not more than 20 mm (frequently longer than 20 mm in *P. parvus* and *P. formosus*); sole hairy on posterior third. TL 112–138, T 50–76, HF 17–20. Weight: 7–10 g.

Distribution: On fine, sandy soils in many parts of the southern half of our state. Widely distributed in arid regions from southern Oregon east to western Utah and Arizona. (See p. 372.)

Food: Seeds of many desert plants, including grasses, goosefoot (*Chenopodium* spp.), and the Desert Trumpet (*Eriogonum inflatum*). It also takes soil-dwelling insects as do other pocket mice.

Reproduction: Rather prolific; litter size ranges from two to eight. Pregnancies occur in spring and fall with a summer lull.

The Little Pocket Mouse remains underground in the winter. It seems not to accumulate fat in the winter and probably subsists on food stored during the summer and autumn.

Great Basin Pocket Mouse [pl. 13d]
(*Perognathus parvus*)

Description: A buff or ashy pocket mouse with soft, silky fur; sole of hind foot furred. Tail long and with a slight apical crest, bicolored. Ear small with an inner lobe at base. TL 160–195, T 85–100, HF 22–27. Weight: 16–28 g.

Distribution: This species occurs in open sagebrush east of the Sierra Nevada. It may sometimes abound on heavily grazed land with sandy soils and also along creeksides. It ranges in the Great Basin north to Canada. (See p. 372.)

Food: Known to take seeds of many forbs and shrubs. Although these seeds are usually rather dry, this pocket mouse is known to eat seeds in the ripening stage when the water content is very high. It is also known to gorge itself on caterpillars.

Reproduction: Breeding seems to occur irregularly throughout the year, except in winter, but the number of broods is not known. Young number from two to eight in a litter.

Two described species, *Perognathus alticola* and *P. xanthonotus*, are rather similar to *P. parvus* and are apparently extremely rare. *P. alticola* is known from Kern and San Bernardino counties; *P. xanthonotus* is known from Walker Pass.

This pocket mouse becomes torpid, at least sporadically, in the winter; it is not active aboveground in the coldest weather.

Desert Pocket Mouse (*Perognathus penicillatus*) [pl. 13b]

Description: A medium-sized pocket mouse with yellow-gray pelage of a rather coarse but not spiny texture. Tail faintly bicolored, often annulated, with a conspicuous tuft and crest. Ears relatively pointed. Sole of hind foot naked. Auditory bullae distinctly separated anteriorly. TL 153–221, T 91–121, HF 21–27. Weight: 14–20 g.

Distribution: Occurs on sandy soils, sometimes on sand and small stones. Common in open areas with scattered, low bushes. Deserts of southern California and also east of Yosemite. Eastward to Texas and south to central Mexico, including Baja California. (See p. 373.)

Food: Seeds of a wide variety of desert plants. Known to store seeds of mesquite, Creosote Bush, and snakeweed. It has been captured in live traps placed 24 cm aboveground in bushes; may climb in search of insects.

Reproduction: Litters vary in size from two to six. Breeds from late winter to spring, with a summer lull, and sometimes again in the autumn. In moist, streamside habitats reproduction may continue all summer.

In contrast to many pocket mice, this species is active in the coldest nights in the winter. Apparently it does not hibernate.

Spiny Pocket Mouse (*Perognathus spinatus*) [pl. 13i]

Description: A rather large, yellow-brown pocket mouse. Tail bicolored and crested. Ears rather small (5–7 mm). Rump with whitish spiny hairs. Sole of hind foot naked. Auditory bullae well separated anteriorly. TL 164–225, T 89–128, HF 20–28. Weight: 19–29 g.

Distribution: Deserts of southwestern part of state. South through Baja California. (See p. 372.)

Food: Not definitely known but presumably seeds of grasses and forbs.

Reproduction: Known to breed from April to July with at least two broods a year. Usually from two to three young, rarely as many as five.

Beavers (Castoridae)

Large, stocky rodents highly modified for an aquatic life. Fur dense and water-repellent. Eyes small and external ears reduced. Nostrils with valves. Tail dorsoventrally flattened, muscular, and naked. Legs short and hind feet completely webbed. Skull heavy and deep with no postorbital projections (fig. 107b, c).

There is one species of beaver in North America today, and a separate, but similar, species occurs in Eurasia. About 10 to 14 million years ago giant beavers evolved in North America, and some survived until the end of the glacial epochs, perhaps 10,000 years ago. One of these ancient rodents approximated Black Bears in size.

Beavers are at home in water and are seldom seen far from a creek or pond. They dive skillfully and have been reported to remain submerged for up to fifteen minutes, during which time there is a reduction of blood flow and blood pressure caused by a very slow heartbeat (bradycardia).

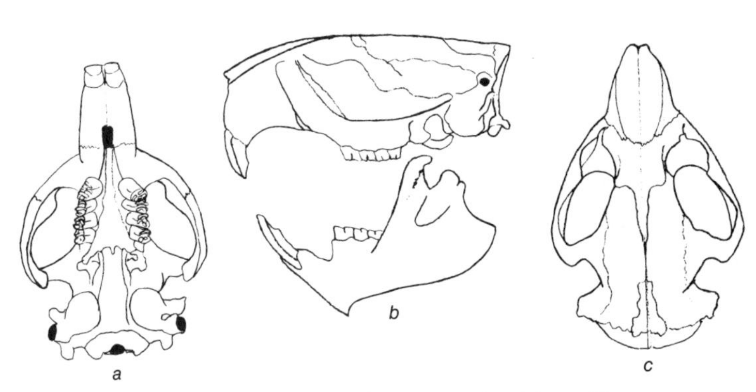

FIG. 107. Skull of the Beaver (*Castor canadensis*): *a,* ventral view; *b,* lateral view; *c,* dorsal view.

FIG. 108. Beaver (*Castor canadensis*).

Beaver (*Castor canadensis*) [fig. 108]

Description: A large rodent, superficially squirrellike, with a large, flattened, and more or less naked tail. Hind feet webbed; second claw divided (the significance of this is not known). Fur very dense; brownish ventrally and dark brown dorsally. TL 1000–1200, T 260–330, HF 150–200, E 23–29. Weight: 11–26 kg. Dentition: 1/1, 0/0, 1/1, 3/3 (see fig. 107).

Distribution: Streams and small lakes throughout the northern two-thirds of the state. Frequently encountered in the Sacramento–San Joaquin Delta, up to 2000 m in the mountains. A pale-colored population occurs along the Colorado River. Throughout most of North America. From northern Mexico north to Newfoundland and Alaska and east to the Atlantic Ocean. Their presence is frequently indicated by their typical gnawing of tree trunks and the construction of dams and beaver "lodges" or dome-shaped stick houses. In the Delta they typically live in burrows in levees. (See p. 373.)

Food: Willows, aspens, and poplars as well as other broad-leaved trees. Also grasses and forbs in the spring. Contains a microbiota (bacteria and protozoa) in the gut for digesting cellulose, an estimated 30 percent of which is absorbed. Reingestion of feces also aids in absorption of food material. Stores food in winter.

Reproduction: Mates in winter (February) and the kits are born 106–110 days later; from two to six young but usually three or four. Weaned by two months. Young disperse at about two years of age, as they approach sexual maturity.

In the past beavers were widely trapped for their fur. They have a pair of musk glands that presumably serve to mark territories, and these glands were once considered to have medicinal value, a myth that gave an added incentive to trap these rodents. Beavers are now common in many parts of North America and can become a pest. Some of the levee breaks in the Delta area are probably at least partly due to the burrowing of beavers. Today their value as a furbearer is low, about $20 for a good skin, and not more than about 2000 are taken annually in California.

Deer Mice, Wood Rats, and Allies (Cricetidae)

Small rodents with a broad zygomatic plate (see fig. 110a) and narrow infraorbital foramen (see fig. 110c). Postorbital process absent (well developed in squirrels, Sciuridae). Color and form variable. Dentition (in California species): 1/1, 0/0, 0/0, 3/3.

This is a large and diverse family that includes most native mice and rats of North America. The species are diverse with respect to development of ears, eye size, tail length, limbs, and color. There is also a variety of adaptations to different habitats.

Cricetid mice and wood rats are among the most familiar of California rodents. They are frequently the most commonly encountered in the field and include those species most likely to enter mountain cabins.

Key to Genera of Cricetidae in California

1. Cheek teeth flat-crowned with enamel patterns of irregular loops (fig. 109a, b); mostly "rat-sized" 2

   Cheek teeth tuberculate (fig. 110a, b); mostly "mouse-sized" 3

2. Venter white or buff, dorsum usually an even shade of brown or gray; ears rather nude; skull weakly ridged (fig.

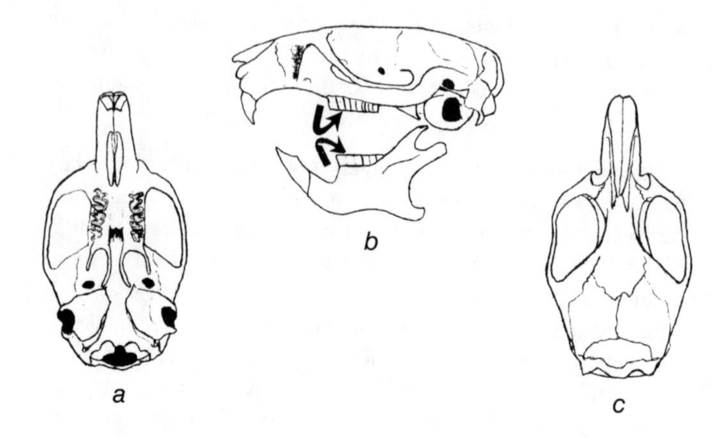

FIG. 109. Skull of the Desert Wood Rat (*Neotoma lepida*): *a,* ventral view; *b,* lateral view (arrows to cheek teeth); *c,* dorsal view.

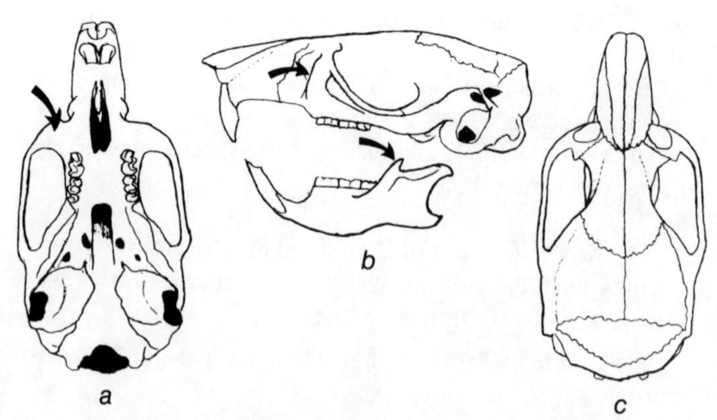

FIG. 110. Skull of the Parasitic Mouse (*Peromyscus californicus*): *a,* ventral view; *b,* lateral view; *c,* dorsal view.

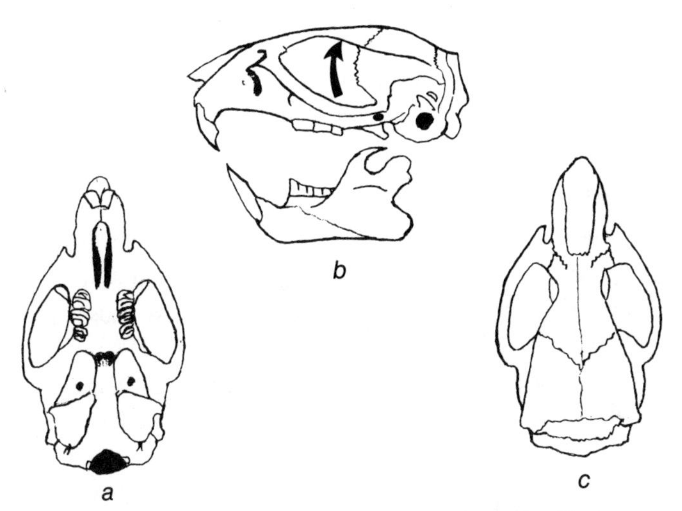

FIG. 111. Skull of the Hispid Cotton Rat (*Sigmodon hispidus*): *a*, ventral view; *b*, lateral view (arrow to ridge over eyes); *c*, dorsal view.

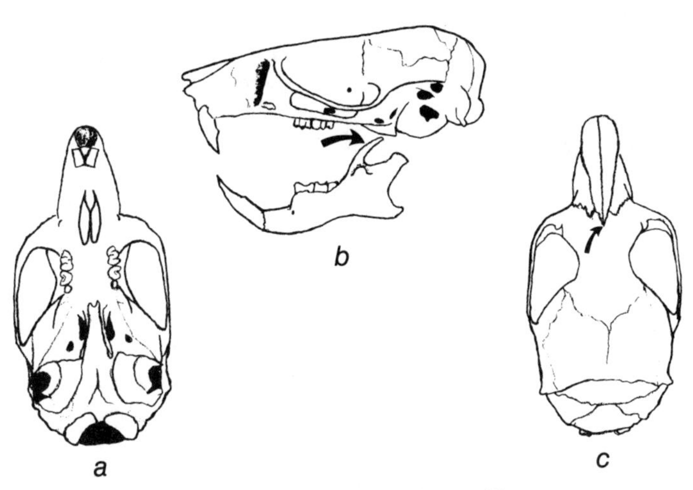

FIG. 112. Skull of the Northern Grasshopper Mouse (*Onychomys leucogaster*): *a*, ventral view; *b*, lateral view (arrow to coronoid process, which is much longer than in species of white-footed mice; see fig. 110): *c*, dorsal view (arrow to nasal bones).

109b) . . . . *Neotoma*

Venter light gray, dorsum a grizzled gray; ears well covered with fur; skull ridged above eyes (fig. 111b) . . . . *Sigmodon*

3. Upper incisors not grooved 4

   Upper incisors grooved . . . . *Reithrodontomys*

4. Tail slender, at least one-third total length; hind foot with five or six plantar tubercles; coronoid process short (fig. 110b) . . . . *Peromyscus*

   Tail thickened, less than one third total length; hind foot with four plantar tubercles; coronoid process long (fig. 112b) . . . . *Onychomys*

Wood Rats (*Neotoma*)

Rat-sized, long-tailed rodents with large naked ears and large eyes. Dorsal color an even gray or brown (not grizzled or agouti). Molars (cheek teeth) flat-crowned (fig. 109b). Skull with weak dorsal (supraorbital) ridges (fig. 109c). Eyes large.

Most areas in California have at least one species of wood rat, and in some regions there are two (usually in different habitats). With its large black eyes, large ears, soft pelage and color, and long tail, a wood rat resembles a giant deer mouse (*Peromyscus*). Its presence is usually revealed by its large house, usually on the ground or partly beneath a rock outcrop, built of twigs, small sticks, cow-pies, and other detritus.

Key to Species of Wood Rats (*Neotoma*) in California

1. Tail with hairs few or moderate but never bushy 2

   Tail bushy; Sierra Nevada and northern California . . . . *cinerea*

2. Tail conspicuously bicolored 3

   Tail very faintly bicolored; statewide except desert regions and High Sierra . . . . *fuscipes*

3. Fur on throat completely white; extreme southwest desert regions . . . . *albigula*

Fur on throat gray at base; southern half of state except San Joaquin Valley . . . . *lepida*

White-throated Wood Rat (*Neotoma albigula*) [pl. 14b]

Description: A grayish wood rat with a pure white throat (hairs on throat white to the base, next to the skin). Tail bicolored. Larger than Desert Wood Rat; also distinguished from that species by the pure white throat patch. TL 282–400, T 76–185, HF 30–39, E 28–30. Weight: 145–200 g.

Distribution: Characteristically in regions of juniper–piñon and at lower life zones in deserts of mesquite and prickly pear. Deserts of extreme southeastern California. From southern Utah and southern Colorado eastward to Texas and south to central Mexico. (See p. 374.)

Food: This rat feeds heavily on succulent sections of cacti and other desert plants. It also takes grasses, forbs, and leaves of *Yucca* spp. in lesser amounts. In some regions the branches of juniper are browsed heavily by this wood rat.

Reproduction: A small litter is born in spring; there may be from one to three young. Studies of this species do not indicate more than a single litter per year.

As this wood rat eats cacti, spines accumulate about its pathways. With time, these spines become hard and form a protective layer which may well deter the entrance of predators into the houses of these rodents. The wood rats, however, run over these trails with no injury. They are seldom found far from prickly pear (*Opuntia* spp.).

Bushy-tailed Wood Rat (*Neotoma cinerea*) [pl. 14c]

Description: A large wood rat with a long bushy tail. Ears large and naked. Venter with a large skin gland, which produces an odor. Dorsum pale gray. Palatine foramina long. TL 335–425, T 140–185, HF 38–44, E 31–33. Weight: 220–435 g.

Distribution: In coniferous forests of northern and eastern parts of state up to 2800 m or above. From British Columbia through the Rocky Mountain states south to northern Arizona and New Mexico. (See p. 374.)

Food: A broad variety of leaves and berries. Also insects and small lizards.

Reproduction: From three to five young are born in the spring.

The Bushy-tailed Wood Rat is one of those species commonly known as "pack rat" or "trade rat" from their habit of stealing small objects about campsites and leaving another "in exchange." This behavior has been traced to its habit of carrying small items about in its mouth. If it encounters something of special interest, it apparently drops what it is carrying and picks up the new object. Thus, it appears to be exchanging one object for another. This also accounts for pieces of campsite garbage, such as bottle caps, around houses of wood rats.

Dusky-footed Wood Rat (*Neotoma fuscipes*) [fig. 113]

Description: A medium-sized wood rat with a grayish brown dorsum and a pale or white venter. Feet brown at base; distal

FIG. 113. Dusky-footed Wood Rat (*Neotoma fuscipes*): note dark marking on top of hind feet only, not including toes.

half white. Tail faintly bicolored and scantily haired. Ear pinnae broad, moderately hairy. TL 260–439, T 130–212, HF 37–44, E 31–34. Weight: 184–358 g.

Distribution: Hardwood forests and brushlands of the state. From Baja California north through western Oregon. (See p. 374.)

Food: Many sorts of leaves, flowers, nuts, and berries. Hypogeous fungi eaten during the spring. It has been shown by radiotelemetry that the Dusky-footed Wood Rat frequently forages high above the ground.

Reproduction: Breeds almost throughout the year; apparently sexually inactive in late autumn and early winter. Usually more than one, but rarely up to five, litters a year. From one to three young in a litter, two being the most common size.

Desert Wood Rat [fig. 114; pl. 14a]
(*Neotoma lepida*)

Description: A relatively small, pale gray wood rat with a distinctly bicolored tail. Underparts pale or white but hairs gray at the base, next to the skin. Ears large and naked. TL

FIG. 114. Desert Wood Rat (*Neotoma lepida*) sleeping in its nest in hot weather.

282–305, T 113–128, HF 29–34, E 23–25. Weight: 100–190 g. (See fig. 109.)

Distribution: Found in a variety of habitats in the southern half of California, often in the vicinity of rocky outcrops. Known also to occupy old burrows of ground squirrels or kangaroo rats. Stick houses often built beneath the cover of a small rock ledge. Also in extreme northeastern part of the state. From southeastern Oregon to Colorado and southward into Baja California. (See p. 375.)

Food: Like other species of wood rats, the Desert Wood Rat eats many sorts of forbs, both leaves and seeds; also browses on the leaves of shrubs and eats berries. It is often associated with cholla and other desert succulents, from which it obtains necessary water. Remains of partly eaten cacti are often strewed about the entrances to their houses.

Reproduction: Usually from two to four young born in late winter or spring within the protection of the stick house. Possibly two litters a year for some females.

This species is a typical inhabitant of juniper–piñon covered hillsides of southeastern California, especially at lower elevations. In the mountains it is generally replaced by the Bushy-tailed Wood Rat. Both species seem to be attracted to rock outcrops. The Desert Wood Rat is typically associated with cacti, for these succulents provide food and probably protection as well.

Grasshopper Mice (*Onychomys*)

Pale stocky mice with short, thick tails. Hind foot with sole mostly furred (usually nude in deer mice, *Peromyscus*). Skull distinctive in the form of the nasal bones, which come to a point posteriorly (see fig. 112a); in related genera the nasals are rounded or truncate posteriorly. The lower jaw is unusual in the long, slender coronoid process (see fig. 112b).

These stout little mice are generally a small part of the rodent fauna in the western United States. By nature they are highly predatory and feed extensively on insects and sometimes on other small mice. They possess delightful vocal talents, and their pleasant "song" is often heard from captives.

Grasshopper mice seem to favor compact soils with a sparse growth of perennial grasses.

Key to Species of Grasshopper Mice (*Onychomys*) in California

1. Tail usually more than 50 percent of body length; desert and arid areas in southern half of state . . . . . . *torridus*

   Tail usually less than 50 percent of body length; Great Basin, roughly the northeastern section of state . . . . *leucogaster*

Northern Grasshopper Mouse [pl. 15k]
(*Onychomys leucogaster*)

Description: A heavy-bodied mouse with a brown dorsum and white venter. Tail relatively thick and less than 50 percent of total length. Coronoid process long (see fig. 112b). TL 120–190, T 29–62, HF 17–25, E 12–17. Weight: 24–38 g. (See fig. 112.)

Distribution: Sagebrush regions of the Great Basin in northeastern California. From southern Canada to northeastern Mexico. (See p. 375.)

Food: Many insects, especially large species such as fleshy larvae of moths and large beetles. Also orthopterans. Some small mice, which they subdue and kill easily. Little vegetable food.

Reproduction: A prolonged breeding season from February to August or even later. At least two broods a year of one to six young each.

This little mouse has a prolonged "song" with elements of both sonic (below 20,000 cps) and ultrasonic (above 20,000 cps) frequencies.

Southern Grasshopper Mouse [pl. 15l]
(*Onychomys torridus*)

Description: A heavy-bodied mouse with a buff to cinnamon dorsum and white belly. Tail thick, usually more than 50 percent of total length. TL 120–165, T 39–52, HF 18–20, E 11–17. Weight: 20–26 g.

Distribution: Desert regions of southern half of state. Especially sandy areas of Mojave and Sonora deserts and parts of the San Joaquin Valley. From central Nevada and southern Utah south through central Mexico, including Baja California. (See p. 375.)

Food: Large insects and small mice; little plant material.

Reproduction: Breeds from May to July. Litters of two to three young, sometimes up to five or six. Usually two litters a year. Young probably do not breed in the year of their birth.

White-footed Mice (*Peromyscus*)

Small, mouse-sized rodents with a well-developed tail. Various shades of brown, buff, or brownish gray dorsally; feet and venter white. Ears moderately developed and projecting out from fur. Upper incisors not grooved. Coronoid process (on lower jaw) rather short (see fig. 110b).

This is a large genus in North America, and most regions have at least one species. In some California habitats up to four species may occur close together. Many species are rather responsive to environmental conditions and form subspecies or geographic races. Biologically they are also adaptable, and the habits and habitat of a given species may vary from time to time and place to place.

Because of their abundance and diversity, several species of deer mice are favorite subjects for both field and laboratory studies. They reproduce easily and are no trouble to care for. They will survive and thrive on a broad variety of foods, including insects, seeds, leaves, and fungi.

Key to Species of White-footed Mice (*Peromyscus*) in California

The following key to the *Peromyscus* in California is designed to assist in the identification of a specimen found in the field. We have omitted details of dental, cranial, and other internal characters that require dissection and microscopic study. Since individual variation increases the difficulty of identification, this key should be used together with the descriptions of species, the color plates, and the geographic ranges of the several species. Immature specimens differ not only in size and color (they are usually grayish) but also in body proportions, and the ratios in the key may not be correct for immature specimens.

1. Tail 55 percent or more of total length 2

   Tail 51 percent or less of total length 4

2. Hind foot usually 17–25 mm; total length usually less than 235 mm; deserts 3

   Hind foot usually 25–29 mm; total length usually 220–266 mm; southern half of state, at low elevations and foothills, except in deserts . . . . *californicus*

3. Tail with long terminal hairs (4–10 mm from tip of tail); Great Basin and deserts from northeastern corner of California to Mojave Desert . . . . *crinitus*

   Tail with short terminal hairs (2–4 mm from tip of tail); deserts of southern California . . . . *eremicus*

4. Tail about 50 percent of total length; hind foot usually 20–27 mm 5

   Tail about 45 percent of total length; hind foot usually 18–20 mm, longer than ear length (measured from notch) . . . . *maniculatus*

5. Ear quite large, 18–27 mm (measured from notch), and about equal to length of hind foot . . . . *truei*

   Ear moderate, 15–20 mm (measured from notch), and less than length of hind foot . . . . *boylii*

Brush Mouse (*Peromyscus boylii*) [pl. 15g]

Description: A moderately large, brown deer mouse with a rather long tail and medium-sized ears. Dorsum light chocolate brown, often with a pinkish area on forearm, and venter white. Distinctive in the long tail (more than 55 percent of total length) and moderate ears (less than length of hind foot). TL 180–238, T 91–123, HF 20–26, E 15–20. Weight: 25–35 g.

Distribution: Usually in brushy areas from sea level to 1700–2000 m, rarely to 3000 m. Common on hillsides of manzanita, ceanothus, and associated shrubs. East to Texas and south through Mexico. (See p. 376.)

Food: A great variety of seeds of forbs and shrubs; leaves and insects; sometimes feeds heavily on hypogeous fungi. Insects usually a major food.

Reproduction: Breeds from April to October; litters of four to six young. Reproduction dependent upon abundance of food.

This is one of the arboreal species of California mice, and it commonly forages in shrubs for such items as manzanita berries and fruits and leaves of the silk-tassel (*Garrya* spp.).

Parasitic Mouse (*Peromyscus californicus*) [pl. 15j]

Description: The largest deer mouse in California, if not in the United States. Distinctive in its large size, relatively long tail (more than 55 percent of total length), and dark brown coloration. TL 220–285, T 117–156, HF 24–31, E 20–26. Weight: 33–55 g. (See fig. 110.)

Distribution: Mostly South Coast Range woodlands, especially in oaks, buckeyes, bay, and other hardwoods. Also foothills of Sierra Nevada south into Baja California. (See p. 376.)

Food: Eats a variety of leafy material and some insects. Known to feed heavily on seeds of the California Bay (*Umbellularia californica*) and also fresh green foliage. Fruits, flowers, and seeds of shrubs.

Reproduction: From one to four young are born after a gestation of twenty-one to twenty-five days. Lactation lasts for thirty-five to forty-five days. Nest sometimes built within the structure of a wood-rat house. This species is unusual among cricetine mice in forming long-term or permanent pair bonds. In addition, the male defends the nest and shares in care of the young.

This species has also been called the "California Mouse," a name which has no special significance and which could be applied with equal logic to a number of other mouselike rodents in our state. It has long been called the Parasitic Mouse, referring to its proclivity for making its home within the houses of wood rats, a habit also seen in *P. truei*. *P. californicus* may actually depend to some extent on these stick houses, for one student observed a decline in abundance of the Parasitic Mouse following the systematic removal of wood-rat houses.

This species is known to climb well and probably does much of its foraging in bushes or trees.

Cañon Mouse (*Peromyscus crinitus*) [pl. 15i]

Description: A rather small deer mouse with soft, loose fur. Tail long (more than 55 percent of total length) and bicolored. Tail with a tuft of hairs projecting some 4 to 10 mm beyond tip. Ear about same length as hind foot. TL 161–192, T 80–118, HF 17–23, E 17–21. Weight: 13–18 g.

Distribution: Desert areas east of Sierra Nevada. Prefers sandy habitat with rocky outcrops; usually not above 1800 m. From western Oregon to western Colorado south to northern Mexico, including Baja California. (See p. 376.)

Food: Seeds of a great variety of forbs, shrubs, and trees. Presumably also insects.

Reproduction: Apparently breeds throughout much of the year. The reproductive activity may depend on sporadic rainfall, as it does with many desert rodents. Litters of three to five young have been found from February to July.

Like some species of *Dipodomys* and *Perognathus*, the Cañon Mouse produces an extremely concentrated urine and is thus able to survive with very little free water in its food. In nature, insects probably provide ample water.

Cactus Mouse (*Peromyscus eremicus*) [pl. 15f]

Description: A rather small deer mouse with loose silky fur; buff in color. Tail usually more than 55 percent of total length;

terminal tuft 1–4 mm long. Hind foot larger than ear. The Cañon Mouse has a longer tuft on the tail (4–10 mm), and the hind foot is longer than the ear in the Cañon Mouse. TL 160–200, T 84–120, HF 18–22, E 13–20. Weight: 18–30 g.

Distribution: Mojave and Sonora deserts and west to coast. Generally sandy areas with some shrubby growth. From southern Nevada and southern Utah south to Texas and Mexico, including Baja California. (See p. 377.)

Food: A varied diet of seeds, leaves, and many insects.

Reproduction: From one to five young are born; breeding extends from February to June. Reproductive activity probably varies according to the duration of winter rains and consequent growth of green plants and insects. Known sometimes to breed in the fall.

The Cactus Mouse is sometimes difficult to find in the summer because it is known to become torpid or estivate in warm weather. This behavior might be induced by a lack of moisture, for in streamside environments they remain active all summer.

Deer Mouse (*Peromyscus maniculatus*) [pl. 15d, e]

Description: One of the smaller members of the genus, distinctive in the relatively short bicolored tail and moderate ears. TL 150–200, T 60–91, HF 18–22, E 14–20. Weight: 14–25 g.

Distribution: Statewide from sea level to above 3000 m. In all kinds of habitat, including forests, brush, grassland, and chaparral. Occurs on Channel Islands. From northern Canada and Newfoundland south to Mexico.

Food: Many kinds of seeds and sometimes leaves. May feed heavily on insects, such as orthopterans and soil-dwelling larvae. May eat hypogeous fungi.

Reproduction: Breeds from April through to November or even December; breeding season is determined largely by the abundance of food. From two to eight young in a litter.

This is the commonest member of the genus and is likely to be encountered almost anywhere. It is the most common native mouse to enter mountain cabins. In the winter it invades these shelters and makes nests of such materials as pillows, mattresses, toilet paper, and tampons; will nibble on virtually anything but glass and metal.

Piñon Mouse (*Peromyscus truei*) [pl. 15h]

Description: A moderately large deer mouse with very large ears (longer than hind foot). Tail about 50 percent of total length. (The Brush Mouse has ears shorter than hind foot, and its tail is more than 55 percent of total length.) TL 177–195, T 87–98, HF 22–24, E 24–27. Weight: 20–29 g.

Distribution: Open woodland and brushy areas, especially piñon and juniper covered hillsides, avoiding the Central Valley, high mountains, and extremely arid regions. From central Oregon and northern Colorado south through Mexico. (See p. 377.)

Food: Eats a variety of seeds, berries, and insects. Fond of pine nuts; the shells can sometimes be found on the ground under the cover of a fallen log or boulder. Also juniper berries.

Reproduction: A litter of three to six young is born from April to June or July, providing time for two or three litters. Sometimes nests within the stick structure of wood-rat houses.

Although this species is commonly associated with the Piñon Pine, it may occur hundreds of miles from this tree. The Piñon Mouse is at home in trees; when released, a captive will frequently climb the nearest tree or shrub.

Harvest Mice (*Reithrodontomys*)

Small, delicate mice with long, nearly nude tails. Ear pinnae large and almost hairless. Color yellow-brown, but immatures are gray. Skull distinctive in the presence of a pronounced groove on upper incisor. Somewhat like House Mouse (*Mus musculus*), which lacks the grooved upper incisor.

These little mice are frequently found in brushy areas which have a dense ground cover of long grass. They are usually smaller than deer mice (*Peromyscus* spp.), but the two may occur together.

Harvest Mouse (*Reithrodontomys megalotis*) [pl. 15c]

Description: A small, delicate buff or brownish mouse with a rather long tail. Distinctive in the grooves on outer surface of upper incisors. Venter white, sometimes with a buffy spot on chest. Tail bicolored. Ears rather large. TL 114–145, T 50–70, HF 15–18, E 12–15. Weight: 9–14 g.

Distribution: Throughout California in lowland and mid-elevations in grassy areas. May occur along margins of cultivated areas if grass and weeds are present. Rather adaptable but avoids forest. From southwest Canada across much of western United States to Mexico.

Food: Seeds of grasses and weeds; some insects, especially cutworms.

Reproduction: The nest is a distinctive ball-shaped structure of fine grass, usually on the surface of ground but sometimes in a dense bush a foot (⅓ m) or more aboveground. Usually three to five young (sometimes up to nine). Breeds in spring and sometimes again in autumn. Young weaned in three and a half weeks and may breed in the year of birth, providing for occasional increases to high densities.

Salt Marsh Harvest Mouse (*Reithrodontomys raviventris*)

Description: Similar to *R. megalotis* but with tail thicker near base (or basal third) and tail one color or only faintly bicolored. Venter brownish or buff; white or nearly so in *megalotis*. TL 118–175, T 56–95, HF 15–21, E 12–14. Weight: 8–12 g.

Distribution: This species is confined to the salt marshes about San Francisco Bay and the Napa, Petaluma, and Suisun salt marshes. Commonly associated with dense growth of

Pickleweed (*Salicornia virginica*). As its habitat is being obliterated, this mouse disappears. It is considered to be an endangered species.

Food: Presumably feeds on seeds or grasses and forbs as well as insects.

Reproduction: Breeds from spring to autumn; litters from one to seven young (usually three or four).

Hispid Cotton Rat (*Sigmodon hispidus*) [pl. 14g]

Description: A small blackish or dark brown rat with grizzled fur. Distinguished from the Roof Rat (*Rattus rattus*) by the grizzled color and shorter tail (tail more than half total length in Roof Rat). The Brown Rat (*Rattus norvegicus*) is larger and browner than the Cotton Rat. The species of *Rattus* are also clearly separable by the three rows of tubercles on the molar teeth; in *Sigmodon* there are two rows of tubercles, and in older specimens these are worn so as to reveal transverse ridge of enamel. TL 224–365, T 81–166, HF 28–41, E 16–24. Weight: 100–225 g. (See fig. 111.)

Distribution: Along margins of watercourses in region of Colorado River. Throughout much of southern half of United States into South America, usually near water. (See p. 380.)

Food: More vegetarian than most native mice. Eats many grasses and forbs, taking few insects.

Reproduction: Breeding season extended, perhaps dependent on presence of fresh vegetation. After gestation of twenty-six to twenty-eight days, three to seven young are born.

The Cotton Rat resembles voles in many ways. Like the vole, it is active both day and night and makes runways through grassy areas.

Voles and Allies (Arvicolidae)

Stocky rodents, mostly small, legs short, eyes small, and ears partly hidden in fur. Colors grizzled brown or blackish, sometimes reddish or buff, but never with belly white. Tail medium or short, less than 50 percent of total length. Skull flat and

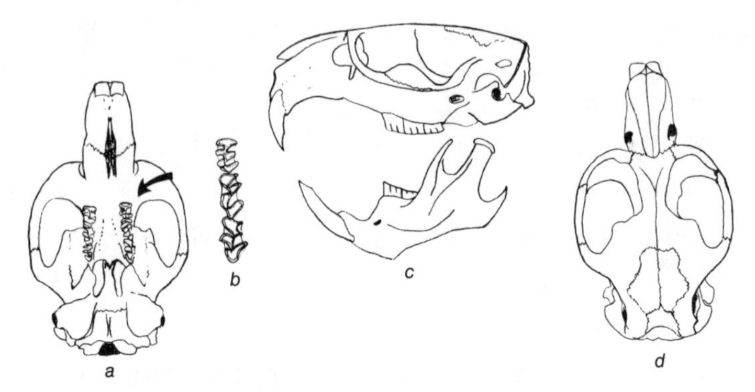

FIG. 115. Skull of the Muskrat (*Ondatra zibethicus*): *a*, ventral view (arrow to upper tooth row); *b*, the upper tooth row shows the typical "loops and triangles" of the cheek teeth of voles; *c*, lateral view; *d*, dorsal view.

broad. Cheek teeth with flat surfaces of distinctive "loops and triangles" which are important in the generic and specific identification (fig. 115b). Dentition: 1/1, 0/0, 0/0, 3/3.

There are many genera and species of voles, and their identification is sometimes difficult. Because an area usually supports only a few different kinds, the locality of capture greatly aids in their recognition. Most species feed on vegetative parts of grasses and forbs, but they sometimes eat cambium of trees and the underground parts of both woody and herbaceous plants. Insects are not generally a major item in the diet of voles.

Voles tend to be prolific. They mature at an early age—three weeks in some species—and may have several large litters a year. Sporadically they become very abundant. At such times they attract many predatory birds and mammals and are also capable of doing substantial damage to field crops. The largest California vole is the Muskrat, the most important furbearer in the United States.

Key to Genera of Arvicolidae in California

1. Tail round in cross section; hind feet without a lateral fringe of stiff hairs 2

   Tail compressed with dorsal and ventral ridges; hind toes and feet with lateral fringes of stiff hairs . . . . *Ondatra zibethicus*

2. Cheek teeth rooted 3

   Cheek teeth rootless 5

3. Lower cheek teeth with inner angles deeper than outer (fig. 116b) 4

   Lower cheek teeth with inner angles not conspicuously deeper than outer (fig. 117b) . . . . *Clethrionomys californicus*

4. Tail less than 20 percent of total length; four pairs of nipples; hip glands present . . . . *Phenacomys*

   Tail more than 30 percent of total length; two (rarely three) pairs of nipples; hip glands absent . . . . . . *Arborimus*

5. Soles of feet with dense hairs; tail scarcely longer than hind foot; external ears mostly concealed in fur . . . *Lagurus*

   Soles of feet hairless; tail considerably longer than hind foot in most species; external ears projecting slightly from fur . . . . *Microtus*

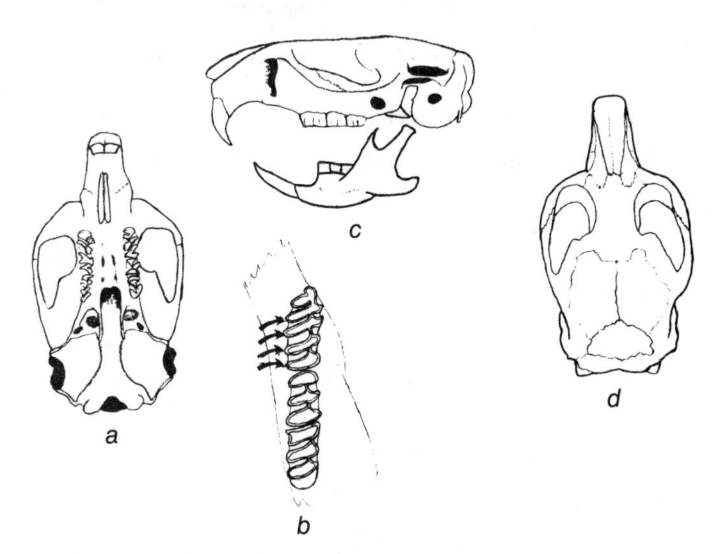

FIG. 116. Skull of the Heather Vole (*Phenacomys intermedius*): *a,* ventral view; *b,* lower right tooth row (arrows to long inner loops of the first lower cheek tooth); *c,* lateral view; *d,* dorsal view.

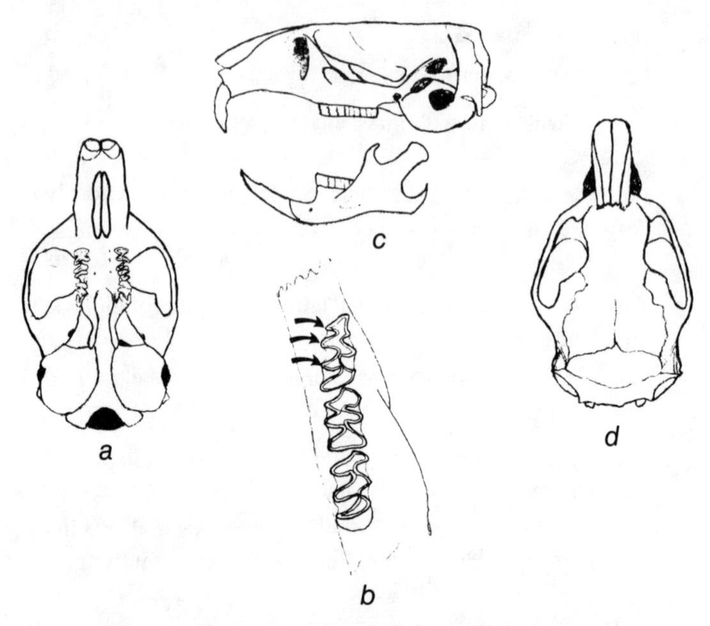

FIG. 117. Skull of the California Red-backed Vole (*Clethrionomys californicus*): *a*, ventral view; *b*, right lower tooth row (arrows to inner angles); *c*, lateral view; *d*, dorsal view.

Climbing Voles (*Arborimus*)

Rather long-tailed voles with rooted molars. Females usually with two pairs of nipples (indicating two pairs of mammae). Hip glands (visible externally on rear flank) absent (present on *Phenacomys*). Tail more than 30 percent of total length. Restricted to coastal forests of northwestern California and Oregon.

Key to Species of *Arborimus* in California

1. Color reddish; tail hairy . . . . *longicaudus*

   Color brown or gray; tail scantily haired . . . . *albipes*

White-footed Vole (*Arborimus albipes*)

Description: A small, dark brown vole with a long, clearly bicolored tail. Venter gray, sometimes with a pinkish cast. Snout darker than rest of head. Claws straight. TL 165–181, T 62–71, HF 19–20, E 10–12. Weight: 17–28 g.

Distribution: Known from streamside thickets in redwood forests in northwestern California. Also in the coastal forests of Oregon. (See p. 378.)

Food: Climbs well and browses on leaves of Red Alder (*Alnus rubra*) and willows. Also feeds on leaves of many forbs.

Reproduction: Usually two or three young. Nipples four in most individuals, but there may be up to six.

Like the following species, the White-footed Vole climbs well and feeds on the foliage of trees. These two species seem to be the only voles that climb more than several inches above the ground.

Red Tree Vole (*Arborimus longicaudus*) [pl. 11a]

Description: A bright chestnut-red or brick-red vole with a long, well-furred tail. Claws curved. Ears partly concealed in fur. This species might possibly be confused with the California Red-backed Vole (*Clethrionomys californicus*), which is a much darker red and has a shorter tail. TL 158–186, T 60–76, HF 19–21, E 10–11. Weight: 24–27 g.

Distribution: Coastal forests in the humid fog belt north of San Francisco Bay. Associated with open stands of Douglas Fir, its major food. Strictly a tree-dwelling mouse. (See p. 378.)

Food: Specialized for feeding only on conifer leaves. In California it browses almost solely on needles of Douglas Fir, but in Oregon it is known to eat needles of the Coast Hemlock and Tideland Spruce. It shuns grasses, forbs, seeds, and insects.

Reproduction: Nest built of fir needles can be found in the outer branches of a conifer, usually a Douglas Fir, usually from 3 to 20 m above the ground. Litters of one to three young are generally distributed throughout the year.

The Red Tree Vole is one of the most unusual of North American mammals. It is perhaps the most specialized feeder, taking only needles of Douglas Fir and occasionally hemlock and spruce. It is confined to a narrow region of humid coastal forests near the ocean; does not venture into less humid areas of the interior, even where Douglas Fir may abound. Because of its specialized diet, it is not attracted to baits. It can sometimes be captured within its nest.

California Red-backed Vole [pl. 11b]
(*Clethrionomys californicus*)

Description: A dark chestnut-colored vole with a faintly bicolored tail. Dorsal fur with many blackish hairs presenting a dark aspect to the reddish color. Venter buffy or cream-colored. Cheek teeth with inner and outer angles equal (fig. 117b). TL 155–165, T 46–55, HF 17–21, E 10–12. Weight: 17–33 g. (See fig. 117.)

Distribution: Coniferous forests north of San Francisco Bay along the coast and north of Lake Tahoe in the Sierra–Cascade ranges. In western Oregon north to the Columbia River. (See p. 379.)

Food: Seeds and insects of the forest floor. Green plant material in stomachs includes lichens. Quantities of fungi are also eaten.

Reproduction: A rather long breeding season, from March to October, provides for several litters. This vole is not very prolific, however, and usually bears two or three young.

This little vole seems not to be common, but population densities do vary. In years of abundance, it can be found in brushland of manzanita and silktassel at around 1500 m in Plumas County.

Sagebrush Vole (*Lagurus curtatus*) [pl. 11c]

Description: A light grayish vole with a rather short tail. Auditory bullae clearly enlarged (fig. 118a). Soles of feet furred. TL 108–140, T 15–26, HF 15–18. Weight: 20–30 g.

Distribution: The Sagebrush Vole, true to its name, is typically found in the high arid sagebrush areas of the Great Basin; in California, it occurs east of the Sierra Nevada both north and south of Lake Tahoe. It is said to favor rather open areas where the sage is sufficiently mature that some branches touch the ground. (See p. 379.)

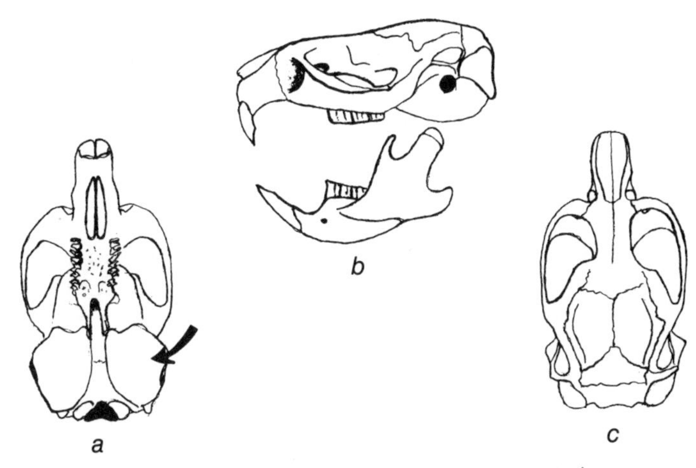

FIG. 118. Skull of the Sagebrush Vole (*Lagurus curtatus*): *a*, ventral view (arrow to auditory bulla); *b*, lateral view; *c*, dorsal view.

Food: Leaves of sage are eaten at all seasons, and they will also eat the cambium layer of the woody growth of sage. Apparently these little rodents do not eat seeds and are not known to eat insects.

Reproduction: From three to six young are born in a concealed nest of grass and shredded bark. Although some desert rodents that feed on seasonally available plants are rather seasonal in their breeding activity, the food of the Sagebrush Vole is available year-round and this species seems not to have a distinct reproductive season.

Meadow Voles (*Microtus*)

Dark blackish voles with grizzled fur, small eyes, and external ears partly concealed in fur. Tail length usually less than 50 percent of total length. Incisors rootless. Outer and inner reentrant angles of molar surfaces about equal (see fig. 123a).

These little rodents, variously known as voles, meadow mice, or field mice, feed very largely on vegetative parts of green plants but may also eat cambium layers of stems and sometimes seeds. They are, therefore, capable of doing tremendous damage to cultivated crops, such as alfalfa and sugar beets, and sometimes have been known to girdle orchard trees. They are generally prolific and attain very high densities. Five species occur in California.

These voles all resemble each other rather closely. They can usually be identified on the basis of (1) relative length of tail and hind foot or total length, (2) shape of incisive foramina, (3) plantar tubercles, and (4) geographic range. The incisive foramina can easily be seen by peeling back the soft flesh in the roof of the mouth from just behind the base of the upper incisors (see figs. 119a–123a).

Key to Species of *Microtus* in California

1. Six plantar tubercles 2

   Five plantar tubercles; hind foot less than 22 mm; side glands (on females) reduced or absent; incisive foramina not constricted (fig. 119a) . . . . *oregoni*

2. Incisive foramina abruptly constricted and narrower posteriorly (figs. 120a and 121a) 3

   Incisive foramina tapered gradually posteriorly or not at all (figs. 122a and 123a) 4

3. Tail more than twice the length of hind foot; coastal marshes of extreme northern California, coastal lowlands of northwest . . . . *townsendii*

   Tail less than twice the length of hind foot; Great Basin . . . . *montanus*

4. Tail more than 33 percent of total length; montane forests . . . . *longicaudus*

   Tail less than 32 percent of total length; usually in lowland meadows . . . . *californicus*

California Meadow Vole [fig. 124; pl. 11e]
(*Microtus californicus*)

Description: A medium-sized vole with six plantar tubercles. Incisive foramina not constricted posteriorly (see fig. 123a). Tail faintly bicolored; more than twice the length of hind

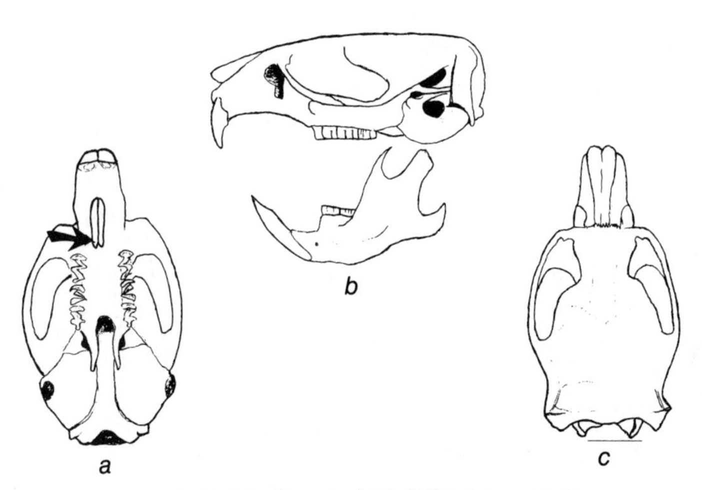

FIG. 119. Skull of the Creeping Vole (*Microtus oregoni*): *a,* ventral view (arrow to palatine foramina, which are of even width throughout their lengths): *b,* lateral view; *c,* dorsal view.

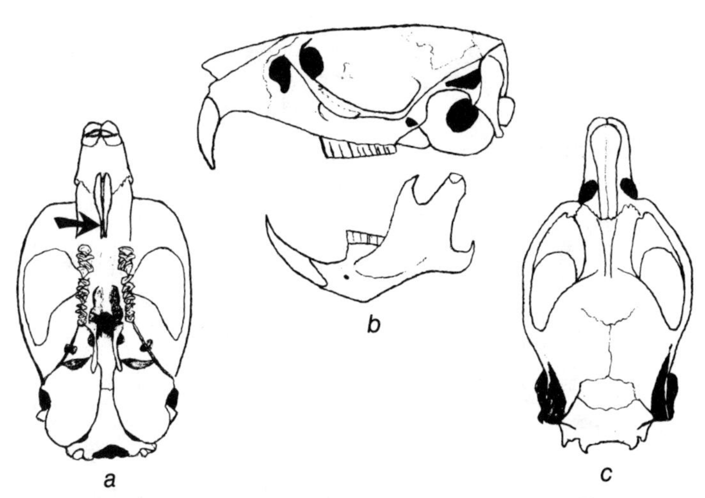

FIG. 120. Skull of Townsend's Vole (*Microtus townsendii*): *a,* ventral view (arrow to palatine foramina, which are constricted posteriorly); *b,* lateral view; *c,* dorsal view.

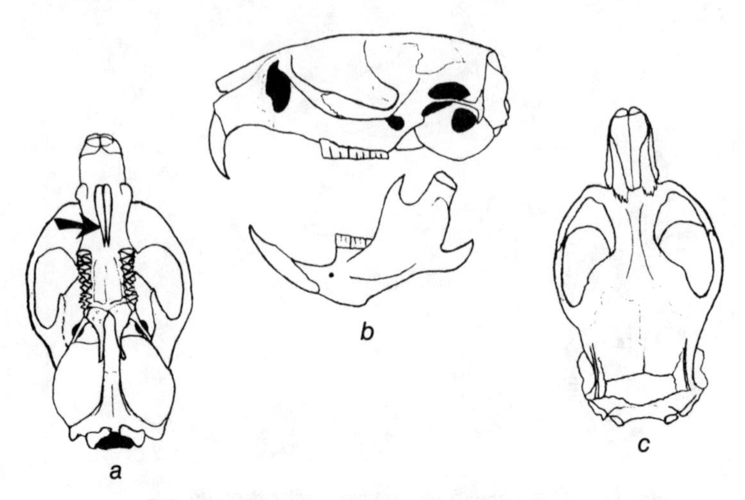

FIG. 121. Skull of the Montane Vole (*Microtus montanus*): *a,* ventral view (arrow to the palatine foramina, which are constricted posteriorly); *b,* lateral view; *c,* dorsal view.

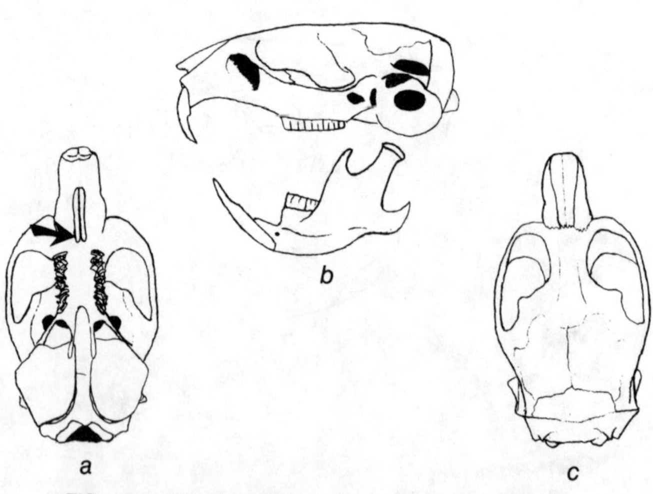

FIG. 122. Skull of the Long-tailed Vole (*Microtus longicaudus*): *a,* ventral view (arrow to palatine foramina, which are of even width throughout their lengths); *b,* lateral view; *c,* dorsal view.

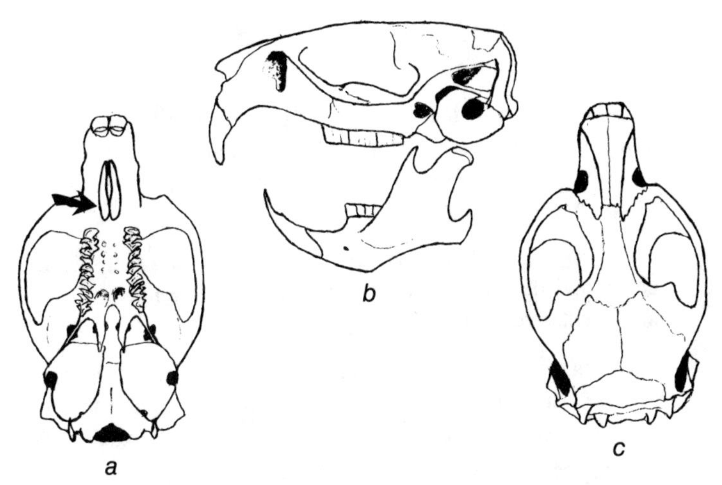

FIG. 123. Skull of the California Meadow Vole (*Microtus californicus*): *a,* ventral view (arrow to rounded posterior ends of palatine foramina); *b,* lateral view; *c,* dorsal view.

FIG. 124. California Meadow Vole (*Microtus californicus*).

foot. TL 157–211, T 39–68, HF 20–25, E 13–16. Weight: 35–72 g.

Distribution: Lowlands and foothills of California up to 1500 m in northern Sierra Nevada. Usually in wet meadows but common in alfalfa plantings. From northern Baja California to Oregon. (See p. 380.)

Food: Many kinds of forbs and grasses, favoring fresh tender leaves and developing seeds. Mature seeds are not preferred and insects are shunned.

Reproduction: May breed throughout the year if fresh green food is abundant. Otherwise breeds when grasses and forbs are sprouting. Usually three to eight young. Females sexually mature between three to four weeks of age.

This is the common meadow vole throughout most of the state except the higher elevations. It is the only vole in the Central Valley. It is common in both alfalfa fields and irrigated pastures and sometimes reaches high densities.

Long-tailed Vole (*Microtus longicaudus*) [pl. 11f]

Description: A moderately large vole with six plantar tubercles and a rather long tail (more than twice the length of hind foot and more than 33 percent of total length). Tail indistinctly bicolored. Incisive foramina of even width from front to back (see fig. 122a). TL 155–221, T 50–93, HF 20–25, E 11–15. Weight: 21–56 g.

Distribution: In woodlands in Sierra Nevada and Cascade ranges, San Bernardino Mountains, and northwestern part of state. Western United States north to Alaska. (See p. 380.)

Food: Grasses and forbs; seeds of many forest shrubs and trees.

Reproduction: From three to eight young in a litter. Two or more litters a year.

Montane Vole (*Microtus montanus*)

Description: A short-tailed vole with a bicolored tail which is usually less than twice the length of hind foot. Six plantar tubercles. Incisive foramina conspicuously narrow posteriorly (see fig. 121a). TL 169–189, T 39–54, HF 20–27, E 11–14. Weight: 30–65 g.

Distribution: Meadows of High Sierra and Great Basin. Western United States north to British Columbia. (See p. 381.)

Food: Almost exclusively green grasses, sedges, and forbs.

Reproduction: Breeding in spring and summer, but seasonality varies from place to place and year to year. Successive litters of three to eight young.

This is a common and widespread vole of the Great Basin. It sometimes attains populations estimated at several thousand per acre; at other times it may become very scarce. Commonly makes discrete runways through grasses and sedges.

Creeping Vole (*Microtus oregoni*) [pl. 11d]

Description: A medium-sized, short-tailed vole with five plantar tubercles. Tail less than twice length of hind foot. Incisive foramina not constricted but rounded posteriorly (see fig. 119a). TL 129–154, T 32–42, HF 16–19, E 9–10. Weight: 18–22 g.

Distribution: Forests of deep soil; generally avoids meadows. In northwestern part of state. North to southwestern British Columbia. (See p. 381.)

Food: Not well known; presumably roots, bulbs, and stems of forest forbs and grasses.

Reproduction: Not well known. Litters of three to six young have been recorded. May breed throughout the year.

Townsend's Vole (*Microtus townsendii*)

Description: A large, dark vole. Six plantar tubercles. Incisive foramina constricted posteriorly (see fig. 120a). Tail faintly bicolored, more than twice length of hind foot. TL 169–222, T 48–70, HF 20–26, E 15–17. Weight: 75–82 g.

Distribution: Marshes of northwestern part of state. North to Vancouver Island. (See p. 381.)

Food: Grasses and forbs. Sometimes a serious pest in fields of clover and alfalfa.

Reproduction: Litter size varies from five to eight. May breed throughout the year on irrigated land, but reproduction is mostly in spring and summer in natural environments.

Muskrat (*Ondatra zibethicus*) [pl. 14h]

Description: A rat-sized vole modified for an aquatic life. Fur chocolate brown; with long glossy guard hairs and dense, water-repellent underfur. Hind feet partly webbed with a fringe of hairs; tail laterally compressed. Eyes small. Conspicuous scent glands near the anal opening give this mammal its name; scent is deposited with feces. Skull heavy and angular; incisive foramina narrowed at each end. (See fig. 115a.) TL 456–553, T 200–250, HF 65–75, E 20–22. Weight: 700–1800 g.

Distribution: Introduced in the Sacramento–San Joaquin Valley, where it is now rather common along watercourses, artificial and natural. From Alaska throughout most of Canada and in the northern two-thirds of the United States. Also introduced in Eurasia.

Food: Feeds on water plants with an obvious preference for cattails. May take some animal food, such as freshwater clams and crayfish.

Reproduction: Breeds from late winter through spring and sometimes summer. Two to ten young after a gestation of twenty-five to twenty-nine days. Young swim in two weeks and are weaned in four weeks. May breed in the year of birth. The nest may be on a platform of dead cattails or in a burrow in bank of levee.

The Muskrat is an important furbearer, but it can also be destructive to irrigation canals. It is usually common and easily trapped and provides supplemental income for a large number of teenagers all across the United States. It may also be a host to such diseases as tularemia and, through its urine, can introduce tularemia into the water in which it swims.

Heather Voles (*Phenacomys*)

A pale brown vole similar to *Arborimus* and resembling some kinds of *Clethrionomys*. The cheek teeth are rooted in the adult, and the lower cheek teeth have inner angles deeper than the outer angles (see fig. 116b). Females with four pairs of nipples. Hip glands present. One species known.

Heather Vole (*Phenacomys intermedius*)

Description: A light gray vole with a short, sharply bicolored tail. Separable from the Red-backed Vole (*Clethrionomys californicus*) and the Long-tailed Vole (*Microtus longicaudus*) by its gray (not reddish or blackish) dorsum and by the sharply bicolored tail. Lower cheek teeth with inner angles deeper than outer. (See fig. 116.) TL 130–153, T 26–41, HF 16–18, E 13–16. Weight: 21–40 g.

Distribution: In the high (2500 m and above) central Sierra Nevada, such as the forests above Lake Tahoe or in the Mt. Shasta area. Sometimes associated with patches of Red Heather (*Phyllodoce breweri*). Widely distributed in coniferous forests across Canada and western United States. (See p. 378.)

Food: Apparently green plants but not well known. They are readily taken in mouse traps baited with walnut meats.

Reproduction: A litter of four to six young born in spring or summer. Probably two broods a year.

The Heather Vole is really not confined to the distribution of Red Heather but occurs in a variety of high-elevation woodlands. It seems to be rather infrequently captured and represents an opportunity for enterprising students to learn the basic facets of this little-known vole.

Old World Rats and Mice (Muridae)

Small to medium-sized rodents of diverse form and color. Represented in North America only by the familiar commensal species, separable from native but rather similar Cricetidae by the presence of three longitudinal rows of tubercles on the cheek teeth (fig. 125b); in the native cricetid mice the molar

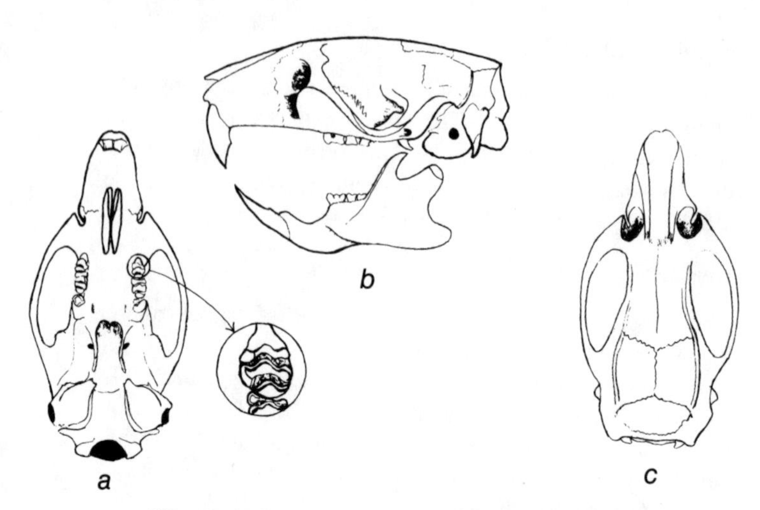

FIG. 125. Skull of the Brown Rat (*Rattus norvegicus*): *a,* ventral view (the cheek teeth, in circle, are characterized by tubercles forming three indistinct longitudinal rows); *b,* lateral view; *c,* dorsal view.

surfaces are either flat or have two rows of tubercles. The murid rodents in North America are also characterized by conspicuous ears and a rather long, nearly hairless scaly tail. Dentition: 1/1, 0/0, 0/0, 3/3 (fig. 125c).

This is the largest family of rats and mice in the Old World, and three species are common in most parts of the United States. The three commensal rats and mice have been carried by ships to all parts of the earth, and their native lands are today not apparent.

Key to Commensal Rats and Mice (Muridae) in California

1. Adult total length more than 320 mm 2
   Adult total length less than 225 mm . . . . *Mus musculus*
2. Tail more than half total length . . . . . . *Rattus rattus*
   Tail less than half total length . . . . *Rattus norvegicus*

House Mouse (*Mus musculus*) [pl. 15a]

Description: A small, dark brown, long-tailed mouse with a light brown or buff-colored venter. Tail almost without hairs and unicolor. Similar in size to harvest mice (*Reithrodontomys* spp.), which have grooved upper incisors, a light dorsum, and a white venter. Species of *Peromyscus* have a white venter and are rarely as dark dorsally as is the House Mouse. Native mice also have two longitudinal rows of tubercles on cheek teeth in contrast to the three rows of the House Mouse. Finally, the House Mouse has an unpleasant odor unlike that of any native species. TL 155–204, T 70–95, HF 17–20, E 11–16. Weight: 12–24 g.

Distribution: Statewide about human habitations and in fields and brushy areas up to at least 2200 m in mountains. Virtually worldwide.

Food: Virtually any edible materials: seeds, leaves, insects, and stored food materials in warehouses and pantries.

Reproduction: Sexually mature between seven and eight weeks; breeds more or less throughout the year. A litter of four to eight (or sometimes more) born after a gestation of three weeks. As many as five litters a year. This pattern allows rapid increase in populations.

These dirty and destructive little mice soil what they do not eat. They destroy stored foods everywhere and pollute what remains. They have been carried throughout the world and are among the most universal mammalian pests. It should be noted, however, that they are among the most common laboratory animals and are the means by which much biological research is conducted.

Brown or Norway Rat (*Rattus norvegicus*) [pl. 14d]

Description: A brownish rat with rather rough or coarse fur. Tail naked and scaly; less than 50 percent of total length. Ears rather large and naked; nose somewhat blunt, a Roman nose. Venter gray. Separable from wood rats (*Neotoma* spp.) by the pigmented venter, which in wood rats is white. Wood rats also have molar teeth with flat surfaces, not tuberculate as in *Rattus* (see fig. 125b). The rather similar Roof Rat (*Rattus rattus*) has a long tail more than 50 percent of total length. TL 300–475, T 120–215, HF 32–44, E 19–24. Weight: 300–525 g.

Distribution: Statewide about buildings and in wild environments up to about 1000 m in the mountains. May be common in rice fields and sometimes in plantings of row crops. Worldwide. Presumably native to Old World tropics.

Food: Omnivorous. Extremely destructive to food products of all sorts.

Reproduction: Prolific, breeding more or less continuously from about twelve weeks of age. From four to ten young in a litter; may mate immediately after giving birth. Thus a female may be simultaneously pregnant and nursing.

The feral Brown Rat must be the most unpleasant mammal in the world. It destroys food both in the field and in storage and damages warehouses and homes when gaining entry. It is a reservoir of plague, some forms of hepatitis, trichina worms, and typhus. It is also the "white rat" of scientific laboratories and, together with the House Mouse, the most valuable experimental mammal in the world.

Roof Rat or Black Rat (*Rattus rattus*) [pl. 14e, f]

Description: A slender rodent varying from various shades of brown to black. Separable from the Brown Rat by the long tail, which is more than half the total length. The Roof Rat may be entirely black or nearly so, in which case it is called the Black Rat; or it may be various shades of brown, with the venter pale buffy; brownish forms seem to be more common in California. Separable from wood rats (*Neotoma* spp.) by the tuberculate surface of the molar or cheek teeth; they are flat-surfaced in wood rats. TL 320–435, T 170–240, HF 32–39, E 19–26. Weight: 160–205 g.

Distribution: Widely distributed except above 800 m. Climbs trees, runs along telephone wires, and commonly enters attics. Worldwide in warm and temperate regions. Probably native to tropical orient.

Food: Largely plant material, most commonly fresh vegetables and fruits in California.

Reproduction: Less prolific than the Brown Rat. Sexually mature between three and four months, breeding throughout the year. Litters of five to eight young (sometimes more) with three to five litters a year.

The Roof Rat in California is most commonly yellow-brown with a strong suffusion of black hairs, but some are entirely black. Its name refers to its tendency to climb, and it is more common in our attics than is the Brown Rat. In southern California it may live in groves of avocados, where it is called the "tree rat." While less destructive to supplies of stored food than is the Brown Rat, the Roof Rat is an important reservoir of both plague and typhus fever and has been implicated in serious outbreaks of the latter in the United States.

In foothill areas the Roof Rat may become common in dense blackberry thickets. It is also known to invade orchards and eat almonds.

Jumping Mice (Zapodidae)

Small mice clearly modified for jumping: hind legs and hind feet enlarged and tail greatly elongated, much longer than head and body combined. Hind feet with five toes. Anterior orbital foramina large (fig. 126b); upper incisors grooved (fig. 126a);

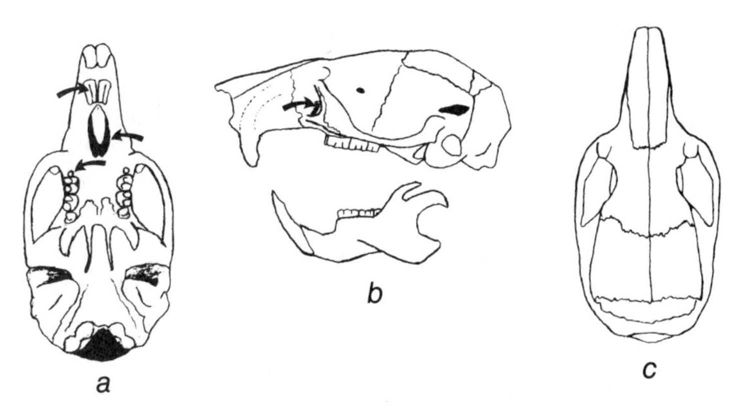

FIG. 126. Skull of the Western Jumping Mouse (*Zapus princeps*): *a,* ventral view (top arrow to grooved incisors, middle arrow to incisive foramina, bottom arrow to peglike premolar); *b,* lateral view (arrow to anterior orbital foramen); *c,* dorsal view.

with a peglike upper premolar (fig. 126a). Internal cheek pouches. Dorsum bicolored.

Jumping mice occur in moist meadows and forests, especially along small creeks. They are easily distinguished from pocket mice by their *internal* cheek pouches (opening within the mouth); the cheek pouches of pocket mice open outside the mouth and are fur-lined. Jumping mice are found in moist habitats; pocket mice occur in dry or arid areas.

Meadow Jumping Mice (*Zapus*)

With the characteristics of the family. A small, peglike upper premolar (see fig. 126a). Two species are currently recognized in our state. They occur in both wooded and grassy habitats but never in arid areas. Dentition: 1/1, 0/0, 1/0, 3/3.

Key to Jumping Mice (*Zapus* spp.) in California

1. Dorsum and sides evenly colored (dark yellow-brown); ear of even pigmentation; coastal redwood forests . . . . *trinotatus*

   Sides a bright straw-yellow set off clearly from darker pigmentation of dorsum; ear with a yellow margin; montane forests . . . . *princeps*

Western Jumping Mouse (*Zapus princeps*)

Description: Dorsum conspicuously darker than the straw-yellow sides. Tail very long and bicolored. Head grayish. Ear with a lightly colored or yellow margin. Upper premolar 0.55 × 0.50 mm (see fig. 126a). TL 215–255, T 121–155, HF 30–37, E 14–18. Weight: 17–28 g.

Distribution: Montane meadows and creeksides in mountains from Klamath Range through northeastern California south to Sierra Nevada. From Washington south in the mountains to California. (See p. 382.)

Food: Largely seeds of grasses and forbs. Also pulpy berries and sometimes insects.

Reproduction: A single litter of four to eight young annually.

These dainty little mice are both diurnal and nocturnal and may occasionally be startled as they forage in tall grass. They escape by leaps of several feet. These mice are profound hibernators and in late summer accumulate quantities of both subcutaneous and visceral fat.

Pacific Jumping Mouse (*Zapus trinotatus*) [pl. 15b]

Description: Sides and back evenly pigmented a dark yellow-brown. Ear evenly dark. Upper premolars 0.70 × 0.75 mm. TL 211–242, T 112–155, HF 30–36, E 13–17. Weight: 16–24 g.

Distribution: Coastal meadows from Marin County northward to Oregon. In coastal forests north to southern British Columbia, not including Vancouver Island. (See p. 382.)

Food: Seeds of forbs and grasses and some leafy material.

Reproduction: One litter of four to eight young is produced annually.

Porcupine (Erethizontidae)

Short, stout rodents with many hairs stiffened and barbed. Modified for tree-climbing; tail prehensile in some species. Skull heavy and deep; no postorbital process. Skull with a large infraorbital foramen (fig. 127b). Incisors deep orange.

This family of porcupines is confined to the New World; another group occurs in Africa. In North America the Porcupine is common and very tame. This family is essentially a South American group which entered North America when the two continents were rejoined, at the close of the Pliocene, some 3 million years ago.

Porcupine (*Erethizon dorsatum*) [fig. 128]

Description: A large, heavyset spiny rodent with a thick tail armed with dozens of heavy quills. Limbs stout. Dorsal pelage with many easily detached spinous hairs, which provide an offensive covering. Color blackish suffused with yellow-buff. Skull with a large infraorbital foramen. (See fig. 127c.)

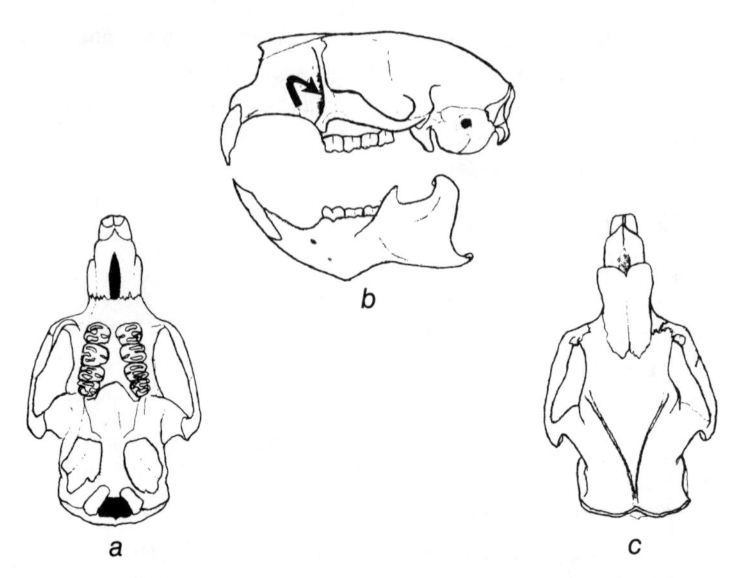

FIG. 127. Skull of the Porcupine (*Erethizon dorsatum*): *a*, ventral view; *b*, lateral view (arrow to infraorbital foramen); *c*, dorsal view.

TL 790–1330, T 145–300, HF 75–101, E 25–42. Weight: 10–18 kg.

Distribution: Virtually all coniferous forests of the state; especially common in cutover or burned areas. From northern Mexico over much of the United States to Alaska. (See p. 382.)

Food: Bark and cambium of coniferous trees, especially pines; rarely hardwoods. Departing from a woody diet in spring and summer, Porcupines may be seen eating grass in mountain meadows. Also roots, stems, berries, leaves, and nuts of many kinds of plants. A rich microbial population in the gut breaks down cellulose to sugars, starches, and vitamins.

Reproduction: Mating in late autumn or early winter. A single young is born in an advanced condition (with dorsal spines soft) some 210 days later. The infant takes solid food within hours and is weaned at an early age.

Like skunks, the Porcupine has few enemies. The barbed quills that cover its back and tail can penetrate not only skin

FIG. 128. Porcupine (*Erethizon dorsatum*).

but also heavy shoe leather. Although few predators attack Porcupines, some large carnivores will kill and eat them under duress. The prey may have its revenge after death, however, for occasionally such carnivores as Bobcats have been found dead with their entire bodies riddled with quills of their last meal.

Although Porcupines are slow-moving and inoffensive, they

can be dangerous and hikers should not molest them. Old tales tell of Porcupines throwing their quills. This does not actually happen, but sometimes they might *seem* to throw their quills. The naturalist Vernon Bailey recounted an attempt to overthrow a Porcupine. After he had repeatedly flipped it over with a stick, many quills became loosened. Finally the animal shook itself vigorously, sending quills out to a distance of 4 or 5 ft (almost 2 m).

LAGOMORPHA

Small herbivores, superficially rodentlike but with two pairs of upper incisors; a small incisor lies behind each large upper incisor (see figs. 129a–133a). Upper rows of cheek teeth farther apart than the lower rows, so that the teeth of only one side at a time can be in opposition to each other. Incisive foramina very large (see fig. 130a). Tail rather short.

This order includes rabbits, hares, and pikas. Pikas are distinctive in their hind limbs, which are scarcely longer than the forelimbs, and in their vestigial tail. Rabbits and hares have hind limbs greatly enlarged for jumping and possess a small but bushy tail.

Hares (*Lepus* spp.) can be distinguished from rabbits (*Sylvilagus* spp.) on the basis of size and cranial features (see the key for rabbits and hares), but the two groups closely resemble one another. The young of hares are precocial (born well furred and capable of walking a short time after birth) in contrast to rabbits, whose young are altricial (born naked, blind, and helpless).

References

Myers, K., and MacInnes, C. D. (Eds.) 1981. *Proceedings of the World Lagomorph Conference (1979)*. Guelph, Ontario: University of Guelph.

Orr, R. T. 1940. *The rabbits of California*. Occasional Papers of the California Academy of Sciences, No. 19. San Francisco: California Academy of Sciences.

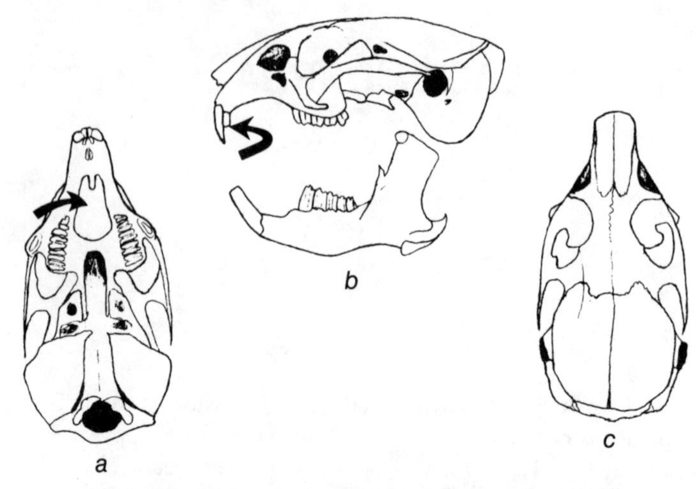

FIG. 129. Skull of the Pika (*Ochotona princeps*): *a,* ventral view (arrow to enlarged incisive foramen); *b,* lateral view (arrow to inner upper incisor); *c,* dorsal view.

Pikas (Ochotonidae)

Small herbivores with rather short rounded ears, short legs, and a minute tail. Hind legs almost the same size as forelegs. With five pairs of upper cheek teeth (fig. 129a, b). Dentition: 2/1, 0/0, 3/2, 2/3.

In North America pikas live on talus slopes of high mountains where they can sometimes be seen perched atop a large boulder. They are diurnal and active throughout the year. In eastern Asia some species of pikas occur deep in coniferous forests where they burrow elaborate systems of tunnels, and some live in open grasslands.

Pika (*Ochotona princeps*) [pl. 16a]

Description: A small, gray-buff mammal with a minute tail. Ears short but broad and projecting above fur. Two pairs of upper incisors distinguish the skull of a Pika from that of a rodent (see fig. 129a). TL 150–210 (tail vestigial), HF 27–35. Weight: 120–130 g.

Distribution: High mountains and talus slopes from southern Sierra Nevada north to Mt. Shasta and east to the Warner

Mountains. From New Mexico and Colorado north to British Columbia. (See p. 383.)

Food: Largely stems and leafy growth of forbs and shrubs. These little mammals gather these materials, dry them in the sun, and then store them beneath the cover of rocks. Presumably these "hay piles" sustain them through the winter.

Reproduction: A single litter, born in June or July, consists of from two to four young.

Pikas are not inconspicuous little mammals and can be readily observed by the patient field zoologist. They frequently sit at the top of a large rock from which they can observe other Pikas as well as enemies. Pikas are alert and shy and are sometimes heard when not seen. In California they seldom occur as low as 1700 m elevation. In the Lava Beds National Monument, Pikas live in short grass over lava outcrops.

Rabbits and Hares (Leporidae)

Herbivores with enlarged hind legs and well-developed external ears, long in most species. The skull has a well-developed supraorbital process (see figs. 130c and 131c). Eyes rather large and placed in the skull so as to provide for a broad field of vision. Dentition: 2/1, 0/0, 3/2, 3/3.

The native species include the various cottontails (*Sylvilagus*) and the hares (*Lepus*). They tend to be abundant and prolific herbivores, and some species are conspicuously cyclic in abundance, attaining population highs approximately every ten years. Like other herbivores, rabbits and hares are especially sensitive to the quality and amount of plant food available; their reproduction is clearly enhanced by a rich food supply. They occur in a variety of habitats, each species being adapted by the nature of limbs and ears for its type of cover and mode of life. Those species, such as the jackrabbits, that live in the open have extremely long ears (for the detection of enemies) and long legs (for escape). In contrast, those that live in dense cover and whose defense lies in concealment have short ears and relatively short legs. The long ears of desert jackrabbits also release excess heat from the animal's body: on a very hot day a jackrabbit can sit in the shade allowing heat to escape from its well-vascularized ears.

Not included in the following section is the European Rabbit (*Oryctolagus cuniculus*), which has been introduced on the Farallons.

Key to Genera and Species of Rabbits and Hares (Leporidae) in California

1. Size large (hind foot 110 mm or longer); interparietal bone not clearly distinguishable in skull (see fig. 130c) . . . . *Lepus* (2)

   Size medium or small (hind foot 105 mm or shorter); interparietal bone outlined by a distinct suture (figs. 131c, 132c, 133c) . . . . *Sylvilagus* (4)

2. Postorbital projection entirely separate from skull (fig. 130c); upper part of tail white 3

   Postorbital projection fused posteriorly to lateral margin of frontal bone; upper part of tail black . . . . *Lepus californicus*

3. Hind foot more than 138 mm; ear from notch more than 100 mm; High Sierra and eastward . . . . *Lepus townsendii*

   Hind foot less than 138 mm; ear from notch less than 100 mm; coniferous forests from 1000 to 2500 m . . . . *Lepus americanus*

4. Tail white below; postorbital projections various but not long and not forming an open rounded concavity with lateral margin of skull 5

   Tail dusky below; postorbital projection short (less than twice as long as antorbital projection) and forming an open rounded concavity with lateral margin of skull . . . . *Sylvilagus idahoensis*

5. Ears sparsely haired on inner surface; vibrissae entirely black or sometimes with a small amount of white on ventral vibrissae 6

   Ears heavily furred on inner surface; vibrissae partly white . . . . *Sylvilagus nuttallii*

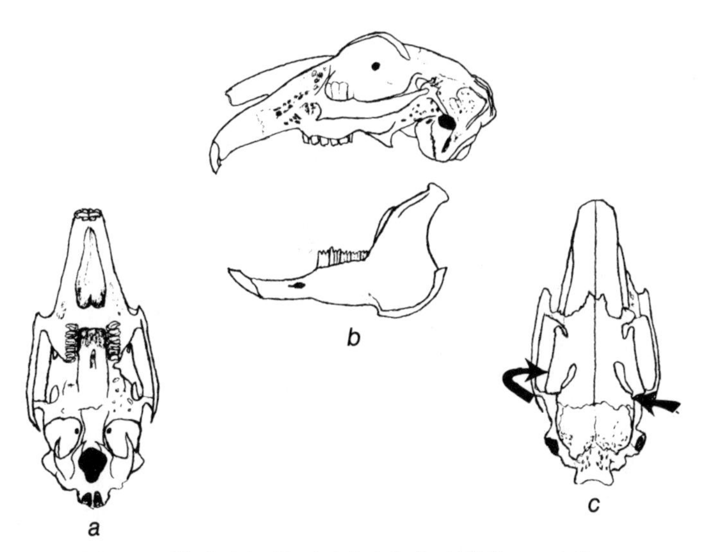

FIG. 130. Skull of the Black-tailed Jackrabbit (*Lepus californicus*): *a,* ventral view; *b,* lateral view; *c,* dorsal view (left arrow to supraorbital process, right arrow to postorbital projection).

6. Ears relatively long, 70–75 mm . . . *Sylvilagus auduboni*
   Ears relatively short, 43–63 mm . . . *Sylvilagus bachmani*

Snowshoe Rabbit (*Lepus americanus*) [pl. 16b]

Description: A medium-sized rabbit with relatively short ears and large feet. White in winter. TL 365–390, T 24–28, HF 112–135, E 68–72. Weight: 900–1100 g.

Distribution: Coniferous forests of Sierra Nevada and Cascades up to 2500 m. Characteristic of dense creekside alder thickets. Although seldom seen itself, the huge tracks of this hare in the snow reveal its abundance. Boreal coniferous forests of the United States, Canada, and Alaska. (See p. 383.)

Food: Willows, alders, and many low shrubs. It causes extreme damage to young pines in montane plantations and is a major obstacle to reforestation of conifers after forest fires.

Reproduction: From three to six young born in June. There is

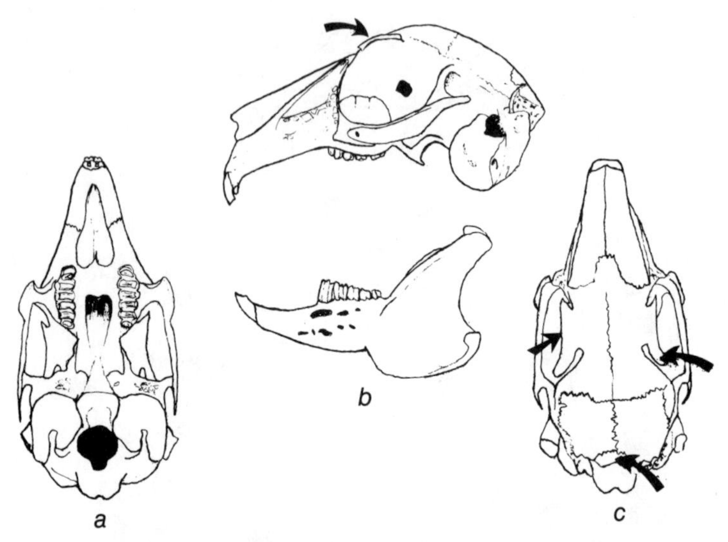

FIG. 131. Skull of Audubon's Cottontail (*Sylvilagus audubonii*): *a,* ventral view; *b,* lateral view (arrow to supraorbital process); *c,* dorsal view (left arrow to supraorbital process, right arrow to postorbital process, bottom arrow to interparietal bone).

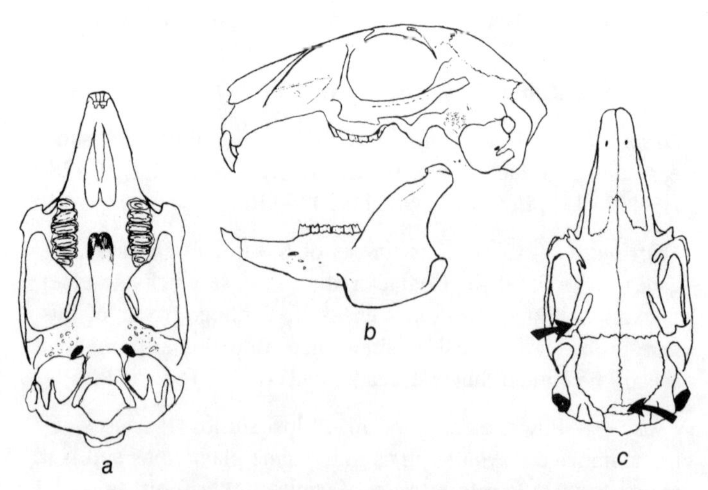

FIG. 132. Skull of the Brush Rabbit (*Sylvilagus bachmani*): *a,* ventral view; *b,* lateral view; *c,* dorsal view (left arrow to postorbital projection, bottom arrow to interparietal bone).

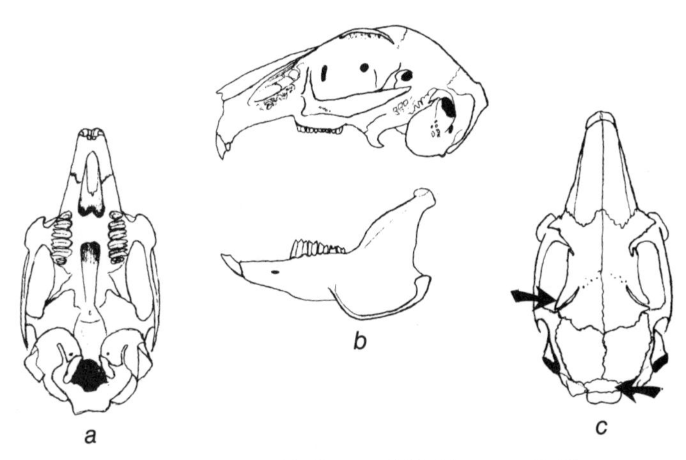

FIG. 133. Skull of Nuttall's Cottontail (*Sylvilagus nuttallii*): *a,* ventral view; *b,* lateral view; *c,* dorsal view (left arrow to postorbital projection, bottom arrow to interparietal bone).

probably a single brood; both a long gestation and an extended period of nursing (which may continue until late August).

This is a common species, though it is seldom seen. Depending on its concealing coloring for protection, it does not readily run from enemies. In the northern Sierra Nevada it is sometimes abundant and thrives in the dense stands of manzanita that grow up after a major forest fire.

Black-tailed Jackrabbit [fig. 134; pl. 16e]
(*Lepus californicus*)

Description: A distinctive, long-legged hare with very long ears and tail black (or partly black) dorsally and grayish ventrally. TL 495–550, T 76–95, HF 117–130, E 105–123. Weight: 1500–2000 g. (See fig. 130.)

Distribution: Widely distributed in the state about forested areas and eastern slopes of the high mountains. Common in deserts, irrigated pastures, and row crops; frequently active in the daytime. May occur up to 2500 m elevation. Much of western United States south to central Mexico and Baja California.

Food: Feeds on many herbs and grasses, including many cultivated crops.

FIG. 134. Black-tailed Jackrabbit (*Lepus californicus*).

Reproduction: May breed at almost any time of the year, probably depending on nature of the food. Nest is a depression in the ground. Usually three or four young in a litter but sometimes as many as seven.

White-tailed Jackrabbit (*Lepus townsendii*) [pl. 16g]

Description: A large jackrabbit with hind foot more than 140 mm. Tail conspicuous and white throughout the year. Possibly confused with the Black-tailed Jackrabbit, which has a blackish tail and a smaller hind foot. Pelage white in winter in the White-tailed Jackrabbit. TL 545–650, T 66–103, HF 145–165, E 98–110. Weight: 2.15–3.44 kg.

Distribution: Along the eastern part of the state; from 1500 m on the western slopes eastward. Much of northwestern United States and south-central Canada. A creature of rather open areas. (See p. 384.)

Food: Browses on many low-growing shrubs.

Reproduction: A single litter of four to six young. As in the Snowshoe Rabbit, lactation may occupy most of the summer.

This is a high-elevation hare not frequently seen. It seems not to occur with either the Black-tailed Jackrabbit (which is found at lower elevations) or the Snowshoe Rabbit (which remains close to cover). The White-tailed Jackrabbit is a little-known species in our state.

Audubon's Cottontail (*Sylvilagus audubonii*) [pl. 16d]

Description: A large, long-legged cottontail with rather short fur. Most nearly resembles Nuttall's Cottontail, but in Audubon's Cottontail ears are sparsely furred and longer than in *nuttallii;* hind feet rather slender and sparsely furred in *audubonii.* TL 370–400, T 40–56, HF 80–95, E 70–75. Weight: 750–900 g (males), 880–1250 g (females). Dorsum generally gray; brownish in *nuttallii.* (See fig. 131.)

Distribution: Most of the shrub-covered part of the state except northern third. In much of the Great Basin, east to Rocky Mountains and south to central Mexico, including Baja California. Where this species occurs together with Nuttall's Cottontail, Audubon's Cottontail tends to occur at lower elevations. (See p. 384.)

Food: Forbs, grasses, and tender branches of shrubs.

Reproduction: Breeds from January to June; litters vary from two to six. Breeds throughout the year on irrigated land.

Audubon's Cottontail is the common cottontail in brushlands and orchards at lower elevations. It may occur together with the Brush Rabbit, but Audubon's Cottontail is far more likely to forage out in the open.

Brush Rabbit (*Sylvilagus bachmani*) [pl. 16c]

Description: A small, short-legged rabbit with moderately pointed (clearly not rounded) ears. Vibrissae black. TL 300–360, T 12–28, HF 70–85, E 43–63. Weight: 560–840 g (females usually larger than males). (See fig. 132.)

Distribution: Throughout chaparral of the western two-thirds of the state from the western slopes of the Sierra Nevada and Cascade ranges to the coast. Favors dense brush and seldom strays far from cover. Also on San José Island and Año Nuevo Island. Throughout Baja California north to Oregon. (See p. 385.)

Food: Feeds largely on grasses and forbs, including sow thistle (*Sonchus* spp.) and Sea Lettuce (*Dudleya farinosa*).

Reproduction: As in other species of *Sylvilagus* the young are born blind and helpless. Breeds from January to June or July; probably two broods a year. Litters range from three to four young. Sexually mature in about ten months.

Pigmy Rabbit (*Sylvilagus idahoensis*) [pl. 16f]

Description: Smallest rabbit in our state (and also in the United States). Very short legs and ears and a buffy underside to the tail (white in other species of *Sylvilagus*). TL 230–295, T 15–24, HF 66–76, E 36–48. Weight: 350–460 g. As in other rabbits, females tend to be larger than males.

Distribution: Eastern margin of the state, especially common in rocky areas dominated by sagebrush. In Great Basin east to Utah and north to eastern Washington and southeast Montana. (See p. 384.)

Food: Sagebrush and grasses.

Reproduction: Breeds in late winter and spring; probably two litters of four to eight young annually. Sexually mature in one year. Nests underground.

These little rabbits are sometimes common and frequently active in the daytime. They are not difficult to see but may not venture far from the protection of dense cover. They build burrow systems, into which they retreat when pursued. They are distinctive in their gait—a mixture of hops and walking. Sagebrush, a favorite food, is said to impart a strong flavor to their flesh.

Nuttall's Cottontail (*Sylvilagus nuttallii*)

Description: A rather large cottontail with moderately short ears, inner surface furred, the tips of which are rounded. Vibrissae partly whitish. Dorsum generally brownish, not gray. TL 335–392, T 30–53, HF 86–100, E 57–85. Weight: 650–980 g. (See fig. 133.)

Distribution: Canyons and creekbeds of the Great Basin, often with a cover of sagebrush. East of the Sierra Nevada south to the Mojave Desert. From Arizona and New Mexico north to Canada, mostly in the intermountain region. (See p. 385.)

Food: A browsing species, especially fond of bitterbrush and sagebrush. Grazes on a broad spectrum of forbs and grasses in the spring.

Reproduction: Breeds from late winter or early spring until early summer. Probably two broods are common, but in some years a female might produce four broods. The young are altricial, born blind and naked. The mother makes a nest in a crevice among rocks, or in a depression in the ground, and lines it with fine hair pulled from her belly. A litter consists of four to eight young (most commonly six).

This cottontail appears to be unique among lagomorphs in its arboreal tendency. During arid periods in midsummer these rabbits may spend much of the early daylight hours up to 3 m above the ground in junipers, apparently licking water from the foliage.

RANGE MAPS

The range maps are guides to the geographic occurrence of California mammals. These maps are based on collection records and, sometimes, also on observations. Such maps are, at best, only approximations of regions in which species may be expected. A species will not occur everywhere throughout its indicated range; usually it will be restricted to a certain habitat. The River Otter and the Water Shrew, for example, remain close to the margins of rivers and creeks within their respective regions of occurrence. There are, moreover, human activities, such as lumbering, mining, and agriculture, that alter the abundance and extent of distribution of many species; the range maps, based partly on past records, may not always indicate accurately the geographic distributions today. Thus, range maps are never precise or entirely accurate. Despite their shortcomings, we believe that range maps are extremely useful when used in conjunction with keys, descriptions, and illustrations.

For species which have experienced drastic reductions in abundance and distribution, such as the Pronghorn and the Wapiti, the current, not the historic, range is shown. We have omitted maps for species that occur throughout the state. We have also left out range maps for introduced mammals; not only are the details of the time and place of introduction sometimes not known accurately, but their occurrence is frequently uncertain and often in a state of flux.

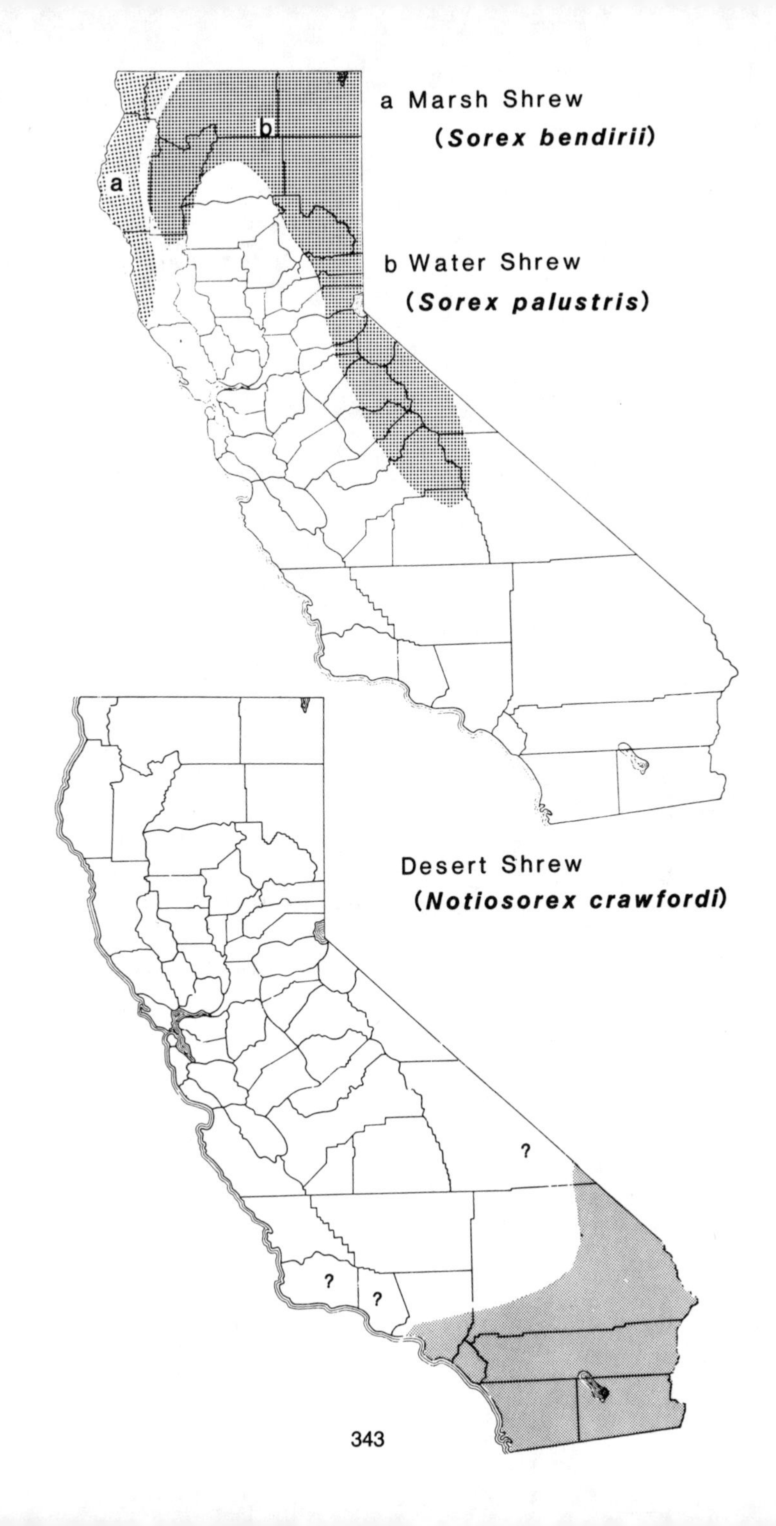
a Marsh Shrew
(*Sorex bendirii*)
b Water Shrew
(*Sorex palustris*)
a
b
Desert Shrew
(*Notiosorex crawfordi*)
?
?
?

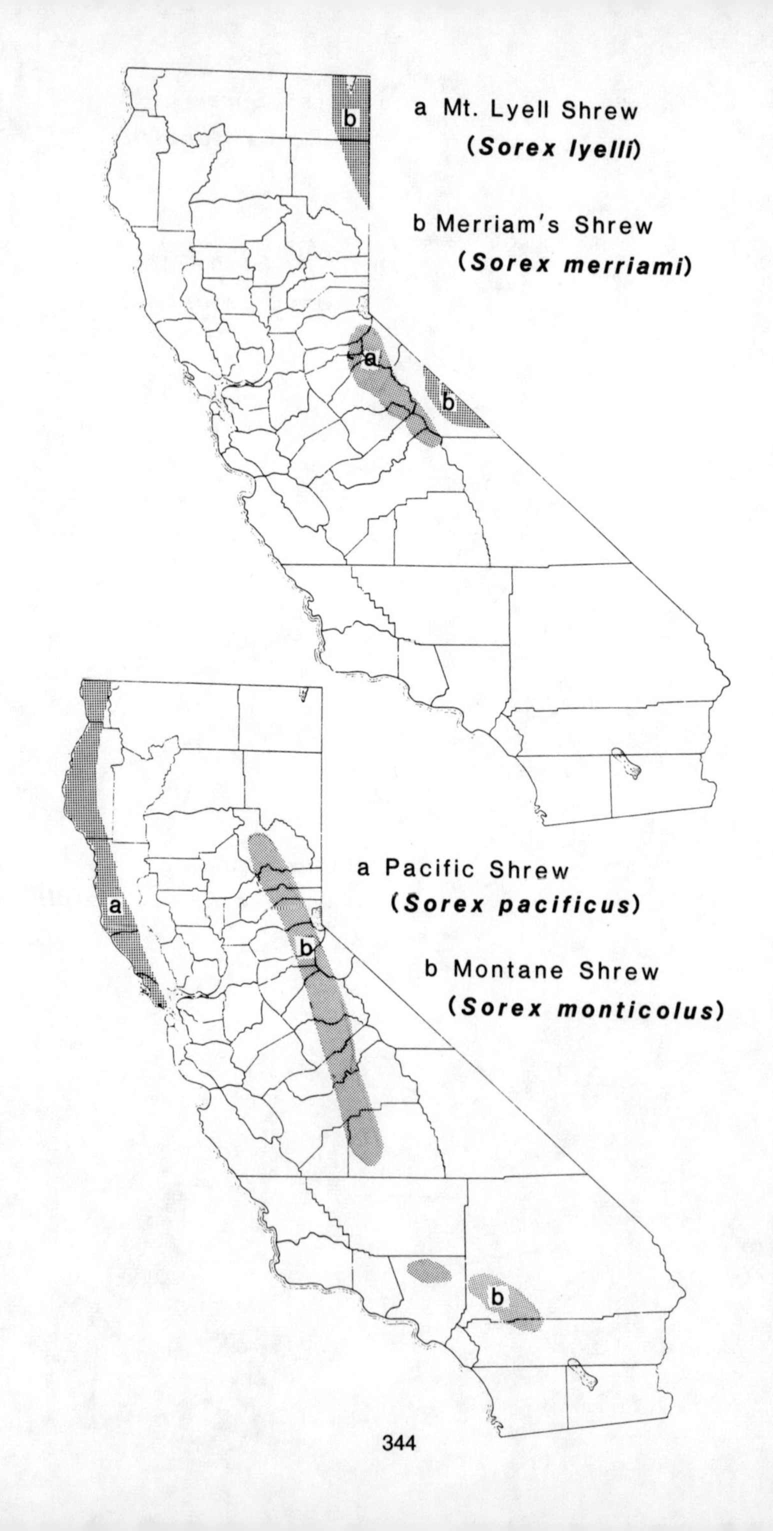
b
a Mt. Lyell Shrew
(Sorex lyelli)
b Merriam's Shrew
(Sorex merriami)
a
b
a
a Pacific Shrew
(Sorex pacificus)
b
b Montane Shrew
(Sorex monticolus)
b

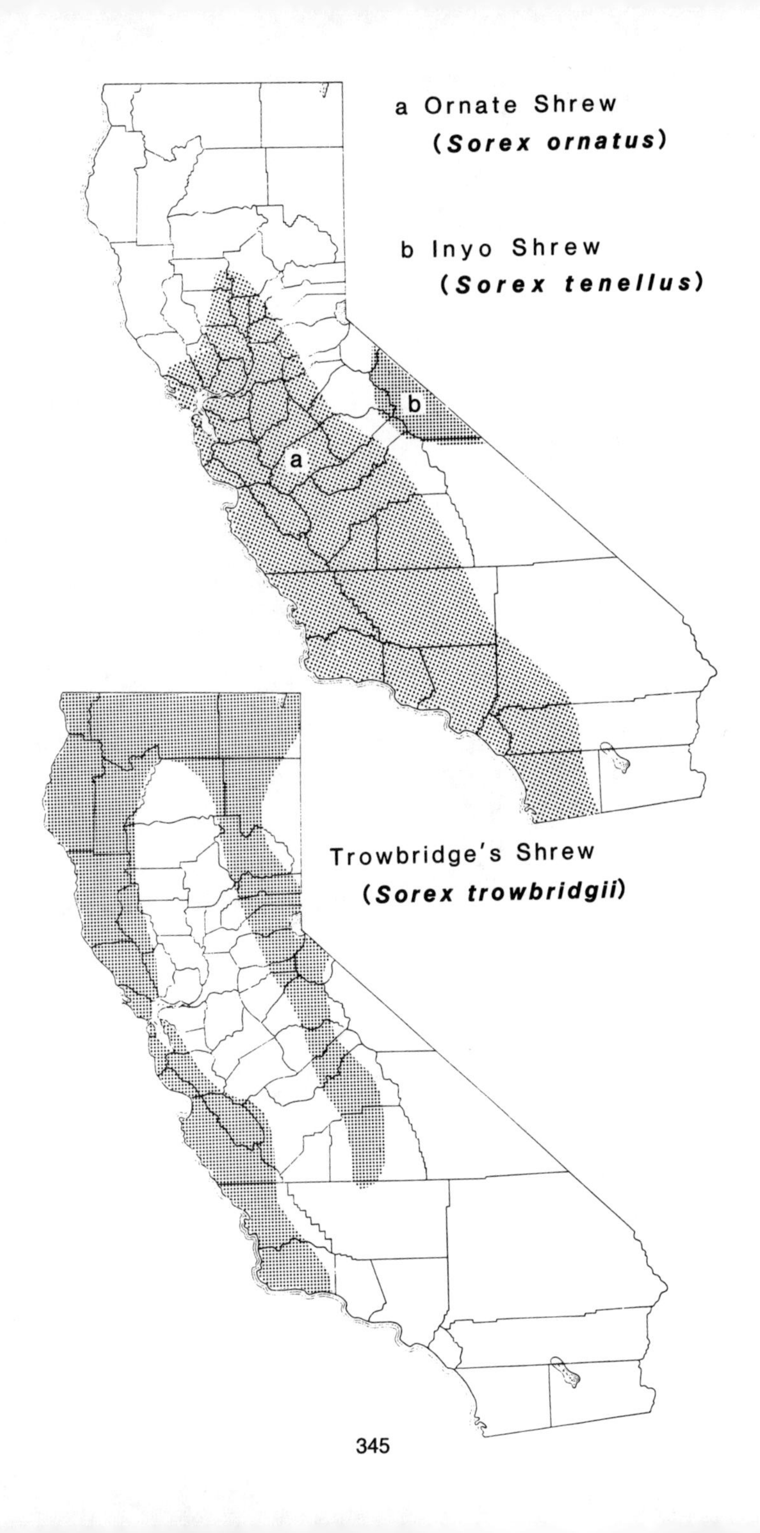
a Ornate Shrew
(Sorex ornatus)
b Inyo Shrew
(Sorex tenellus)
b
a
Trowbridge's Shrew
(Sorex trowbridgii)

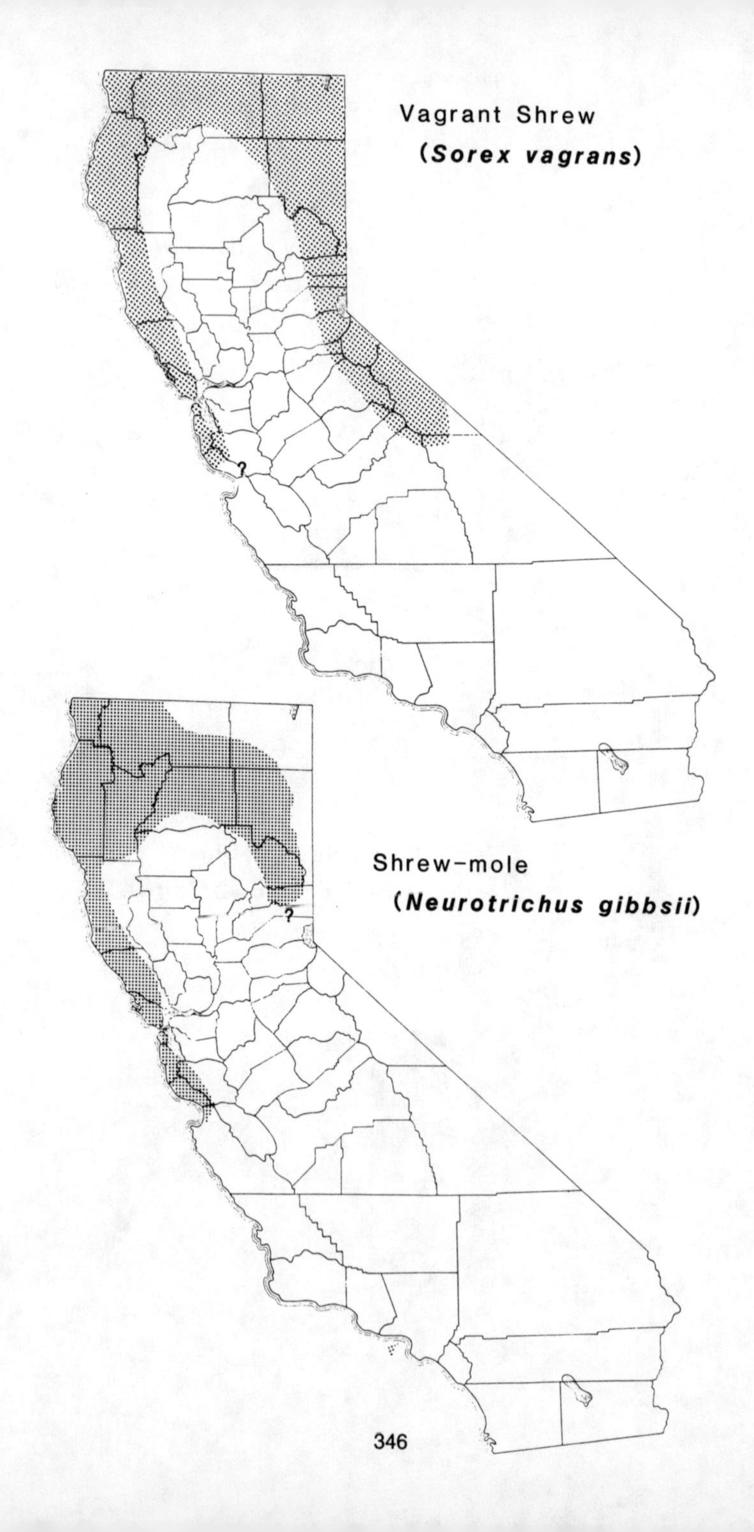
Vagrant Shrew
(*Sorex vagrans*)
?
Shrew-mole
(*Neurotrichus gibbsii*)
?

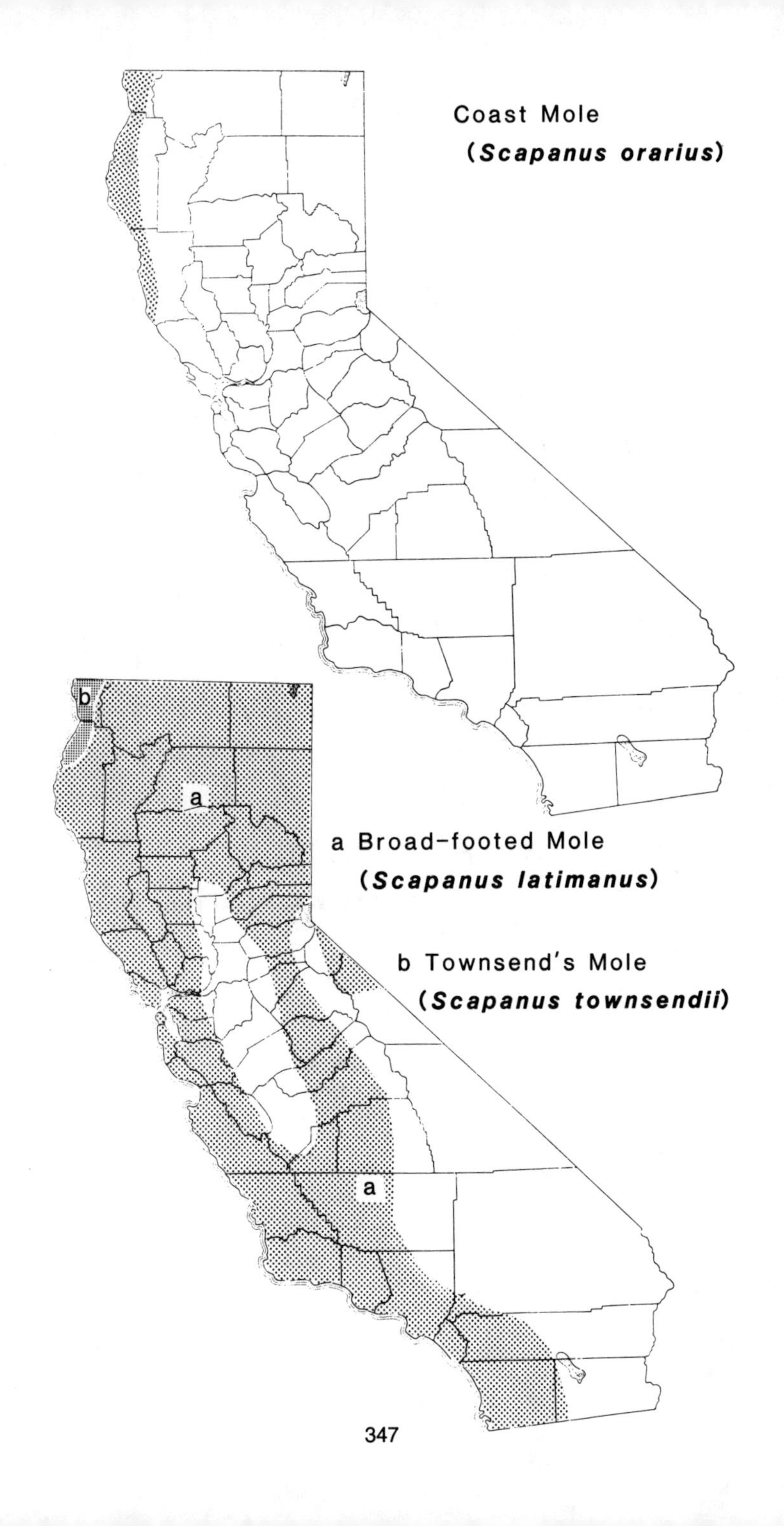
Coast Mole
(Scapanus orarius)
b
a
a Broad-footed Mole
(Scapanus latimanus)
b Townsend's Mole
(Scapanus townsendii)
a

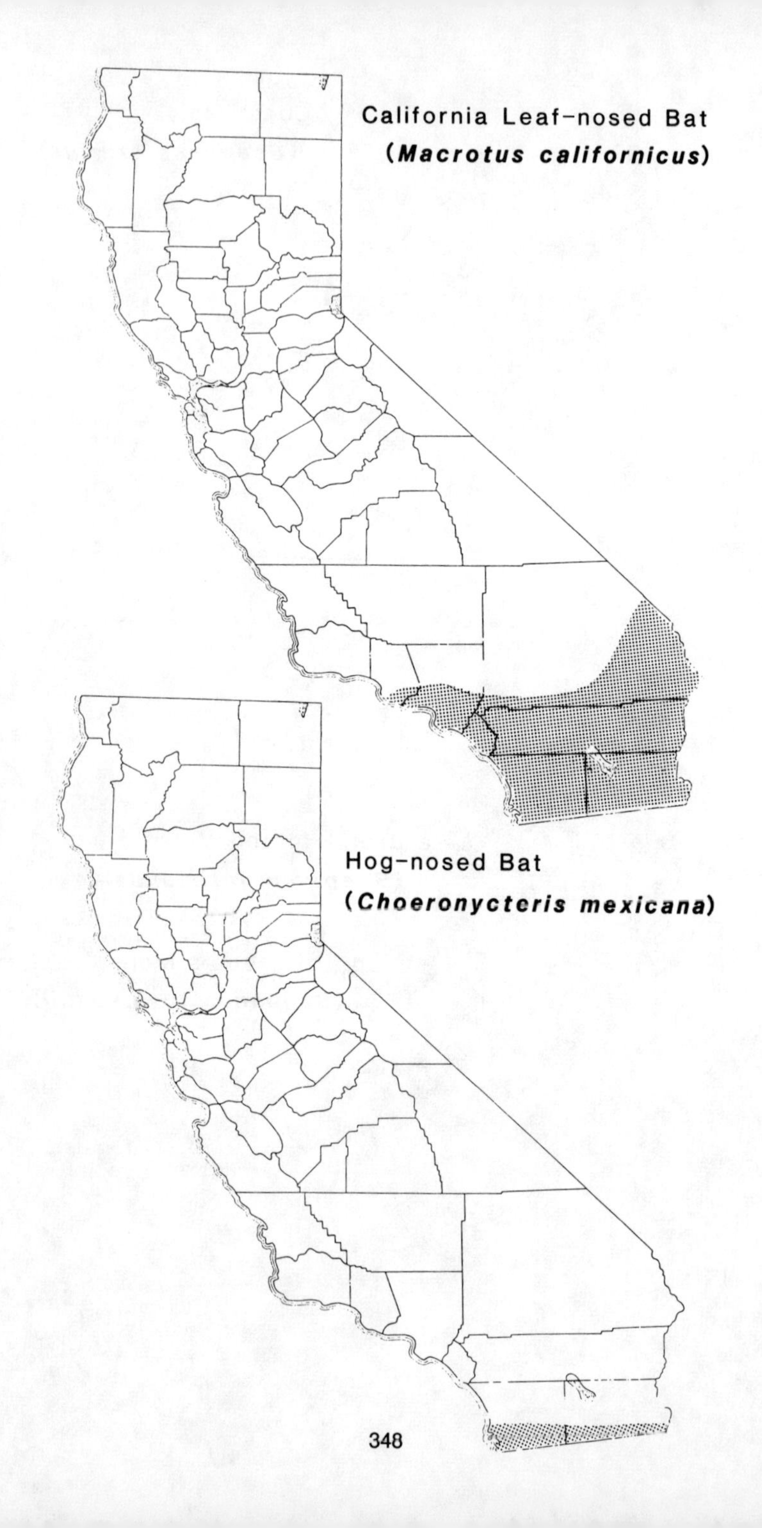
California Leaf-nosed Bat
(*Macrotus californicus*)
Hog-nosed Bat
(*Choeronycteris mexicana*)

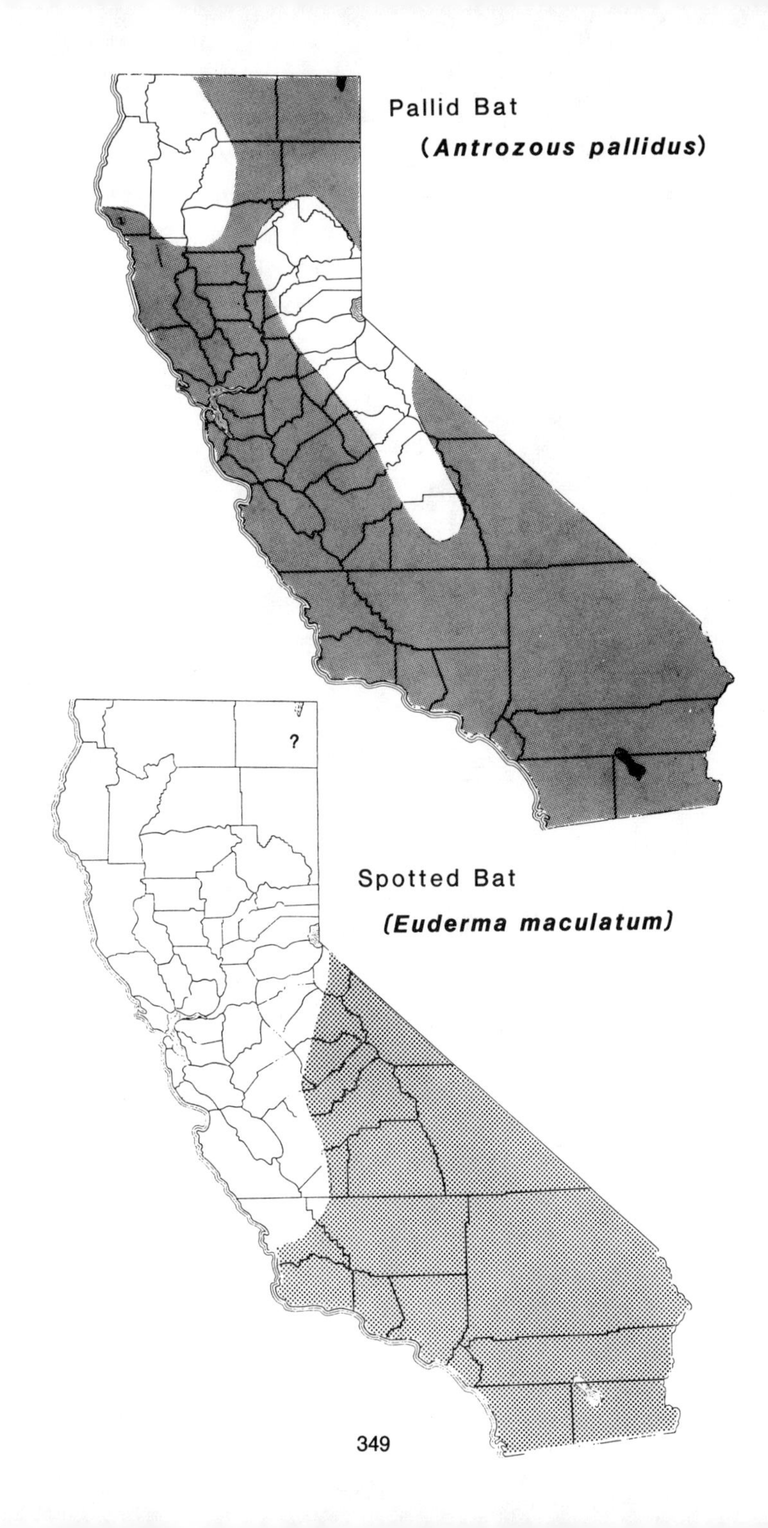
Pallid Bat
(Antrozous pallidus)
?
Spotted Bat
(Euderma maculatum)

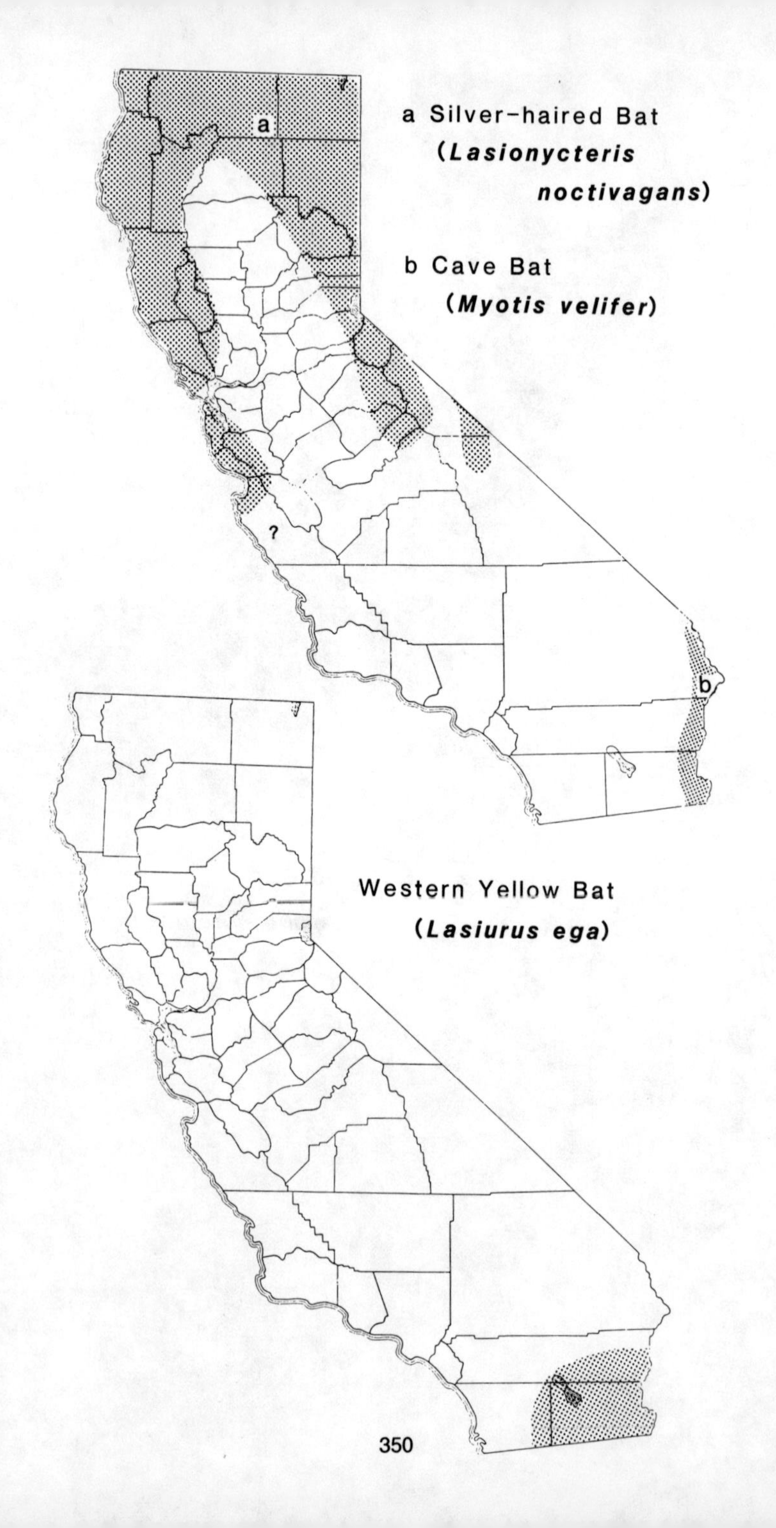
a
a Silver-haired Bat
(*Lasionycteris noctivagans*)
b Cave Bat
(*Myotis velifer*)
?
b
Western Yellow Bat
(*Lasiurus ega*)

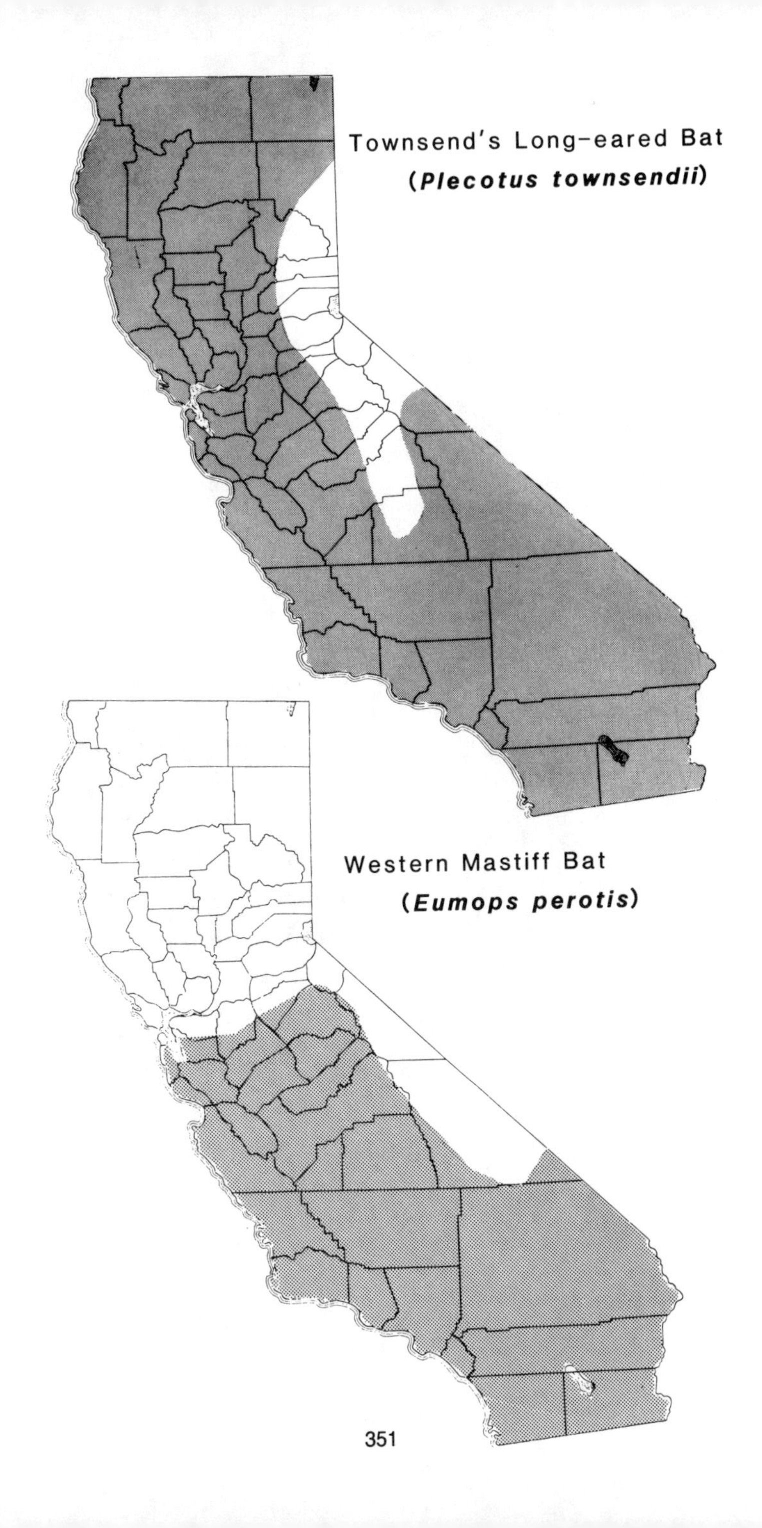
Townsend's Long-eared Bat
(Plecotus townsendii)
Western Mastiff Bat
(Eumops perotis)

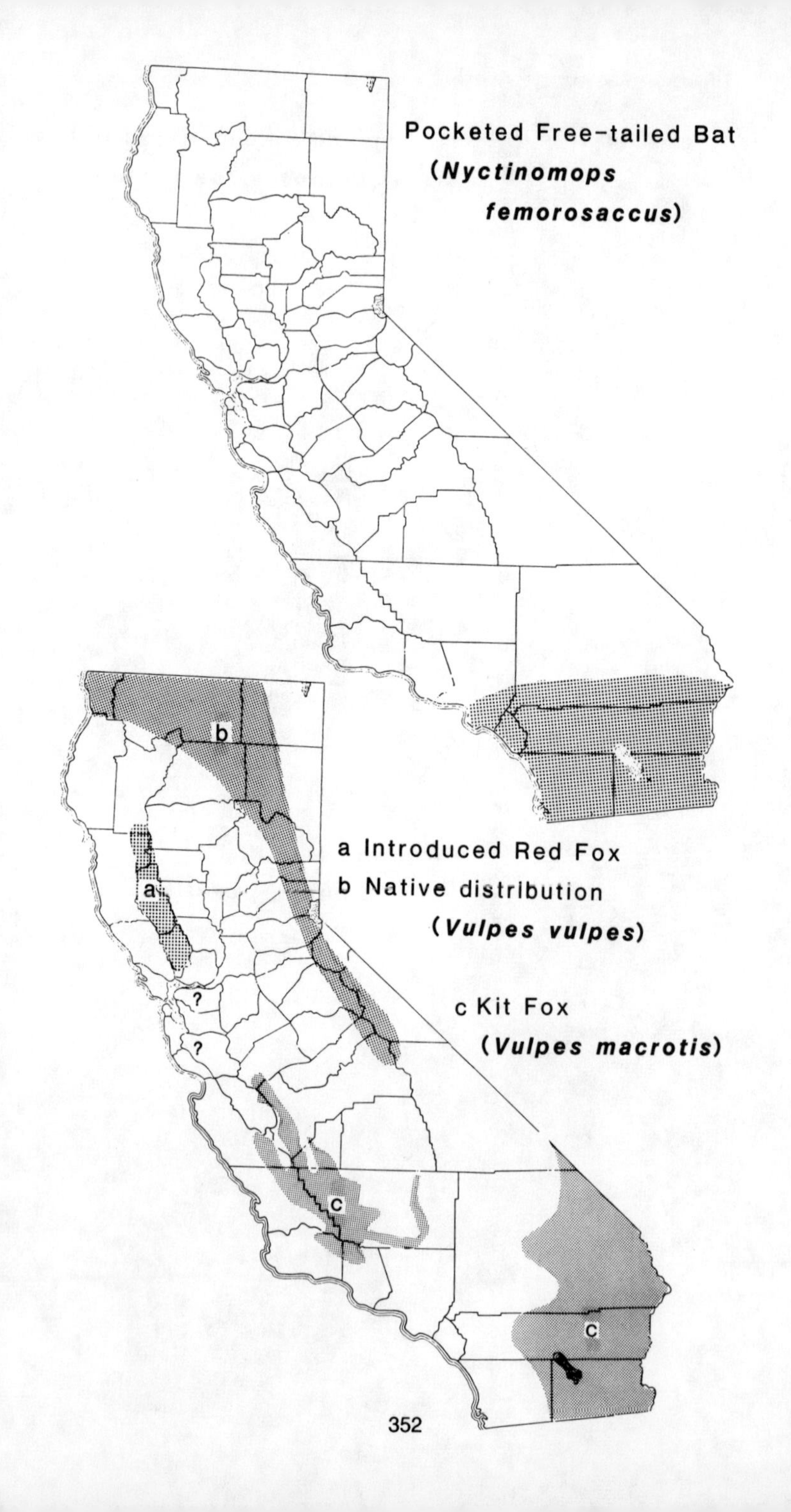

Pocketed Free-tailed Bat
(Nyctinomops
femorosaccus)
a Introduced Red Fox
b Native distribution
(Vulpes vulpes)
c Kit Fox
(Vulpes macrotis)
a
b
c
c
?
?

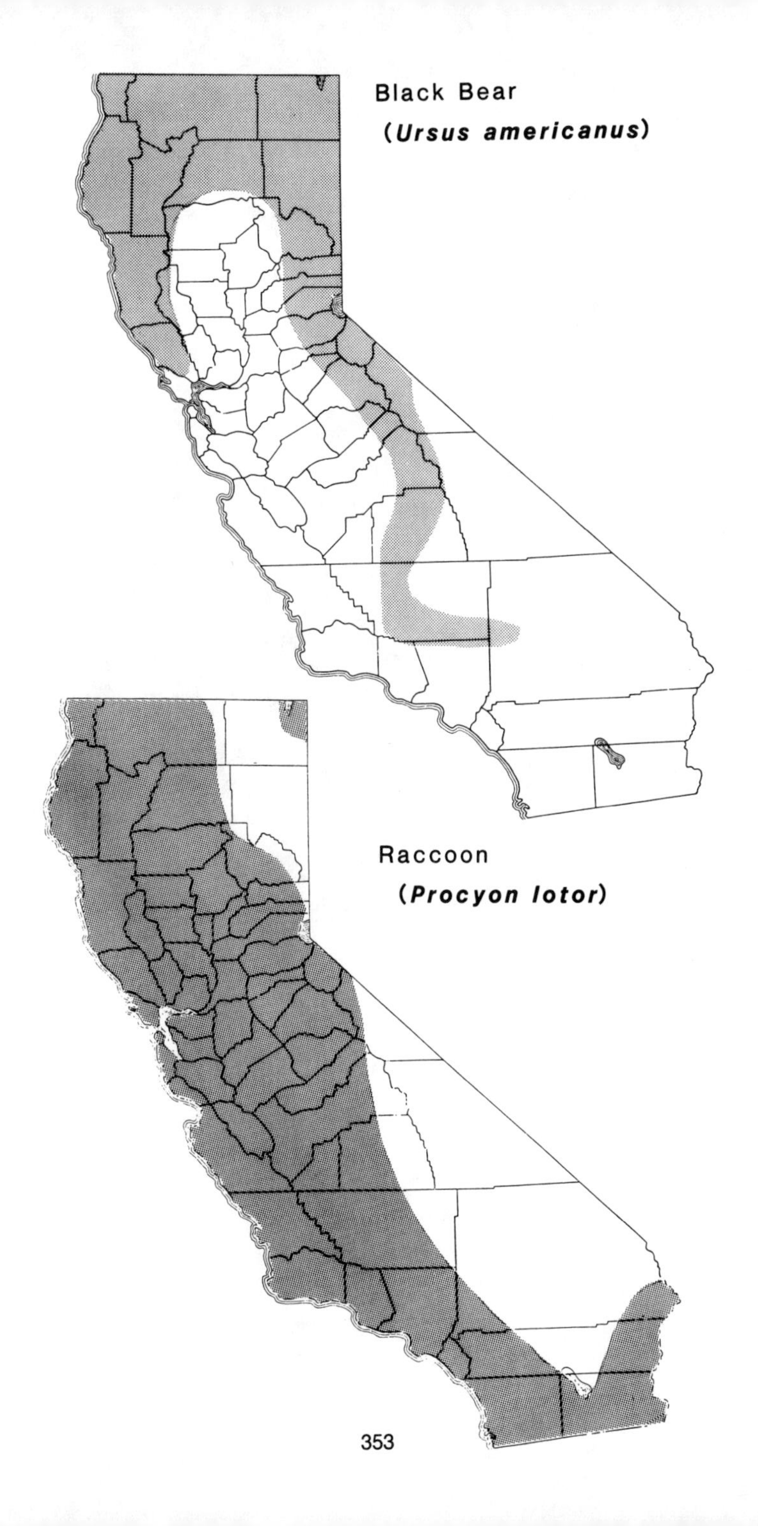
Black Bear
(Ursus americanus)
Raccoon
(Procyon lotor)

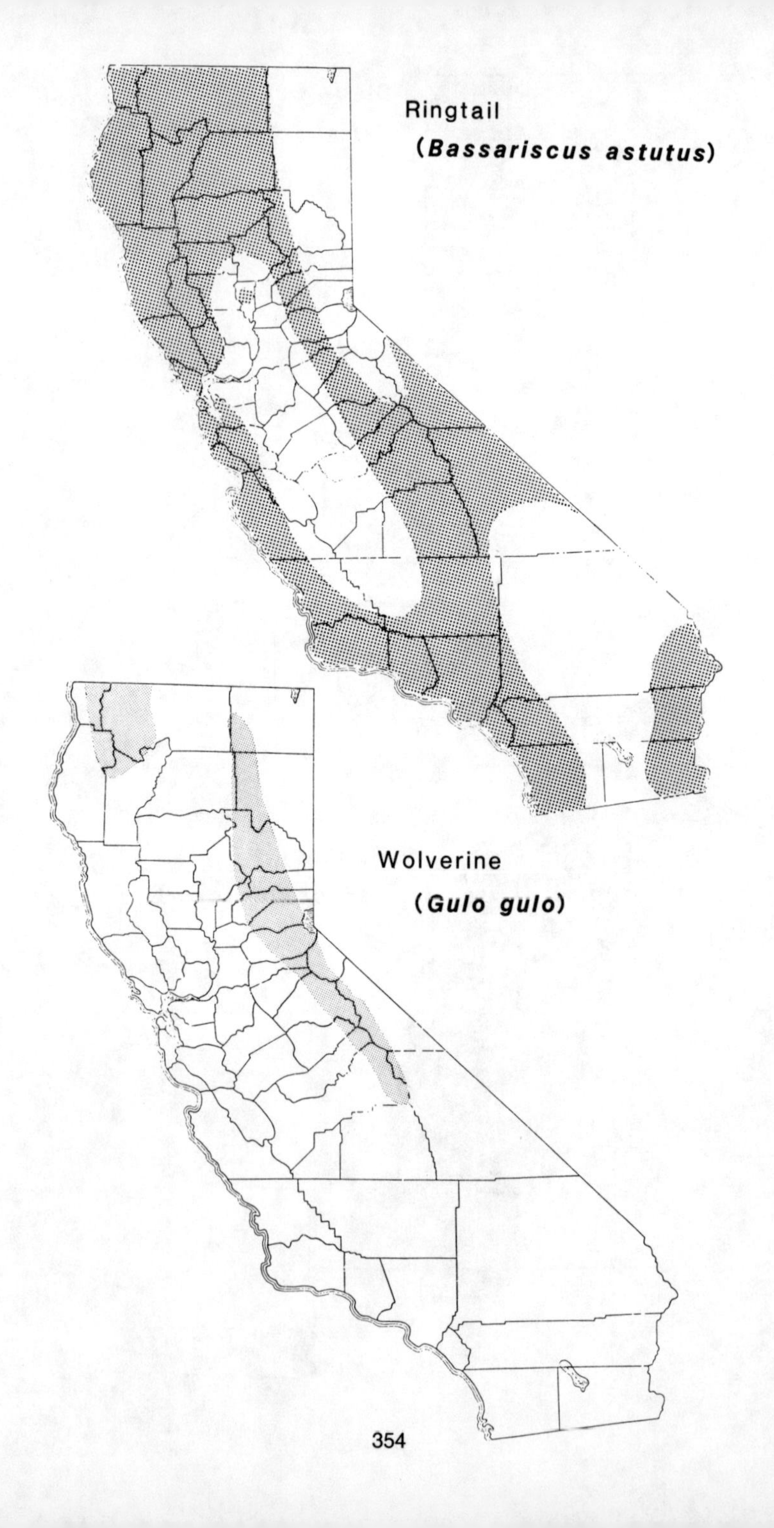
Ringtail
(Bassariscus astutus)
Wolverine
(Gulo gulo)

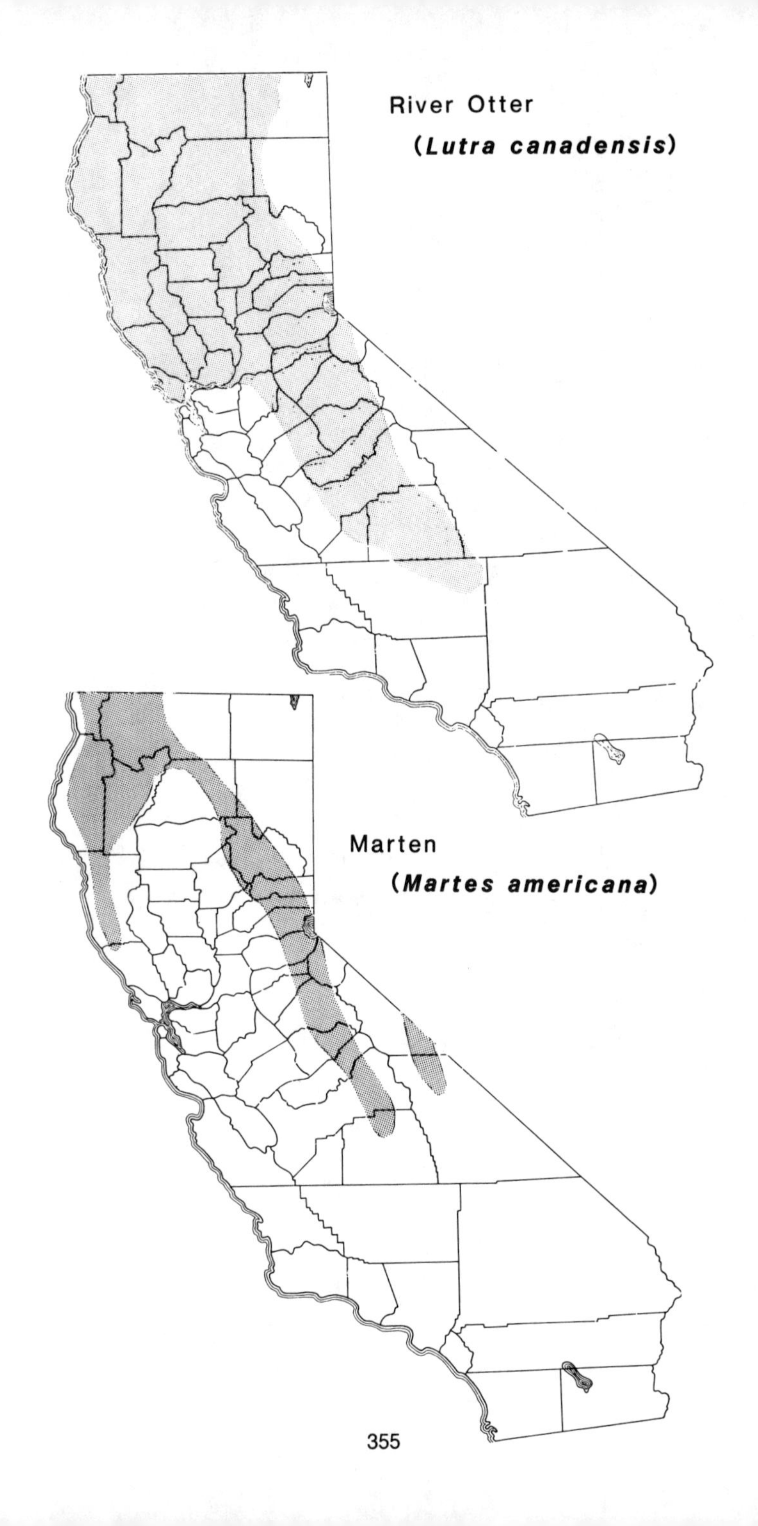
River Otter
(*Lutra canadensis*)
Marten
(*Martes americana*)

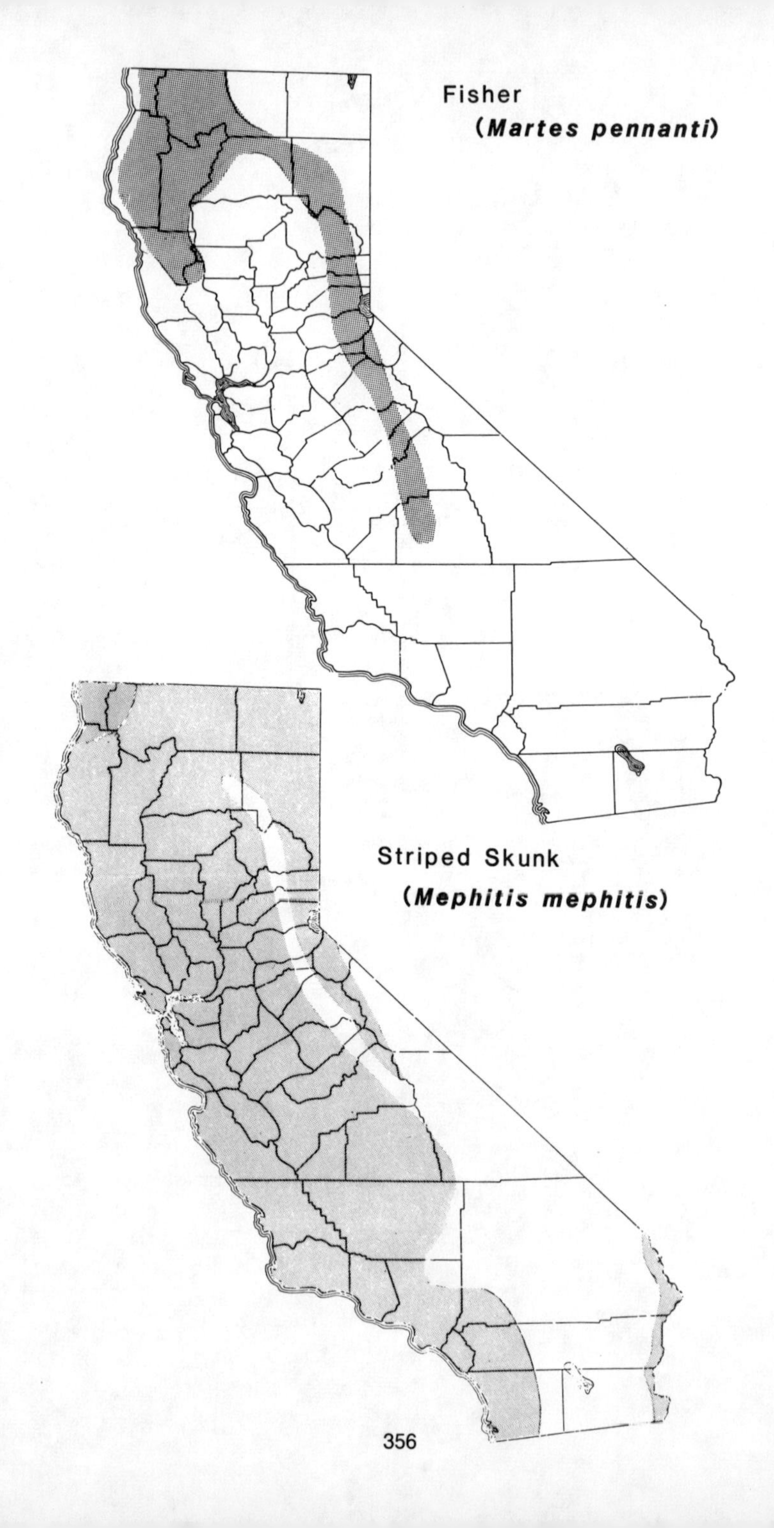
Fisher
(*Martes pennanti*)
Striped Skunk
(*Mephitis mephitis*)

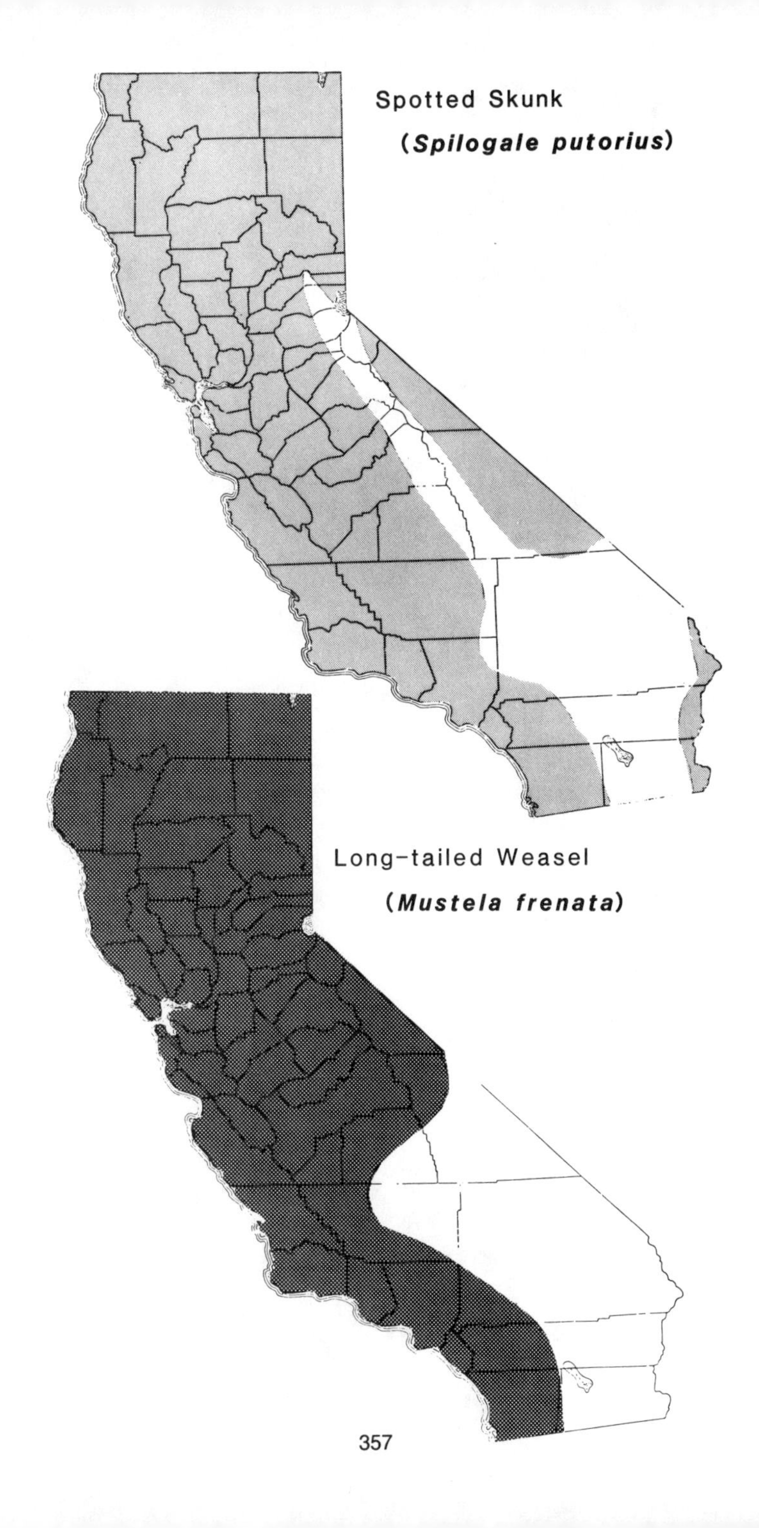
Spotted Skunk
(*Spilogale putorius*)
Long-tailed Weasel
(*Mustela frenata*)

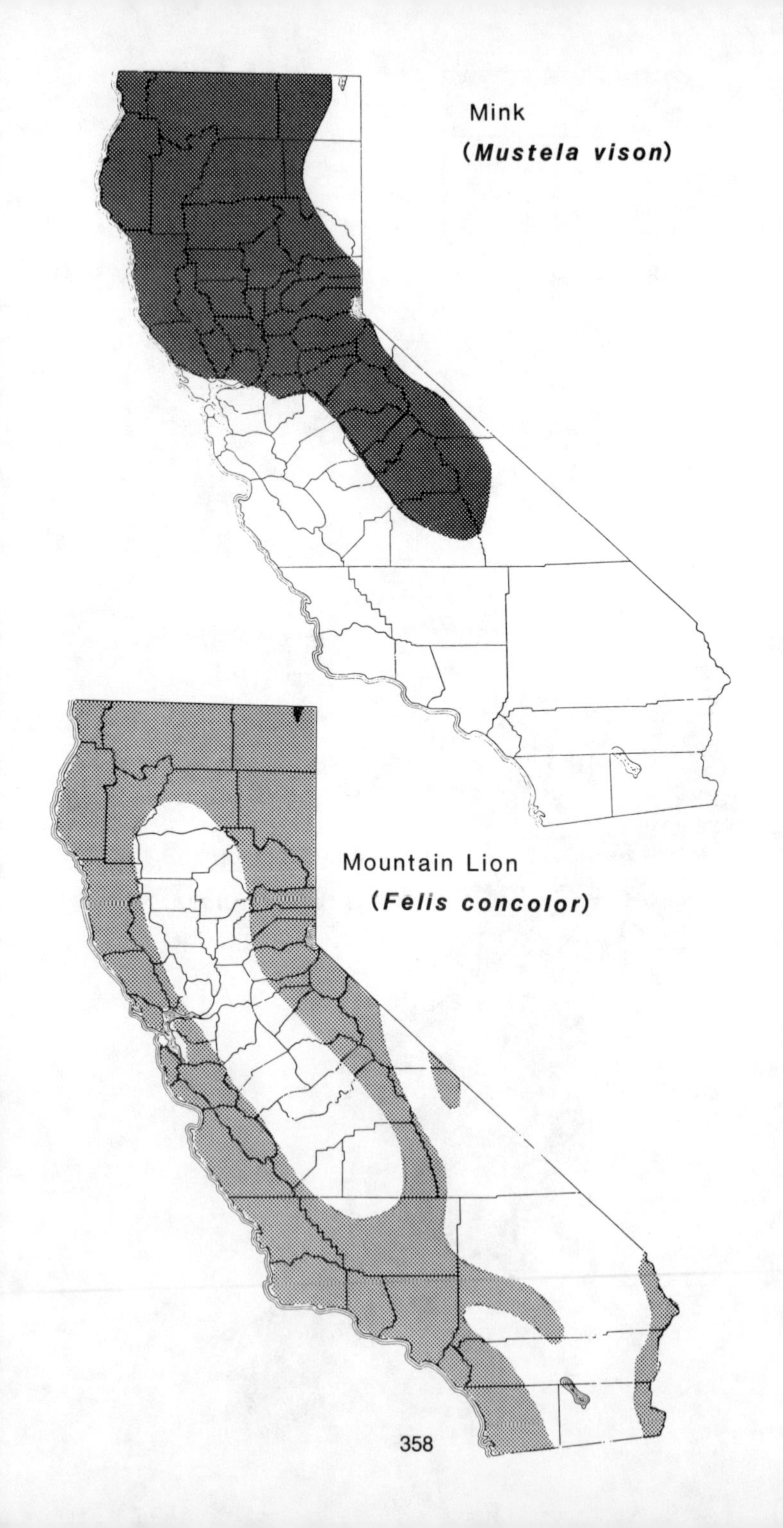

Mink
(*Mustela vison*)
Mountain Lion
(*Felis concolor*)

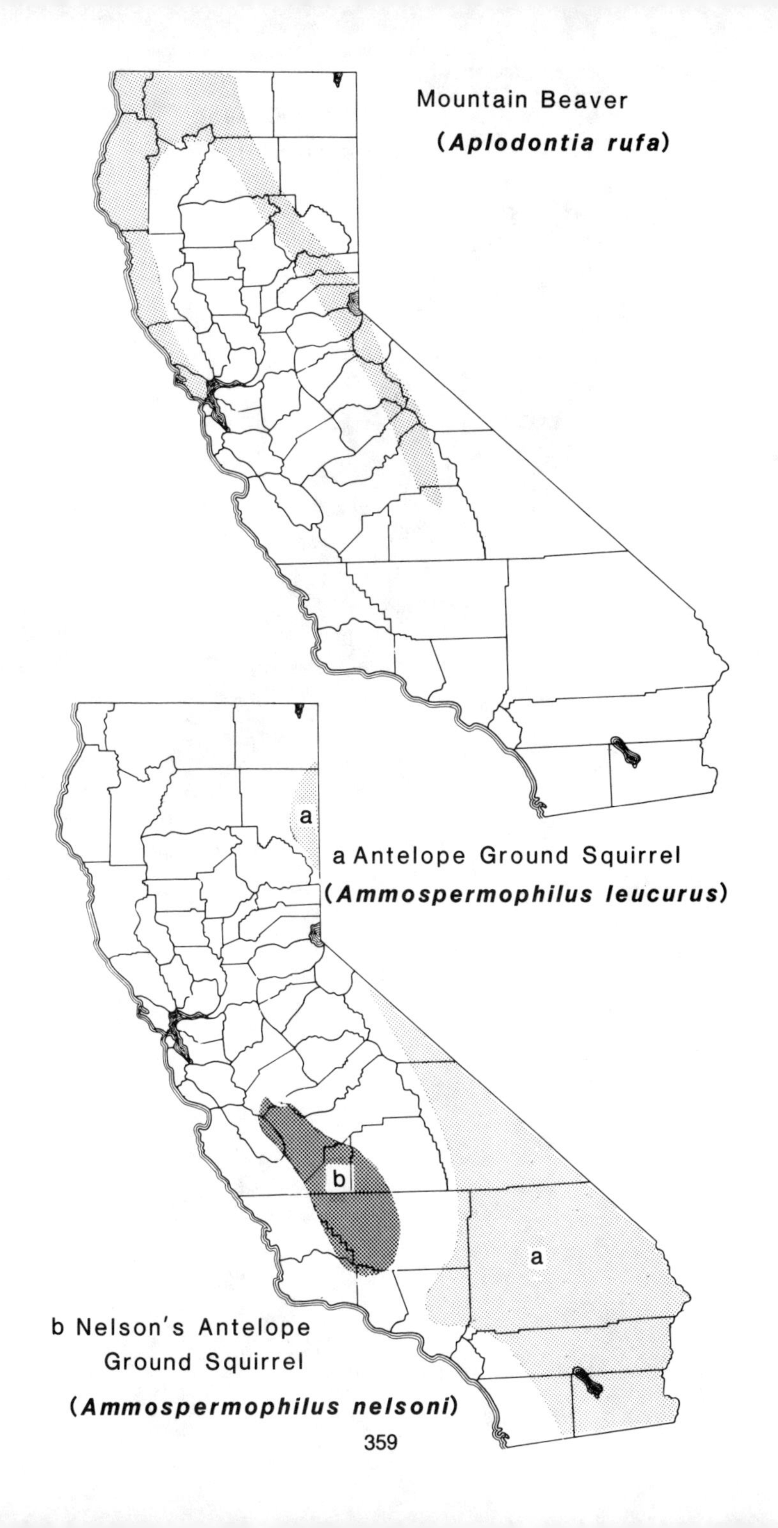
Mountain Beaver
(*Aplodontia rufa*)
a
a Antelope Ground Squirrel
(*Ammospermophilus leucurus*)
b
a
b Nelson's Antelope
Ground Squirrel
(*Ammospermophilus nelsoni*)

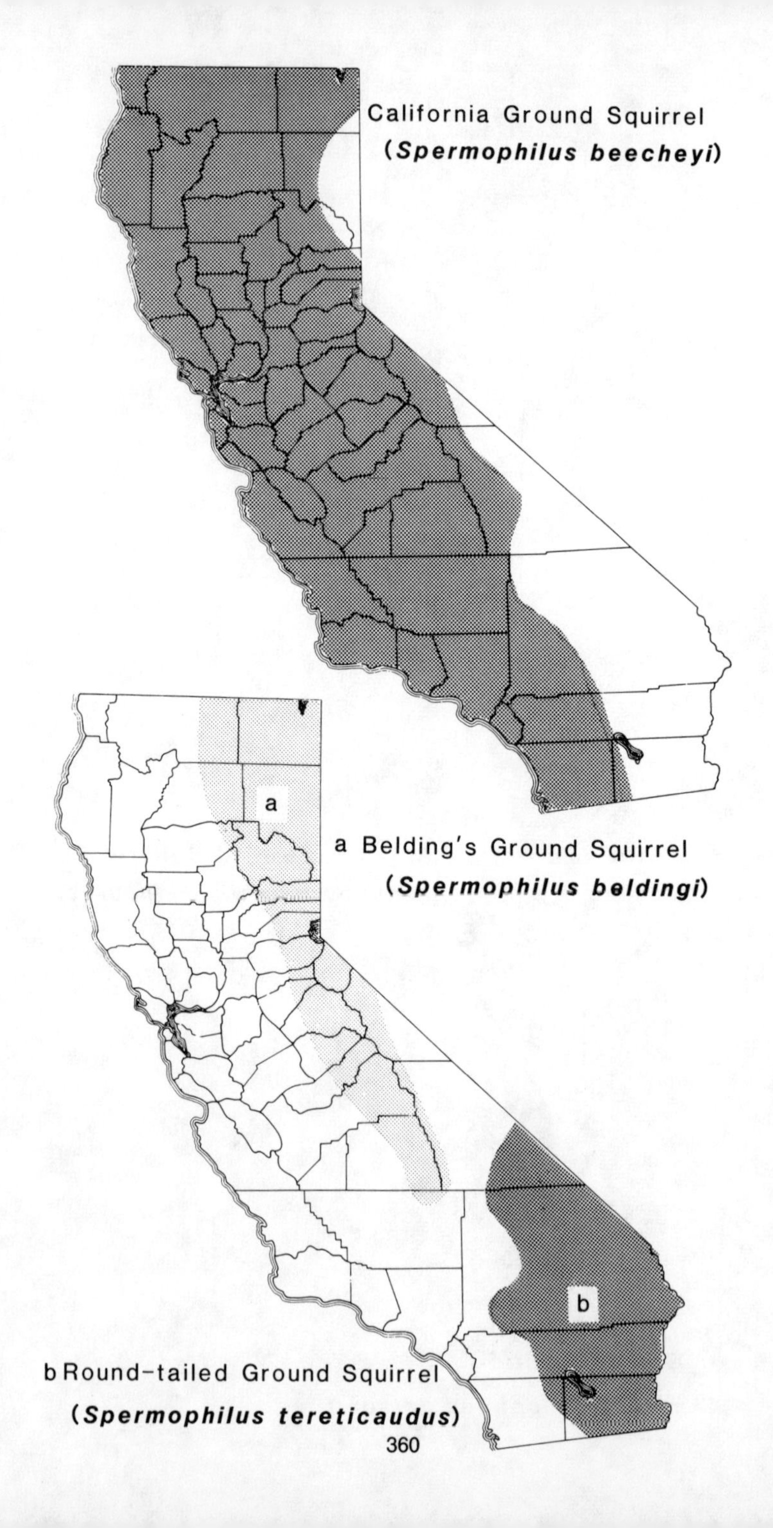
California Ground Squirrel
(*Spermophilus beecheyi*)
a
a Belding's Ground Squirrel
(*Spermophilus beldingi*)
b
b Round-tailed Ground Squirrel
(*Spermophilus tereticaudus*)

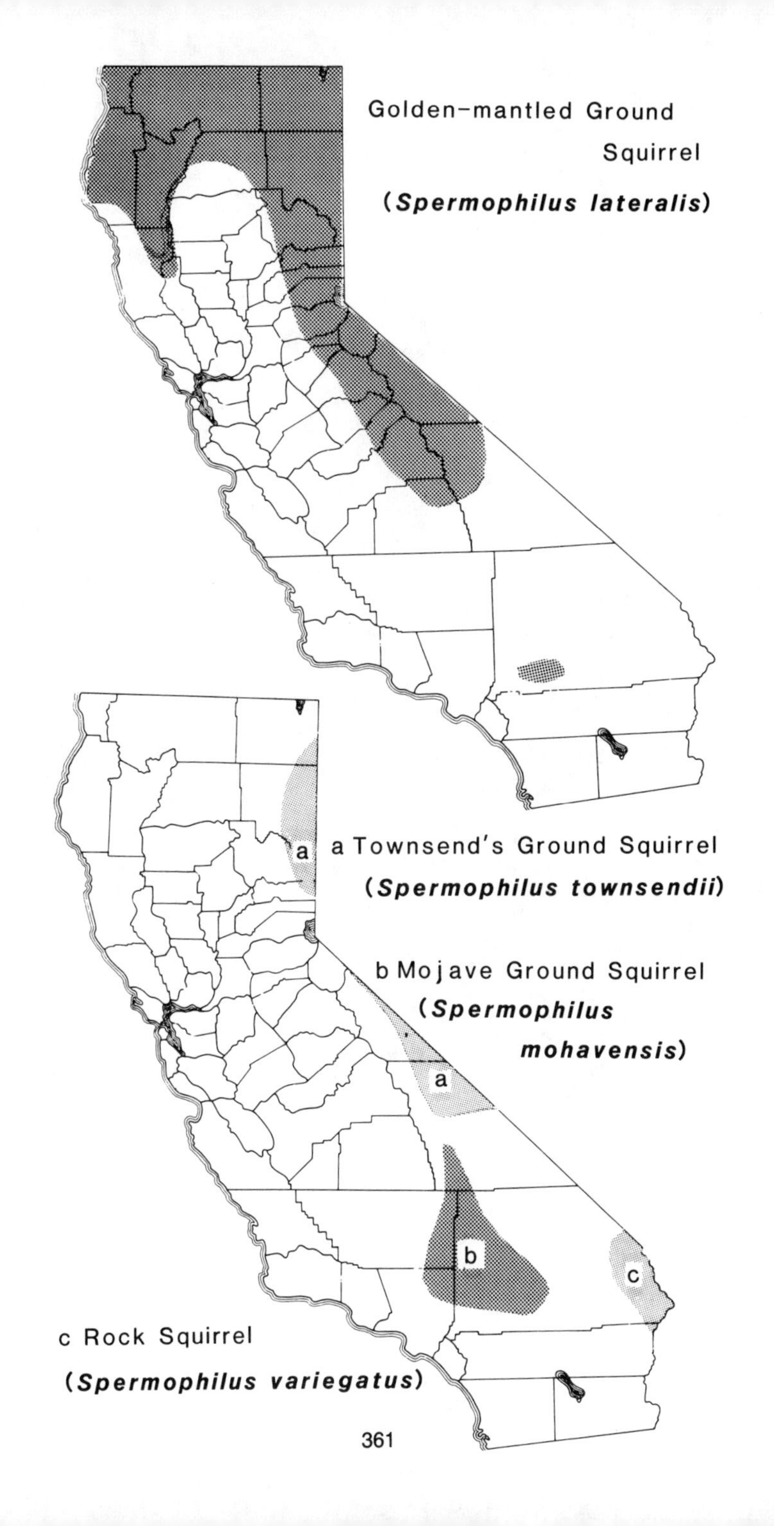
Golden-mantled Ground Squirrel
(*Spermophilus lateralis*)
a
a Townsend's Ground Squirrel
(*Spermophilus townsendii*)
b Mojave Ground Squirrel
(*Spermophilus mohavensis*)
a
b
c
c Rock Squirrel
(*Spermophilus variegatus*)

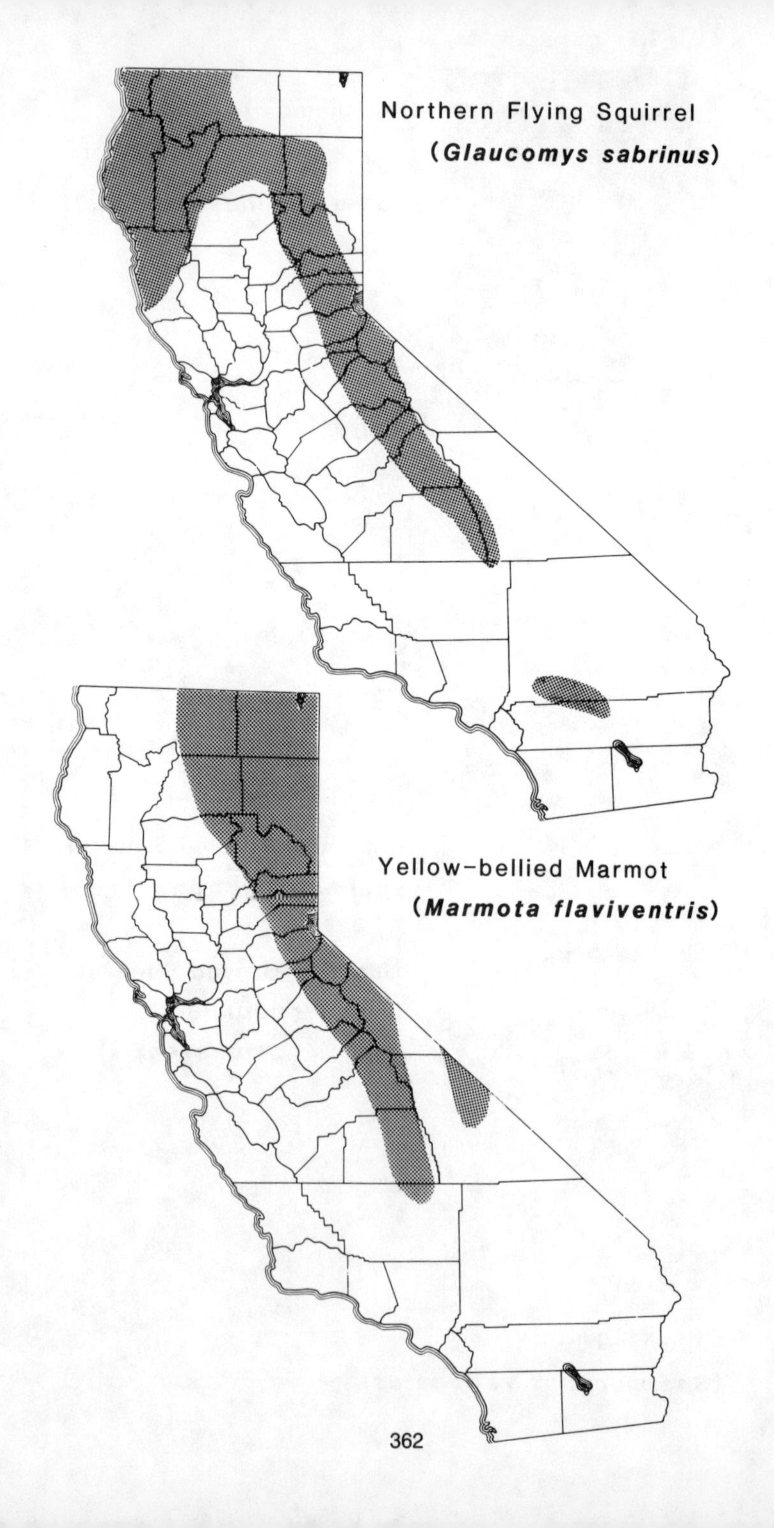
Northern Flying Squirrel
(*Glaucomys sabrinus*)
Yellow-bellied Marmot
(*Marmota flaviventris*)

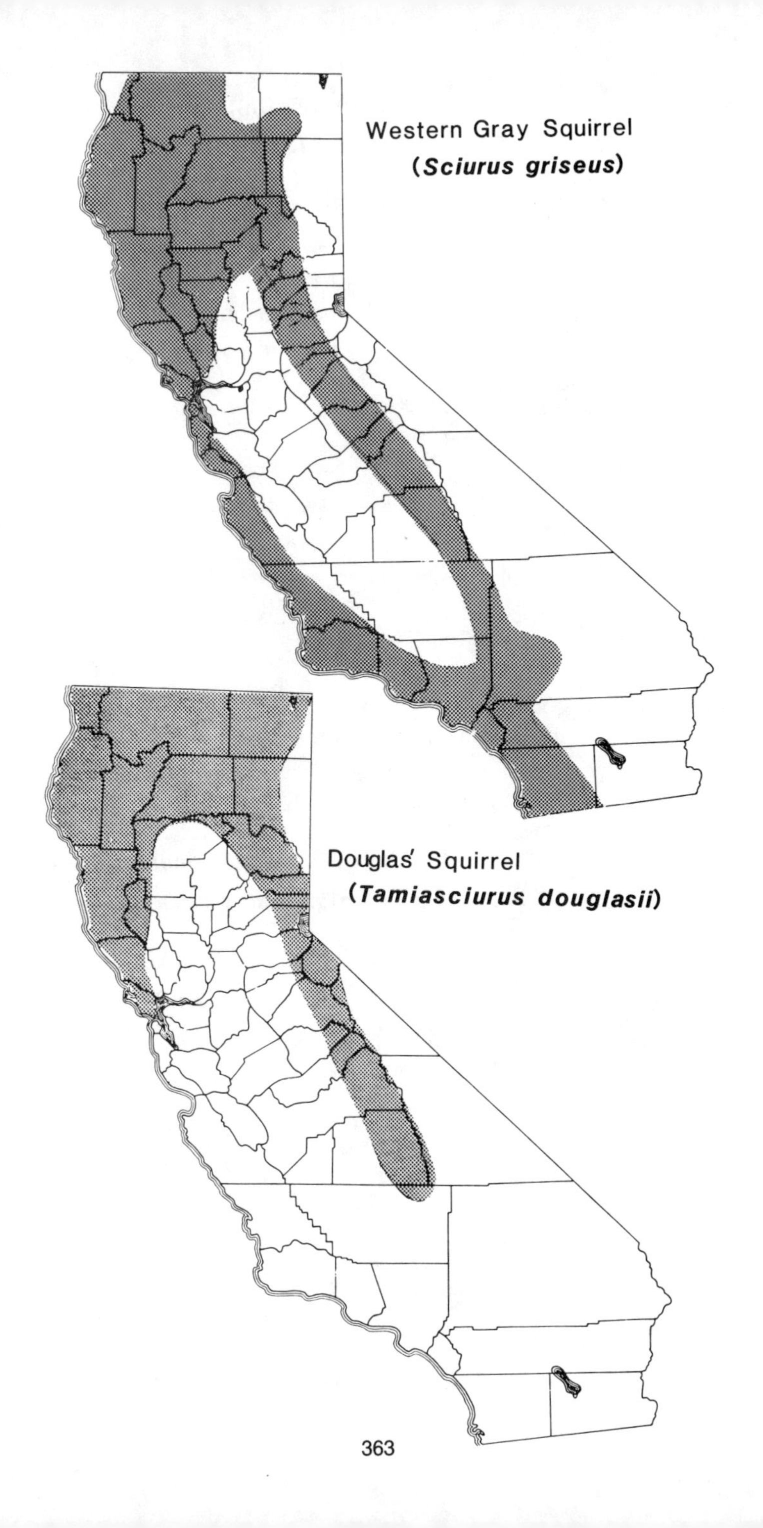

Western Gray Squirrel
(*Sciurus griseus*)
Douglas' Squirrel
(*Tamiasciurus douglasii*)

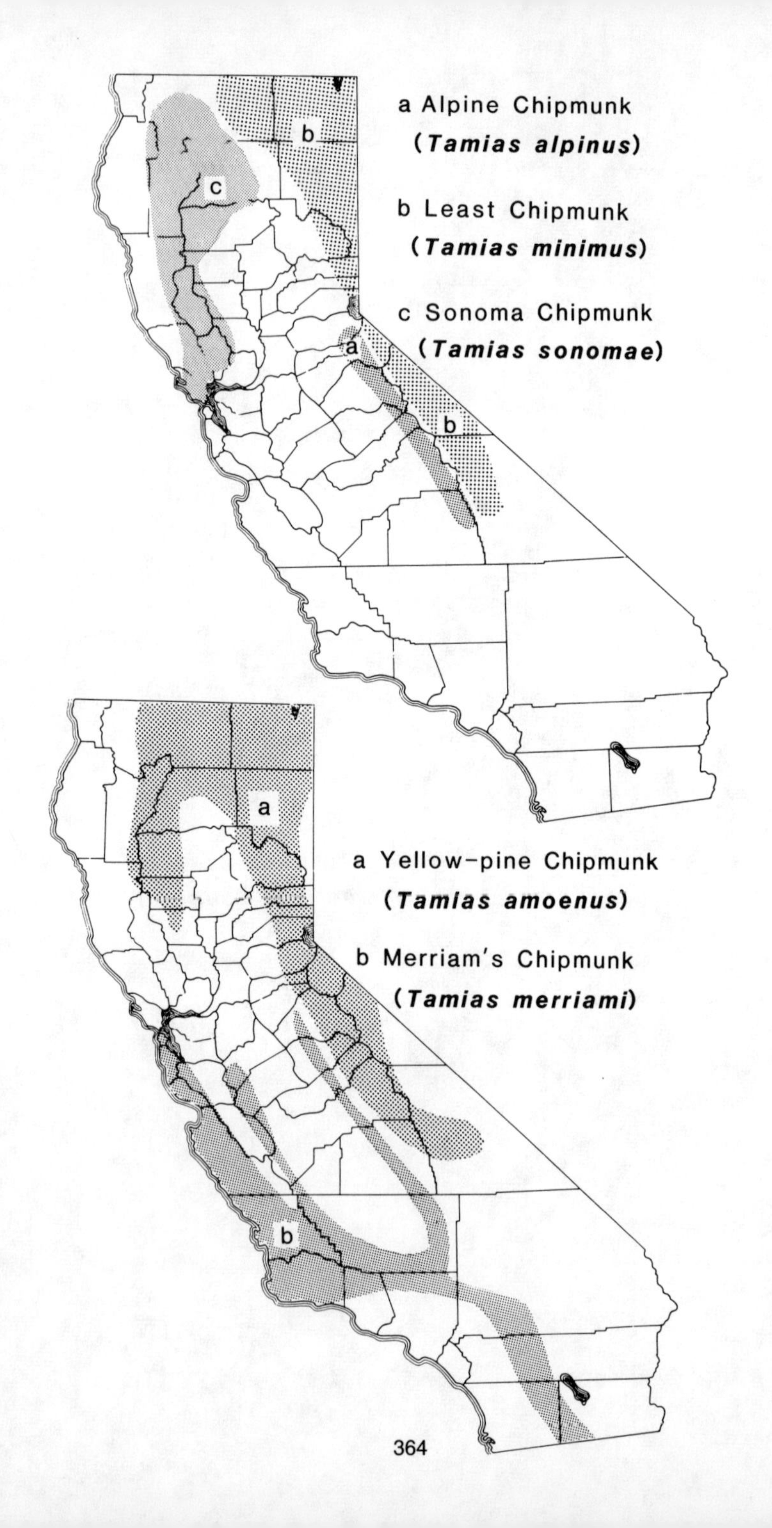
a Alpine Chipmunk
(Tamias alpinus)
b Least Chipmunk
(Tamias minimus)
c Sonoma Chipmunk
(Tamias sonomae)
a
b
c
b
a Yellow-pine Chipmunk
(Tamias amoenus)
b Merriam's Chipmunk
(Tamias merriami)
a
b

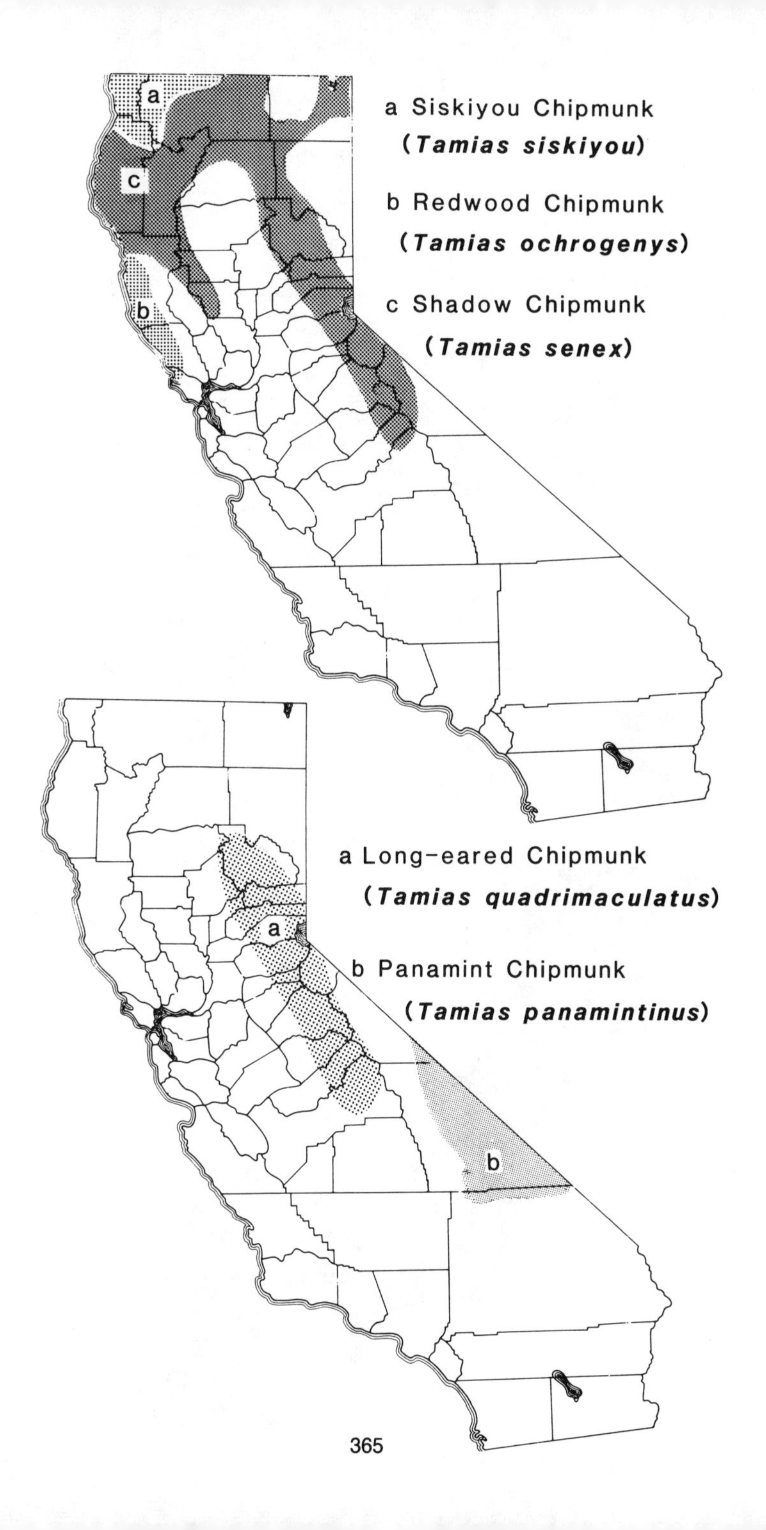
a
c
b
a Siskiyou Chipmunk
(Tamias siskiyou)
b Redwood Chipmunk
(Tamias ochrogenys)
c Shadow Chipmunk
(Tamias senex)
a
b
a Long-eared Chipmunk
(Tamias quadrimaculatus)
b Panamint Chipmunk
(Tamias panamintinus)

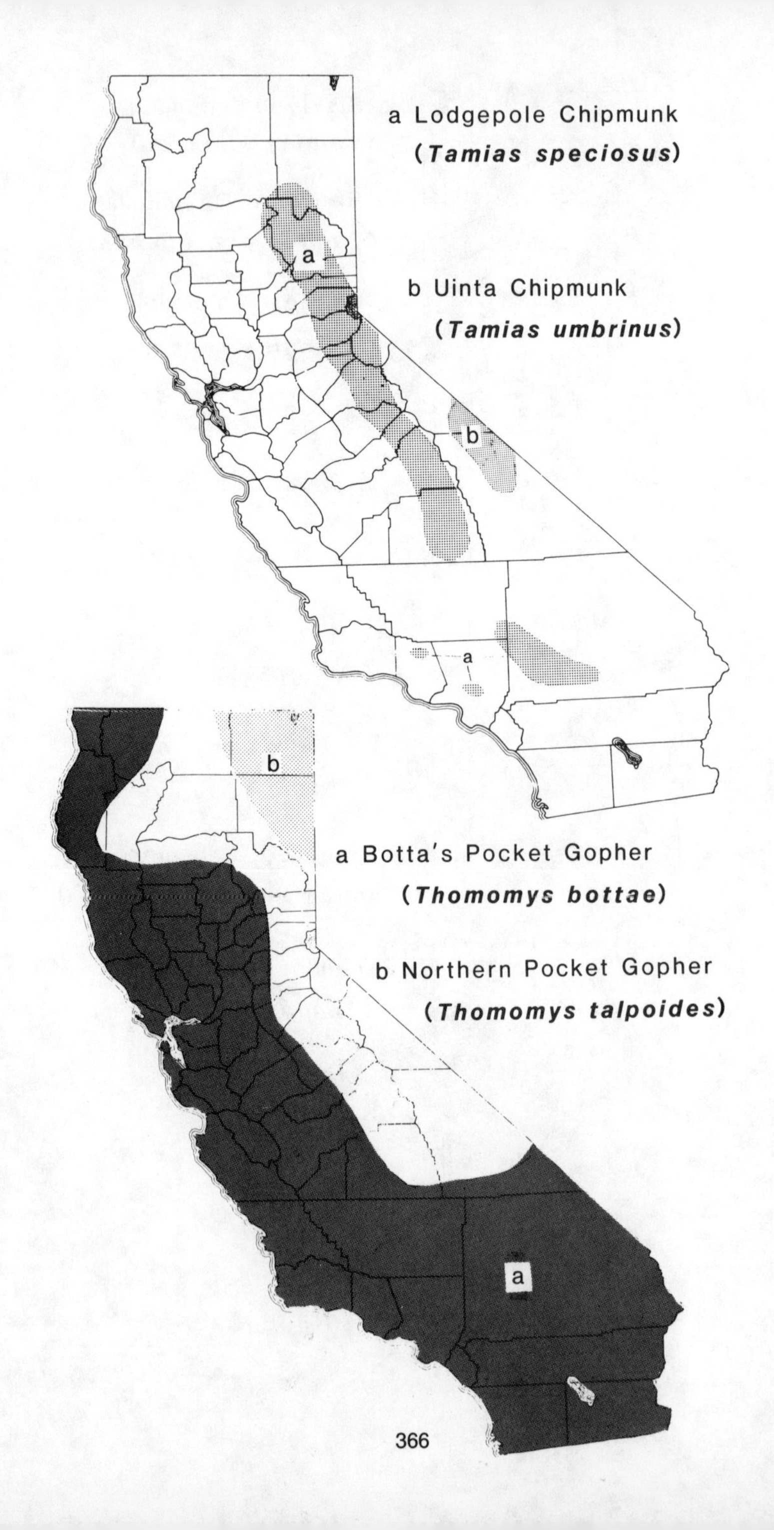
a Lodgepole Chipmunk
(*Tamias speciosus*)
b Uinta Chipmunk
(*Tamias umbrinus*)
a
b
a
a Botta's Pocket Gopher
(*Thomomys bottae*)
b Northern Pocket Gopher
(*Thomomys talpoides*)
b
a

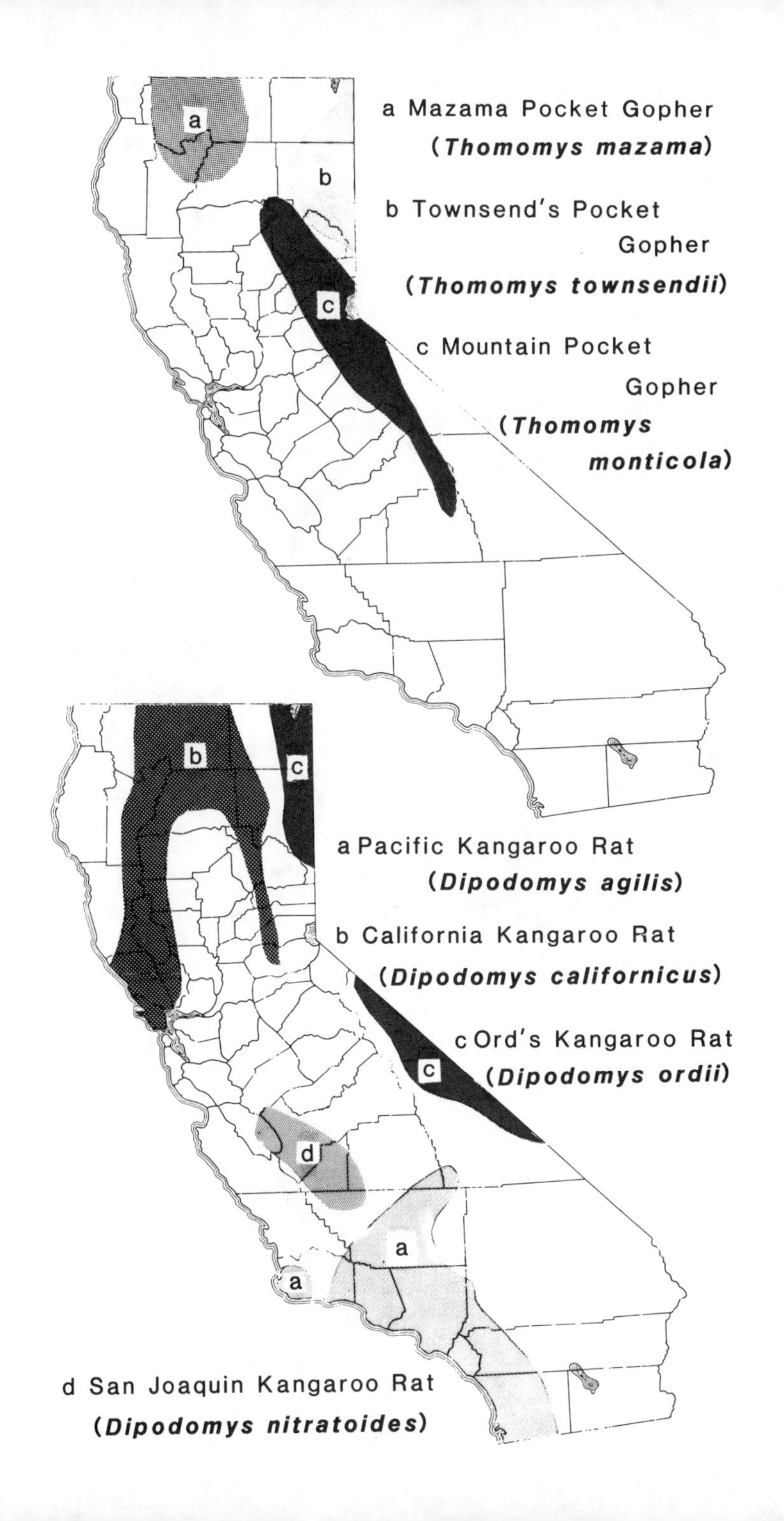

a
b
c
a Mazama Pocket Gopher
(Thomomys mazama)
b Townsend's Pocket Gopher
(Thomomys townsendii)
c Mountain Pocket Gopher
(Thomomys monticola)
b
c
c
d
a
a
a Pacific Kangaroo Rat
(Dipodomys agilis)
b California Kangaroo Rat
(Dipodomys californicus)
c Ord's Kangaroo Rat
(Dipodomys ordii)
d San Joaquin Kangaroo Rat
(Dipodomys nitratoides)

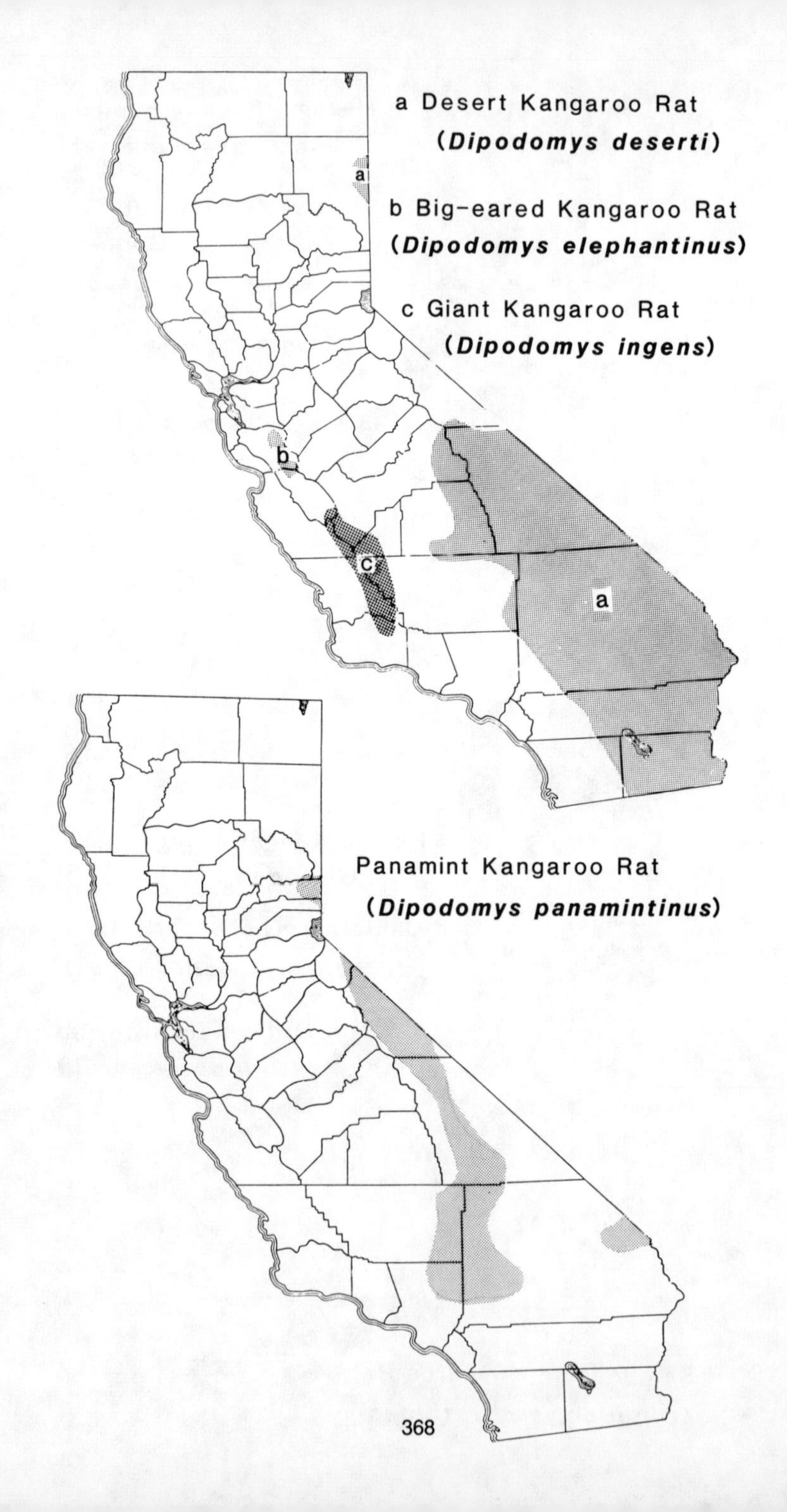
a Desert Kangaroo Rat
(Dipodomys deserti)
b Big-eared Kangaroo Rat
(Dipodomys elephantinus)
c Giant Kangaroo Rat
(Dipodomys ingens)
a
b
c
a
Panamint Kangaroo Rat
(Dipodomys panamintinus)

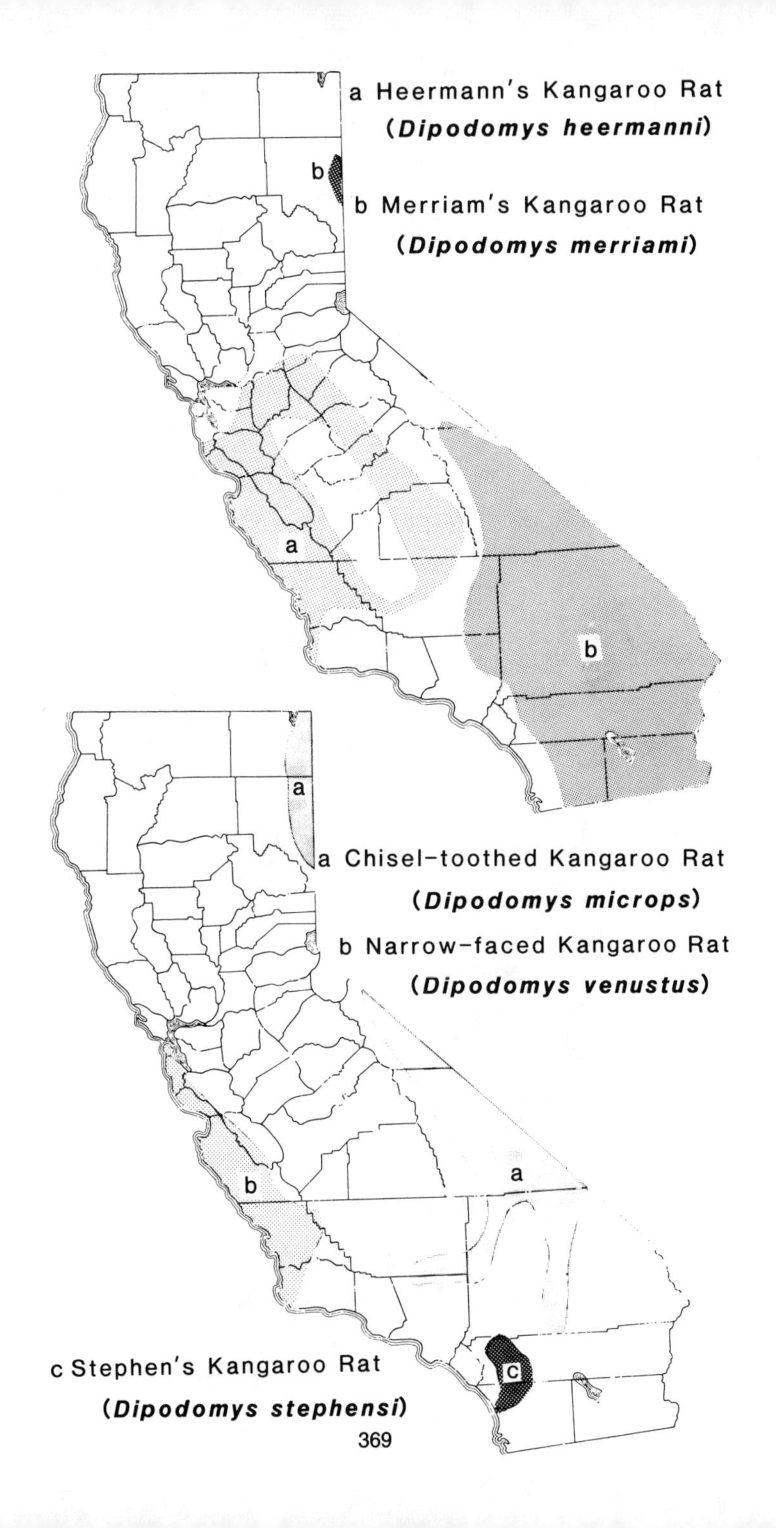
a Heermann's Kangaroo Rat
(*Dipodomys heermanni*)
b Merriam's Kangaroo Rat
(*Dipodomys merriami*)
a
b
b
a Chisel-toothed Kangaroo Rat
(*Dipodomys microps*)
b Narrow-faced Kangaroo Rat
(*Dipodomys venustus*)
a
a
b
c Stephen's Kangaroo Rat
(*Dipodomys stephensi*)
c

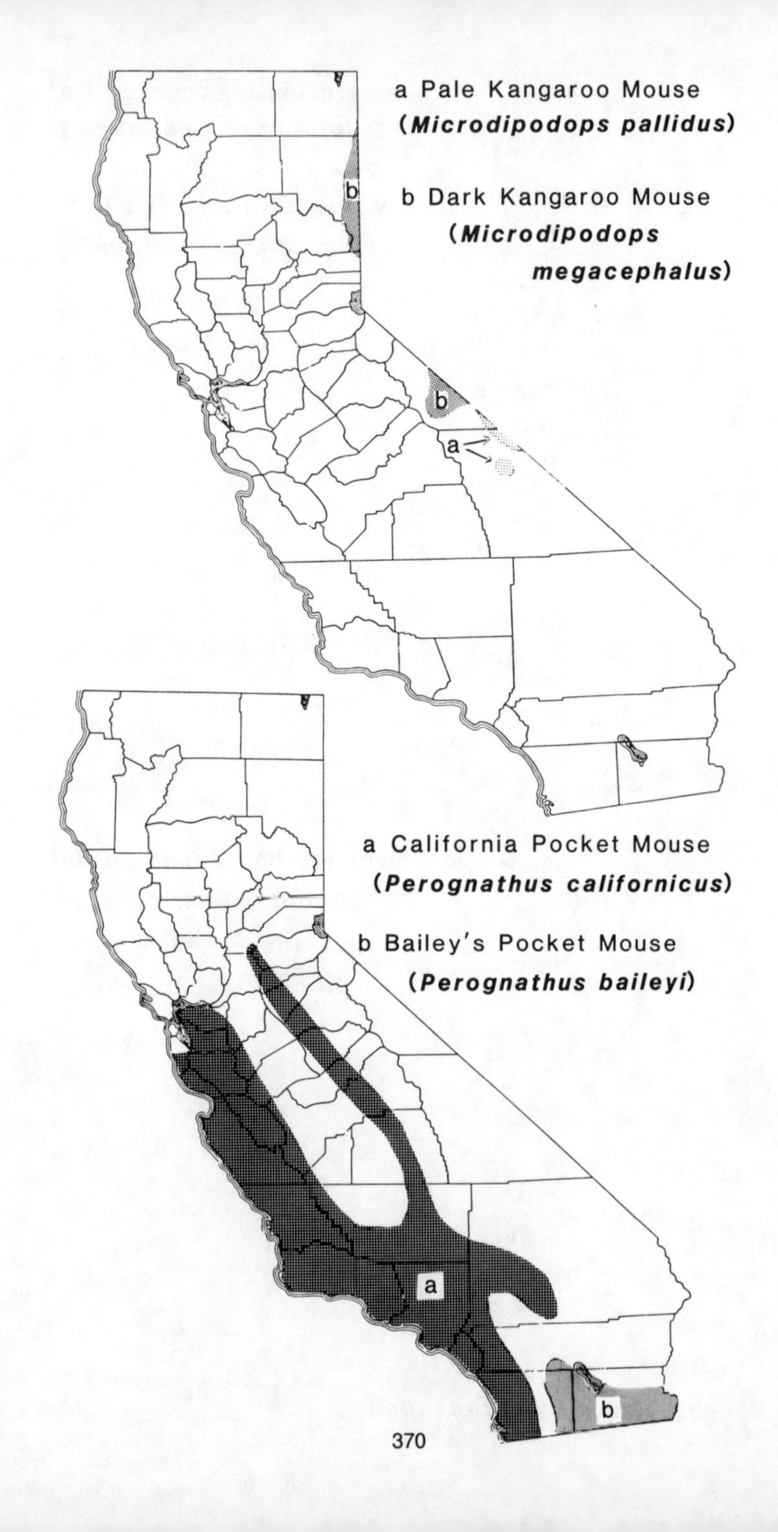
a Pale Kangaroo Mouse
(Microdipodops pallidus)
b Dark Kangaroo Mouse
(Microdipodops megacephalus)
b
b
a
a California Pocket Mouse
(Perognathus californicus)
b Bailey's Pocket Mouse
(Perognathus baileyi)
a
b

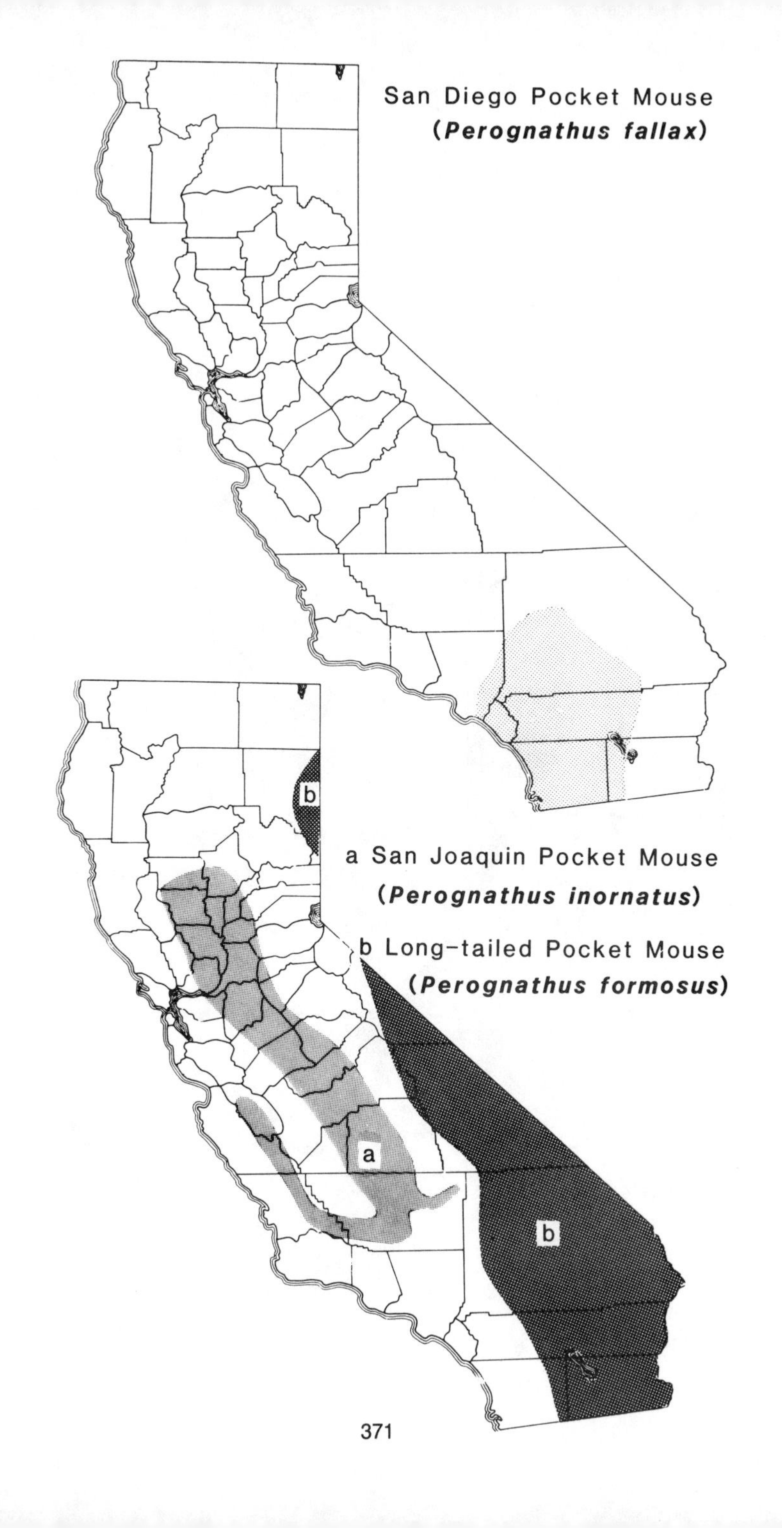
San Diego Pocket Mouse
(*Perognathus fallax*)
b
a San Joaquin Pocket Mouse
(*Perognathus inornatus*)
b Long-tailed Pocket Mouse
(*Perognathus formosus*)
a
b

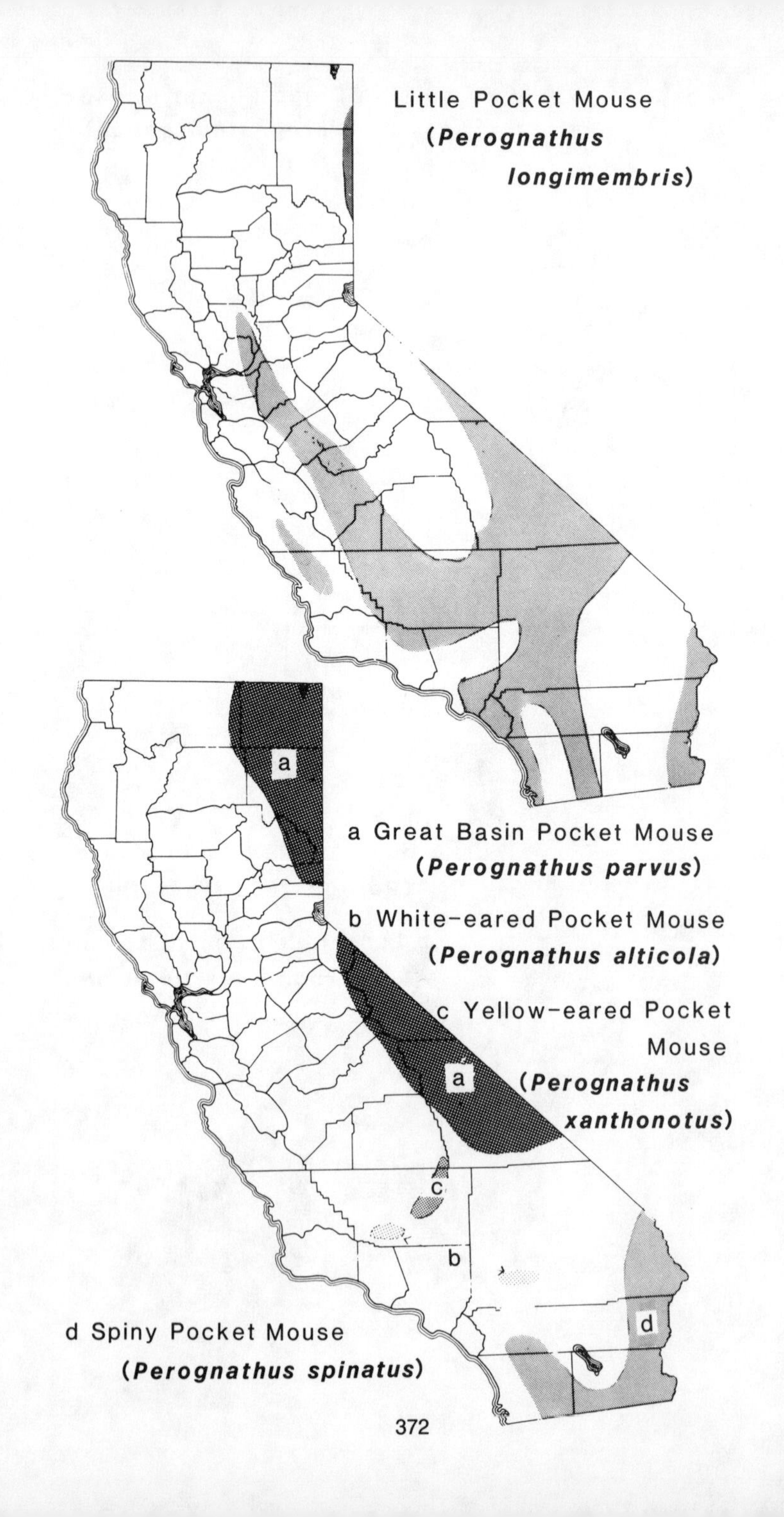
Little Pocket Mouse
(Perognathus
longimembris)
a
a Great Basin Pocket Mouse
(Perognathus parvus)
b White-eared Pocket Mouse
(Perognathus alticola)
c Yellow-eared Pocket
Mouse
a
(Perognathus
xanthonotus)
c
b
d
d Spiny Pocket Mouse
(Perognathus spinatus)

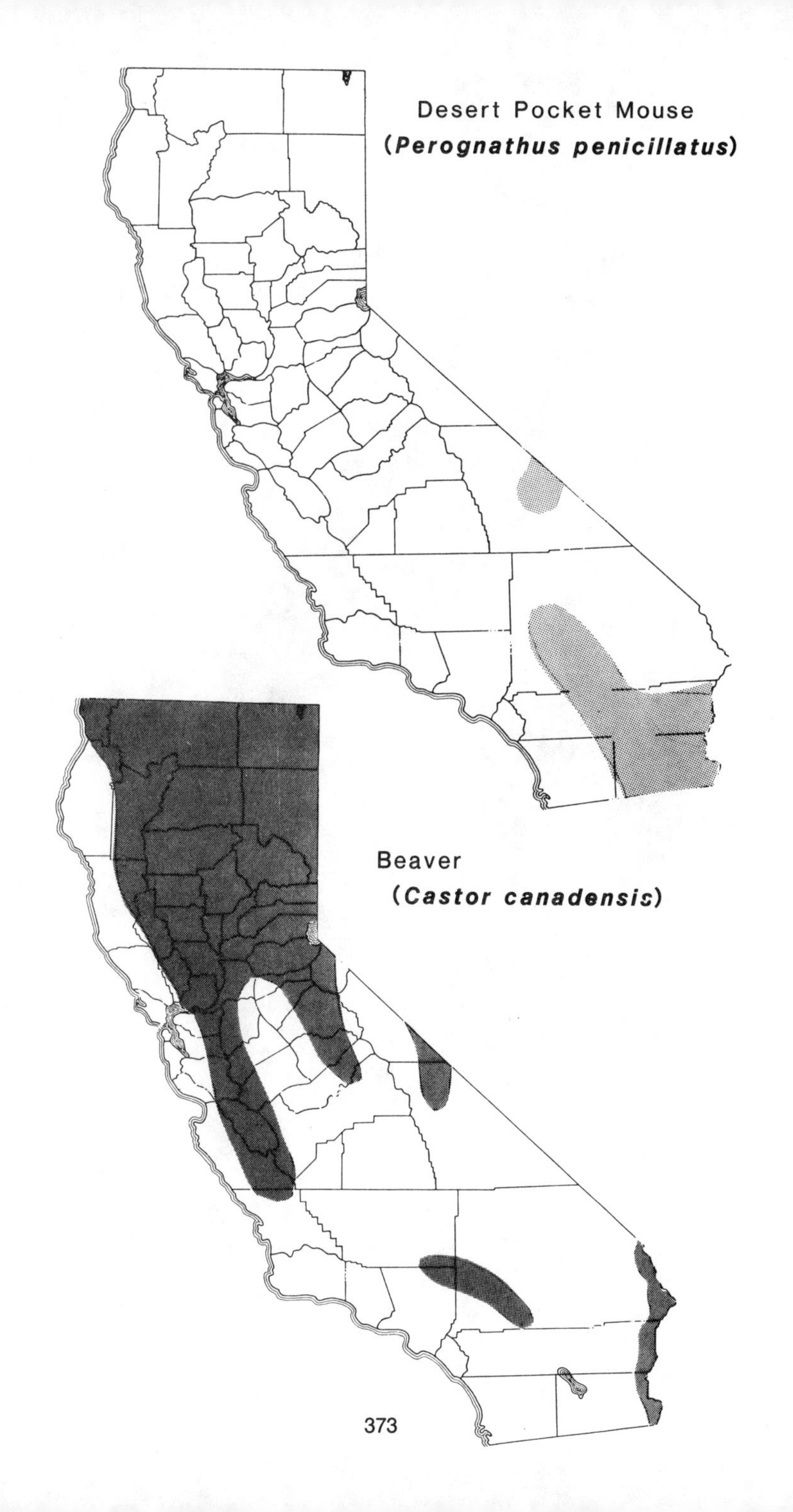
Desert Pocket Mouse
(Perognathus penicillatus)
Beaver
(Castor canadensis)

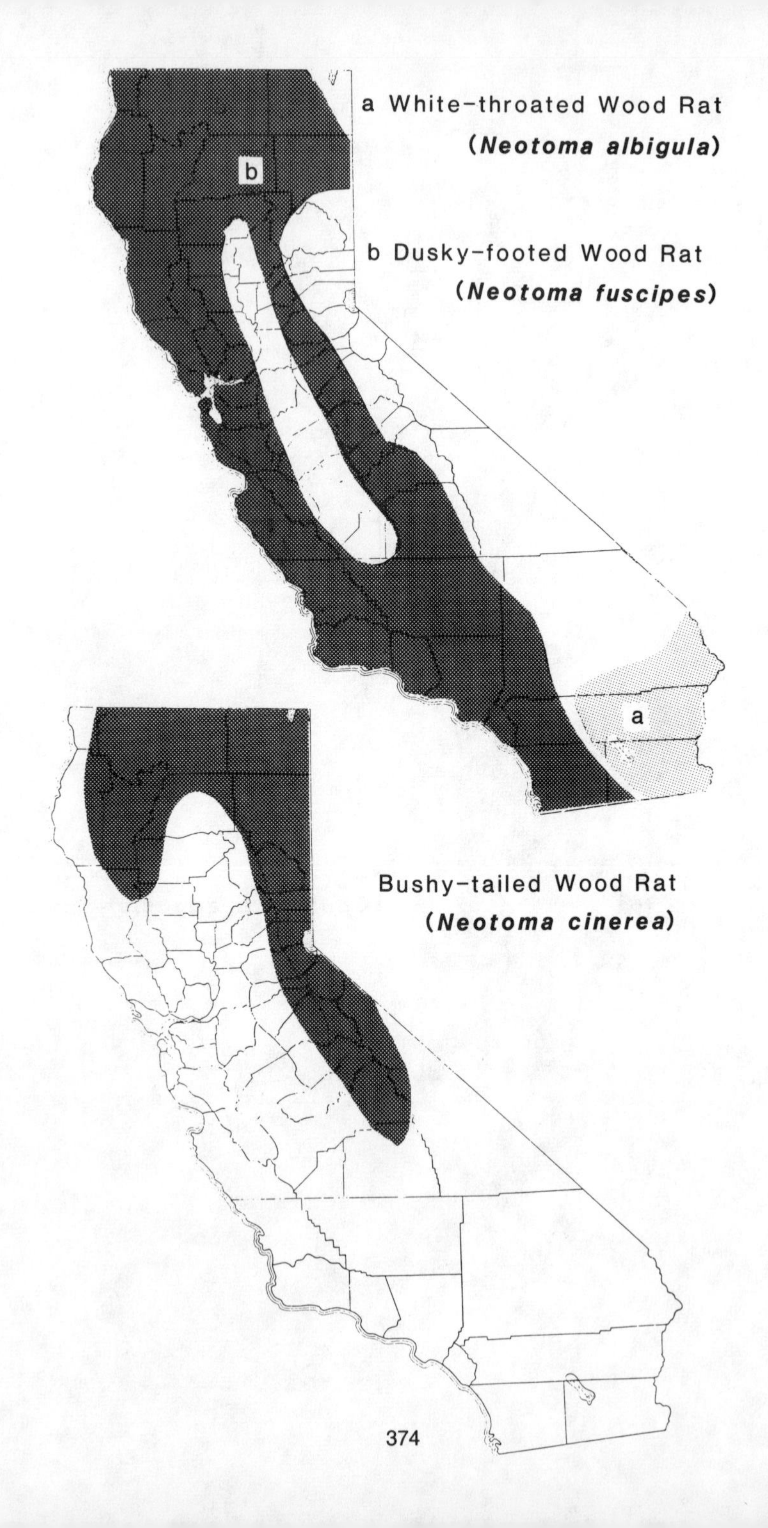
a White-throated Wood Rat
(*Neotoma albigula*)
b Dusky-footed Wood Rat
(*Neotoma fuscipes*)
b
a
Bushy-tailed Wood Rat
(*Neotoma cinerea*)

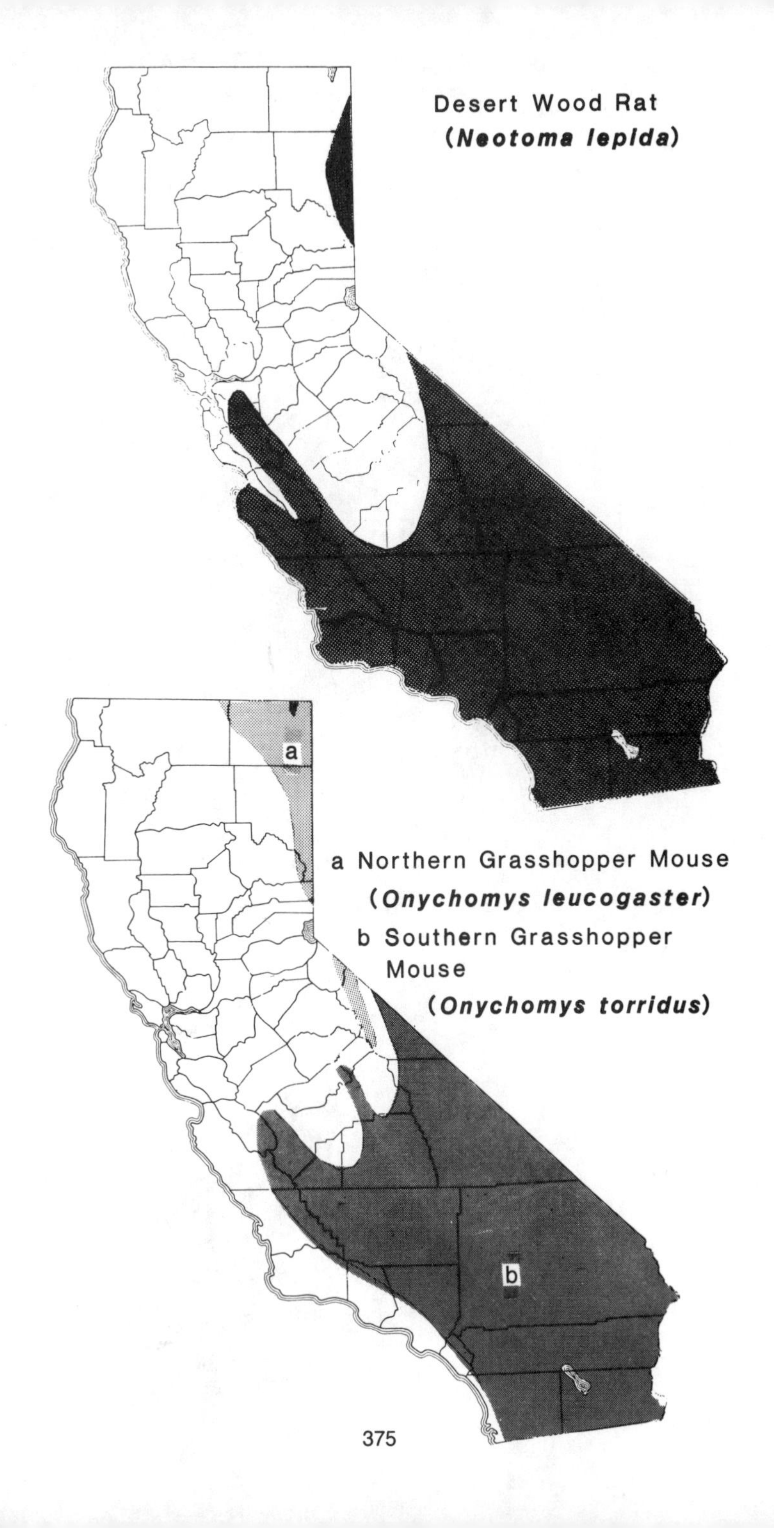
Desert Wood Rat
(Neotoma lepida)
a
a Northern Grasshopper Mouse
(Onychomys leucogaster)
b Southern Grasshopper
Mouse
(Onychomys torridus)
b

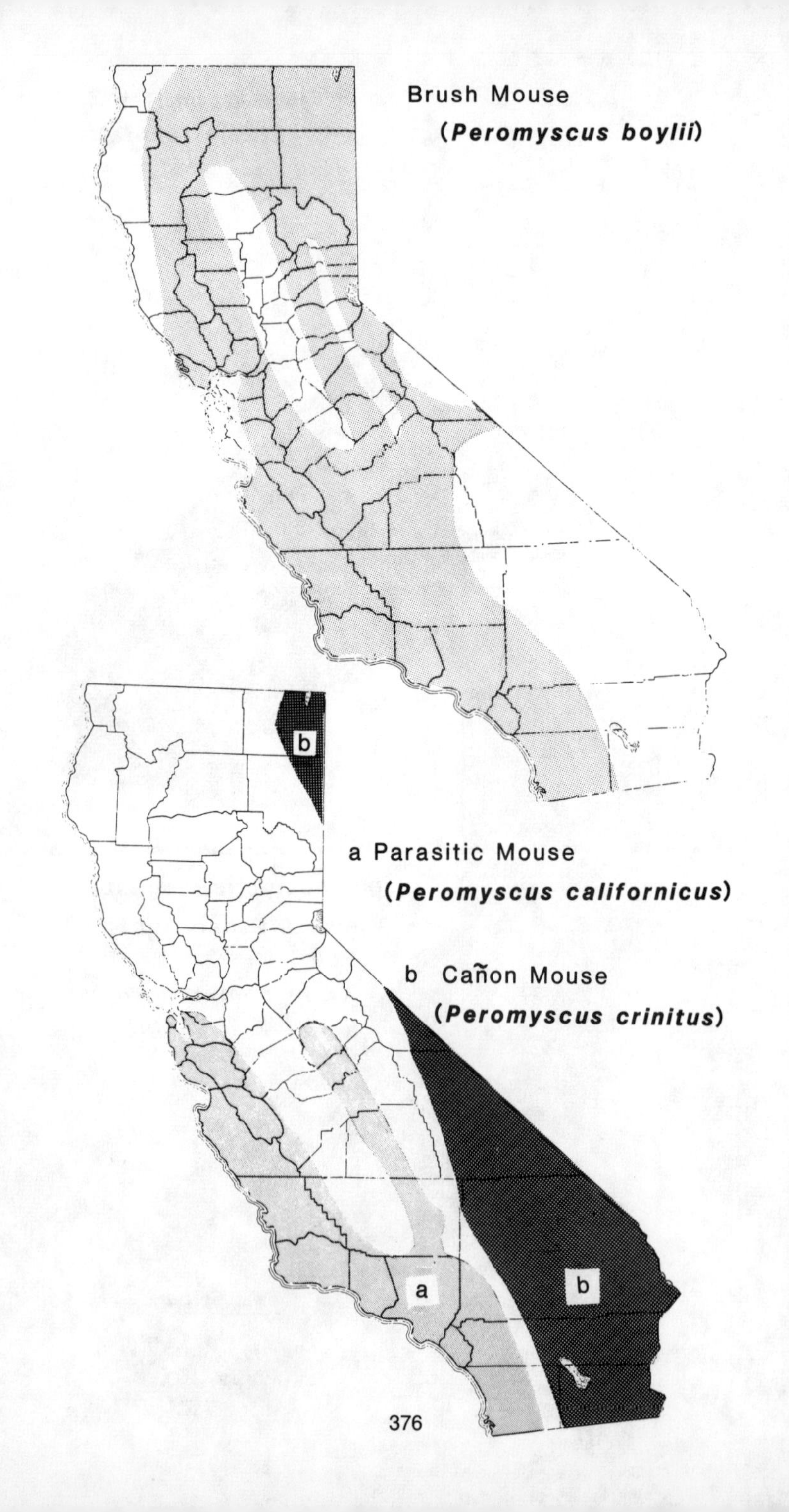
Brush Mouse
(*Peromyscus boylii*)
b
a Parasitic Mouse
(*Peromyscus californicus*)
b Cañon Mouse
(*Peromyscus crinitus*)
a
b

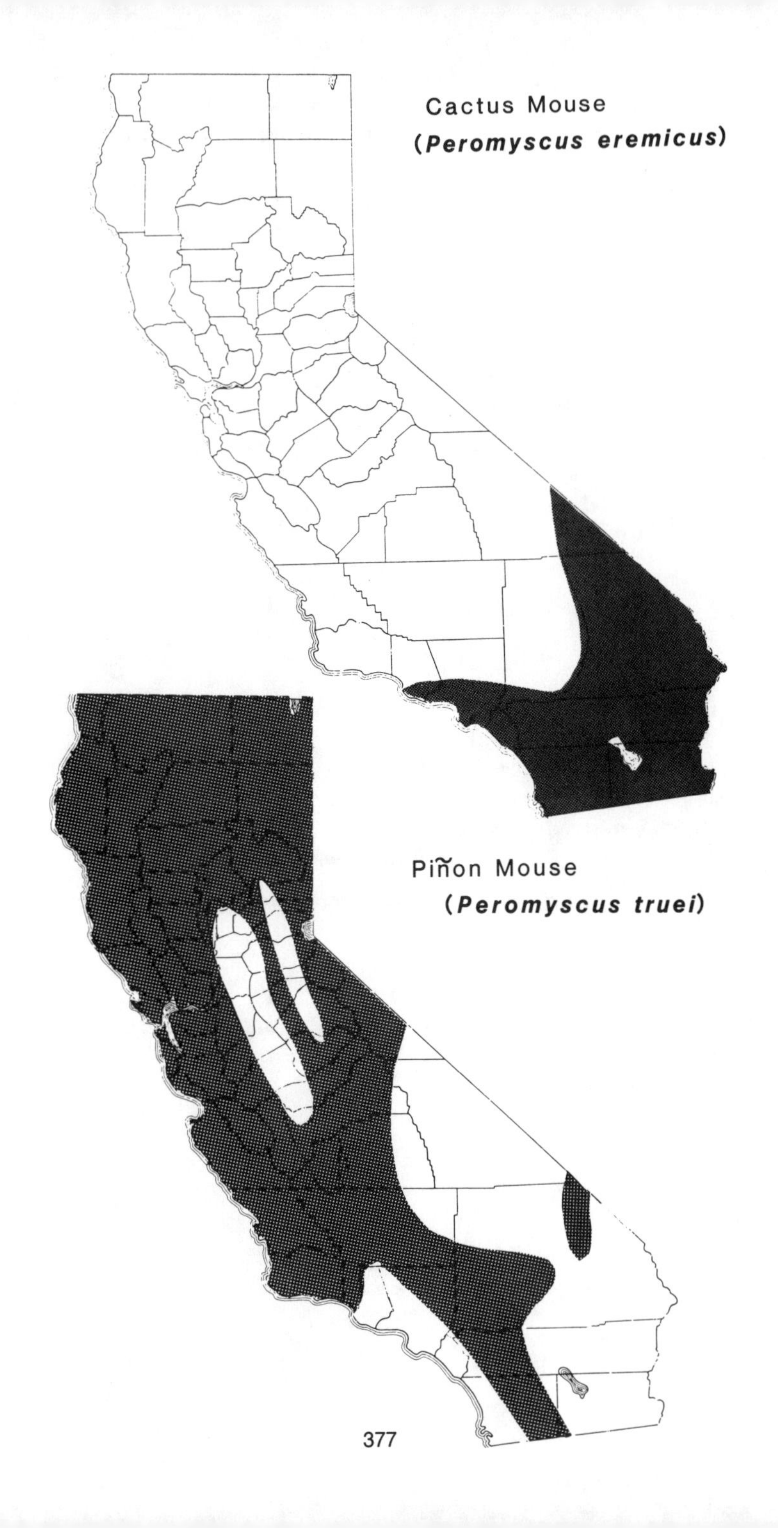
Cactus Mouse
(*Peromyscus eremicus*)
Piñon Mouse
(*Peromyscus truei*)

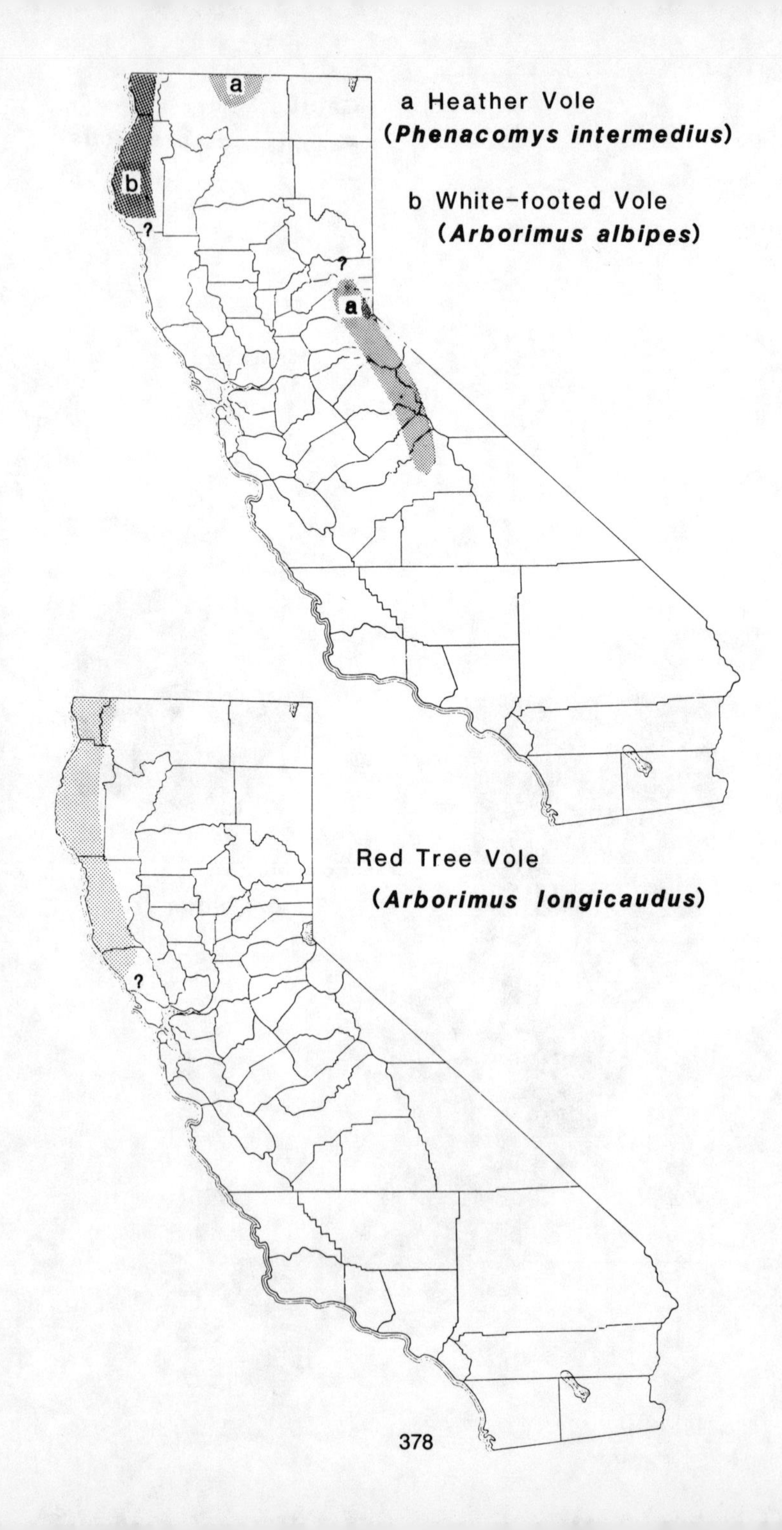
a
b
?
?
a
a Heather Vole
(Phenacomys intermedius)
b White-footed Vole
(Arborimus albipes)
?
Red Tree Vole
(Arborimus longicaudus)

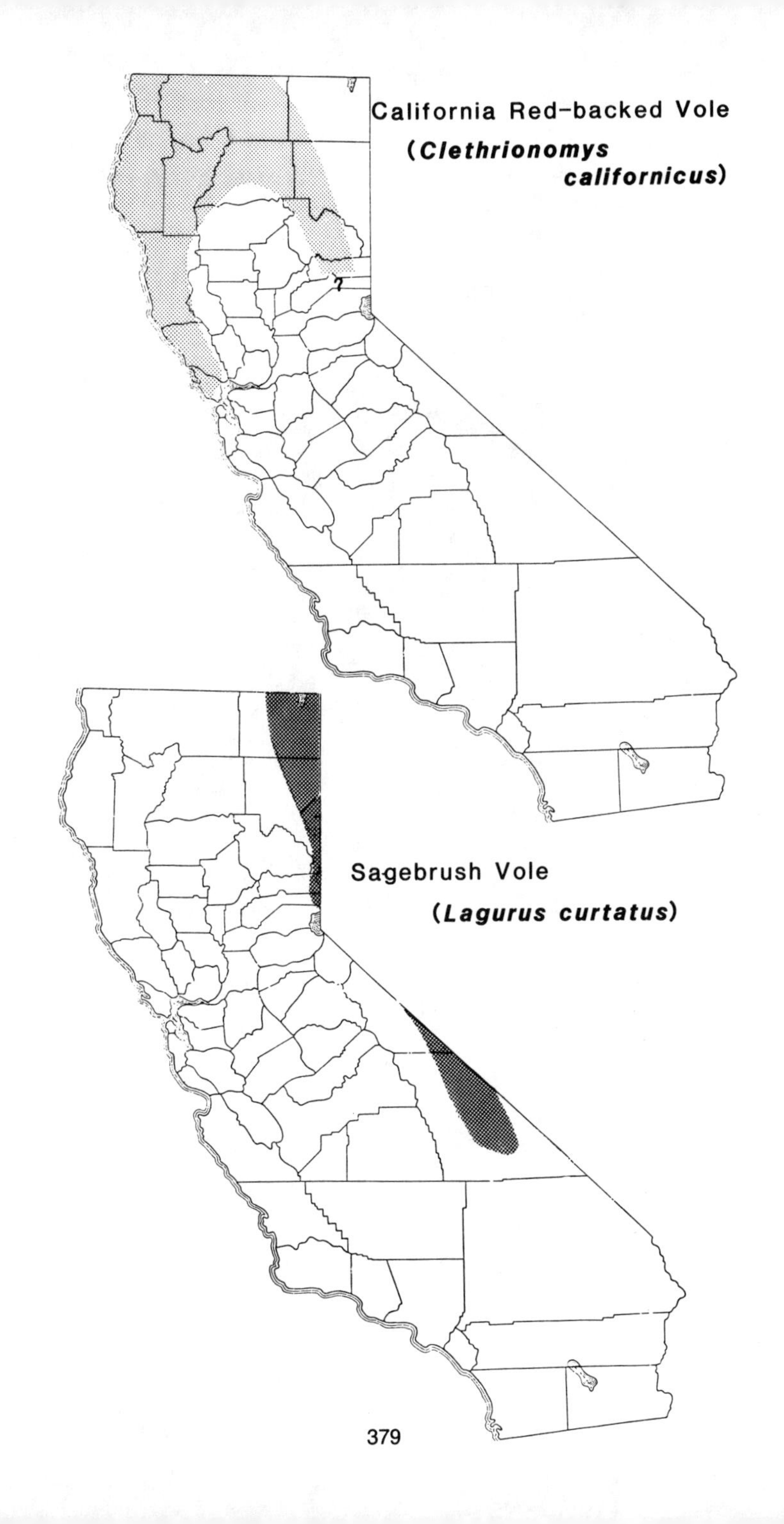
California Red-backed Vole
(*Clethrionomys californicus*)
?
Sagebrush Vole
(*Lagurus curtatus*)

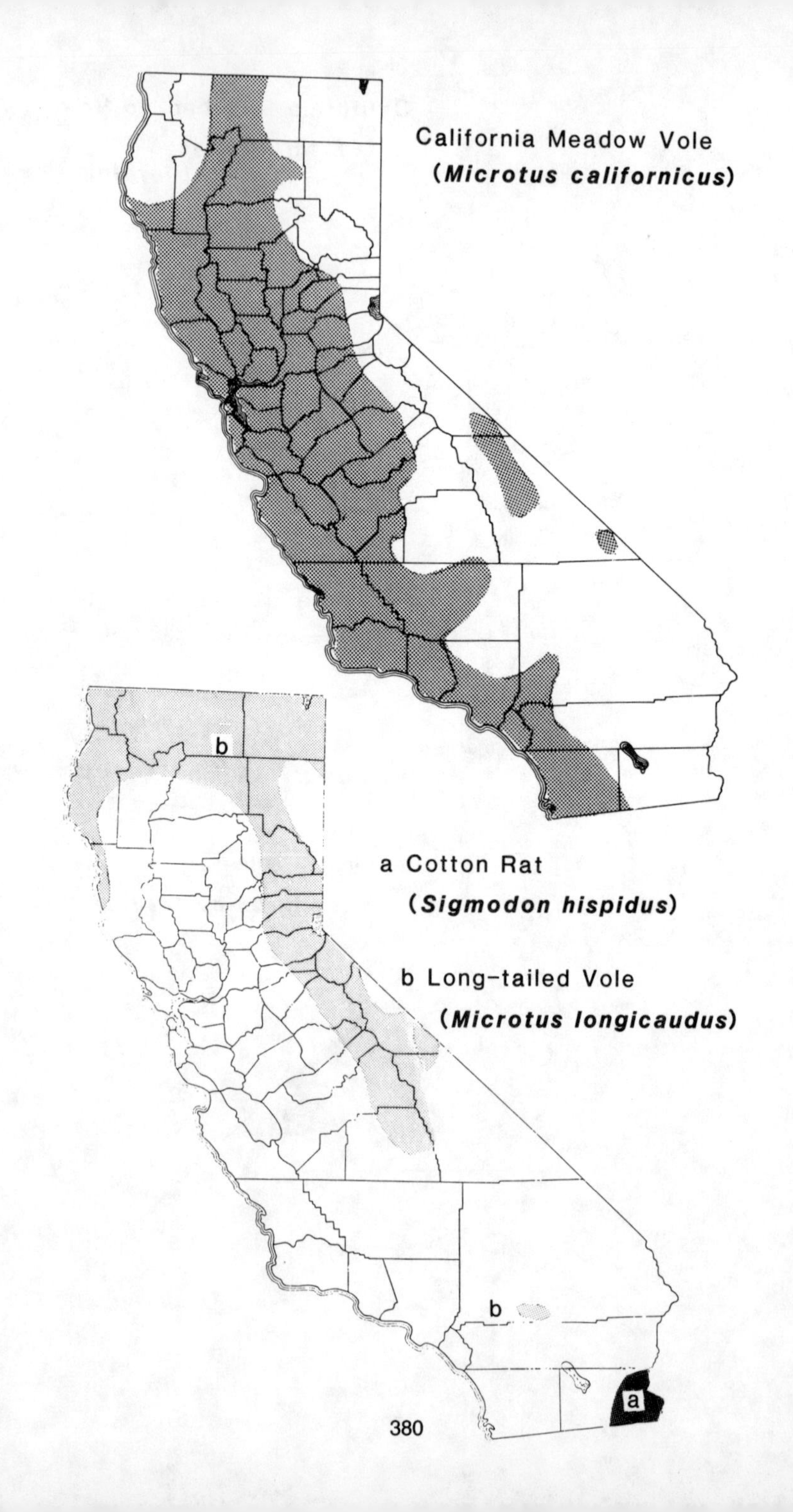
California Meadow Vole
(*Microtus californicus*)
b
a Cotton Rat
(*Sigmodon hispidus*)
b Long-tailed Vole
(*Microtus longicaudus*)
b
a

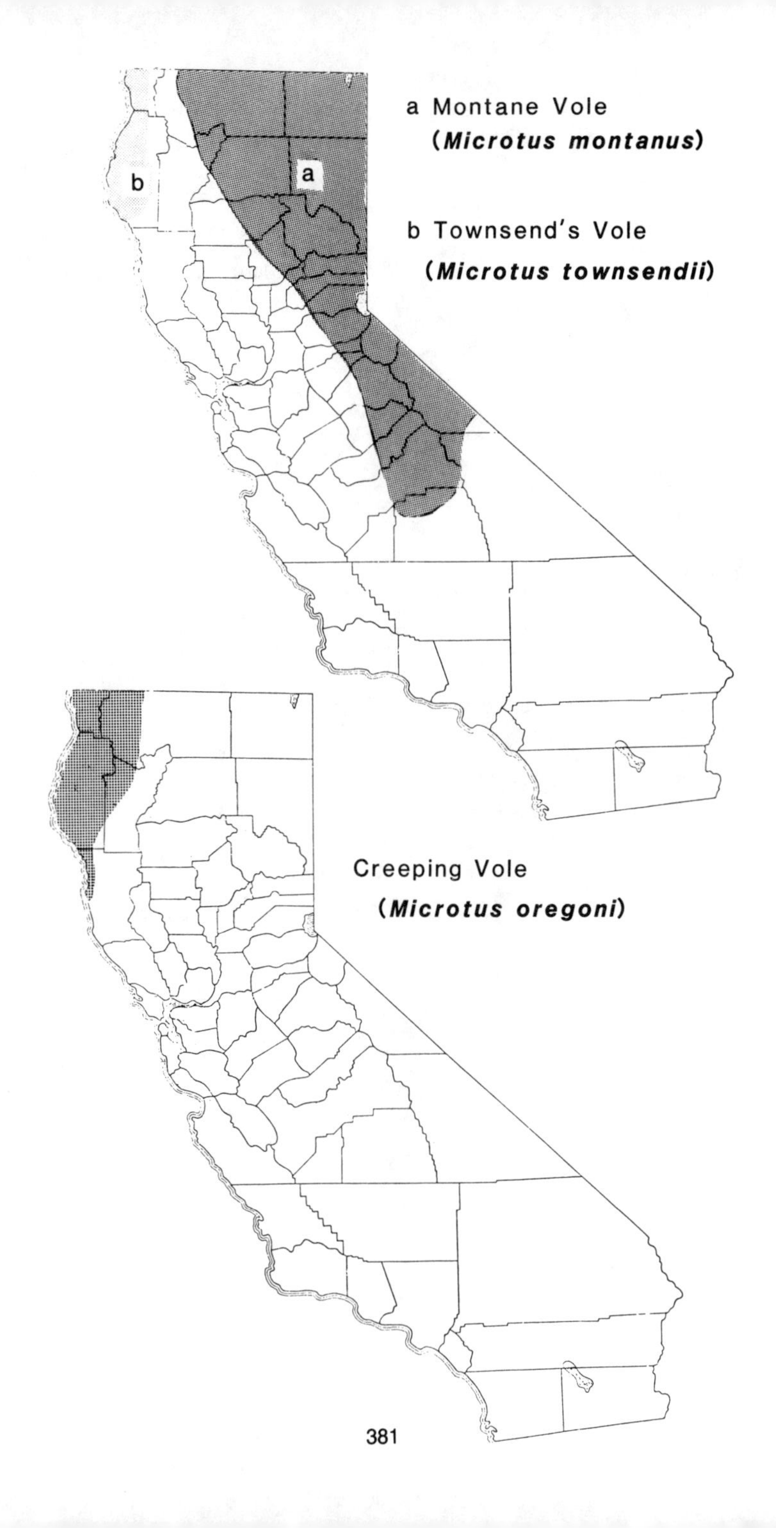
a Montane Vole
(*Microtus montanus*)
b Townsend's Vole
(*Microtus townsendii*)
a
b
Creeping Vole
(*Microtus oregoni*)

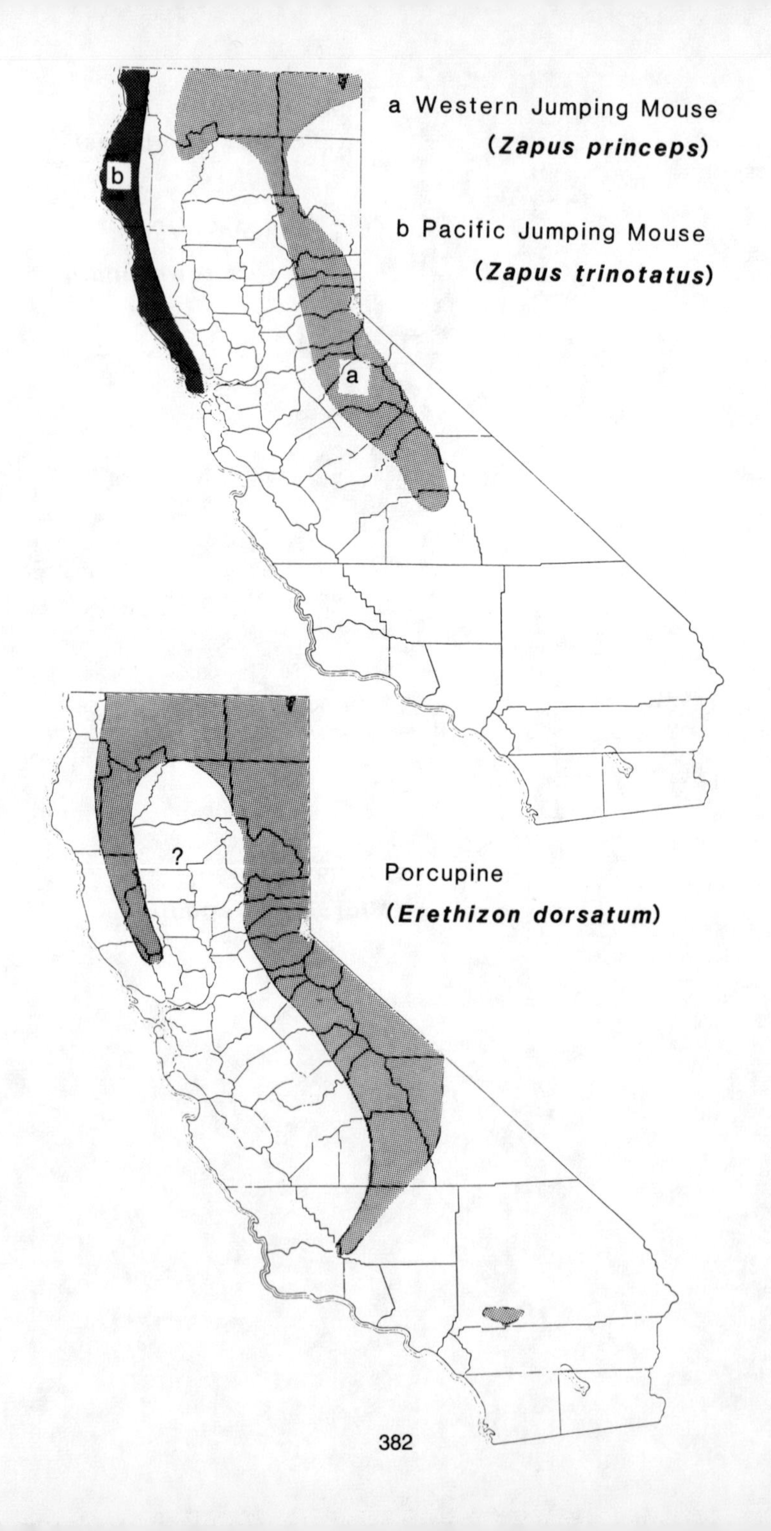

a Western Jumping Mouse
(*Zapus princeps*)
b Pacific Jumping Mouse
(*Zapus trinotatus*)
b
a
Porcupine
(*Erethizon dorsatum*)
?

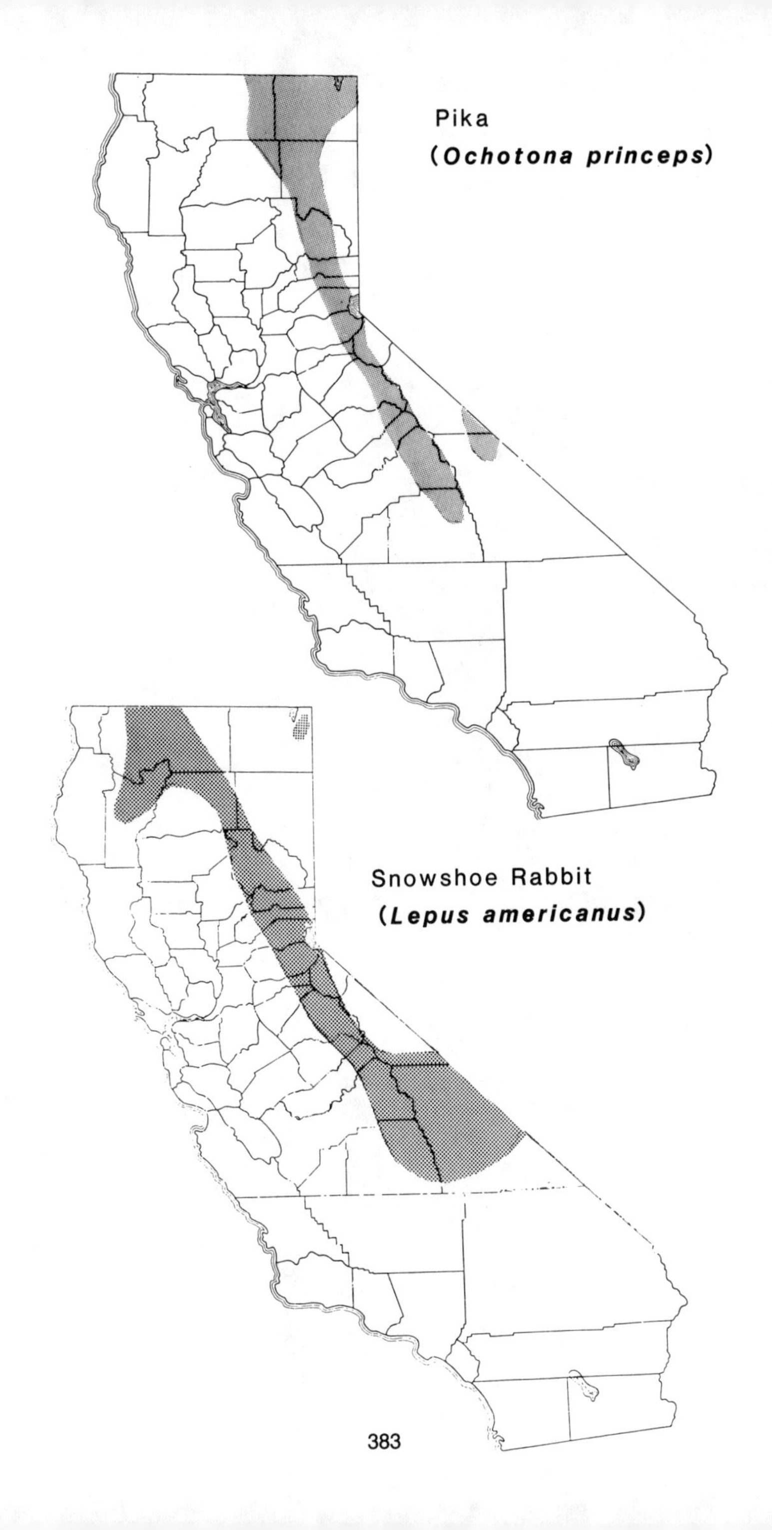
Pika
(Ochotona princeps)
Snowshoe Rabbit
(Lepus americanus)

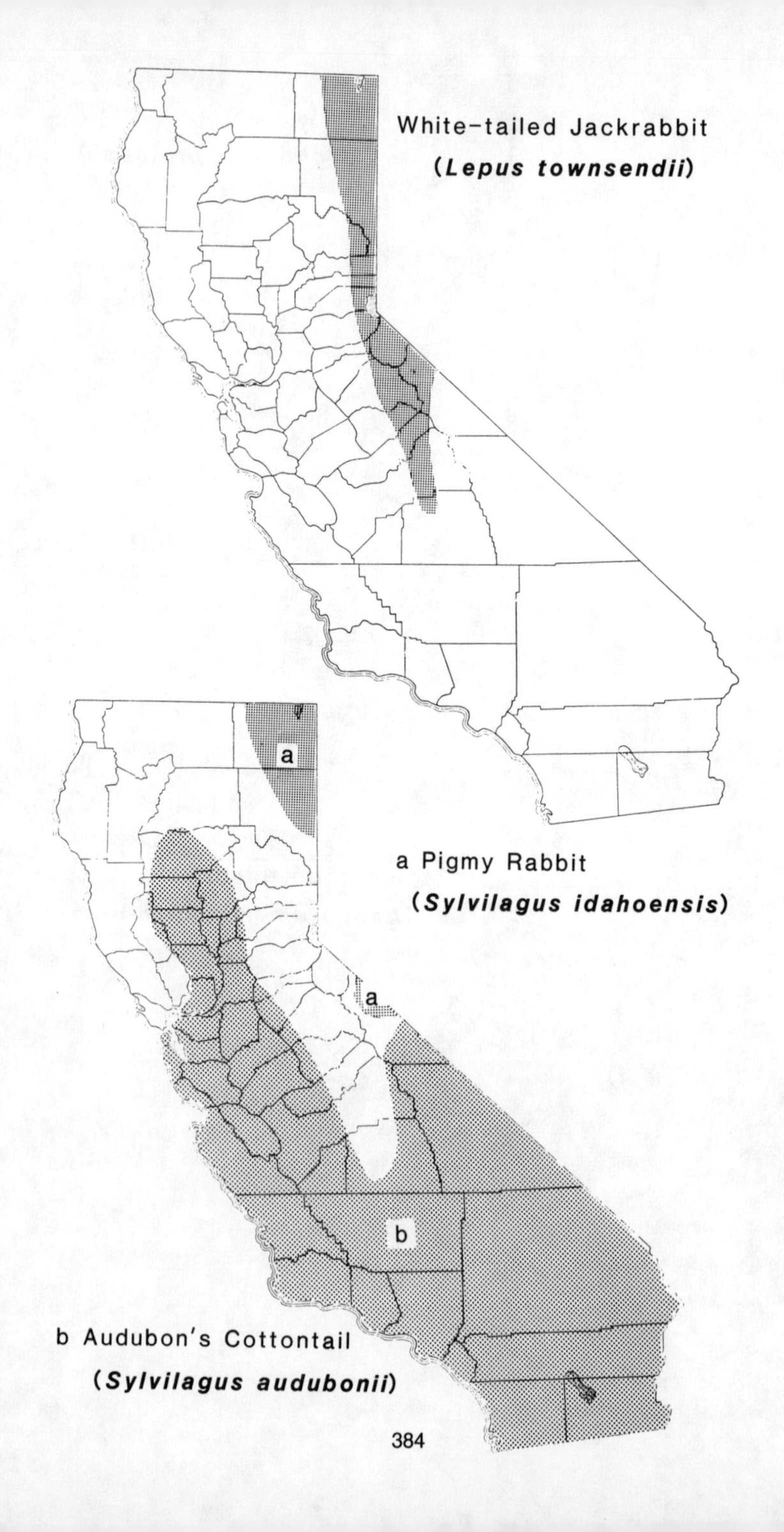
White-tailed Jackrabbit
(Lepus townsendii)
a
a Pigmy Rabbit
(Sylvilagus idahoensis)
a
b
b Audubon's Cottontail
(Sylvilagus audubonii)

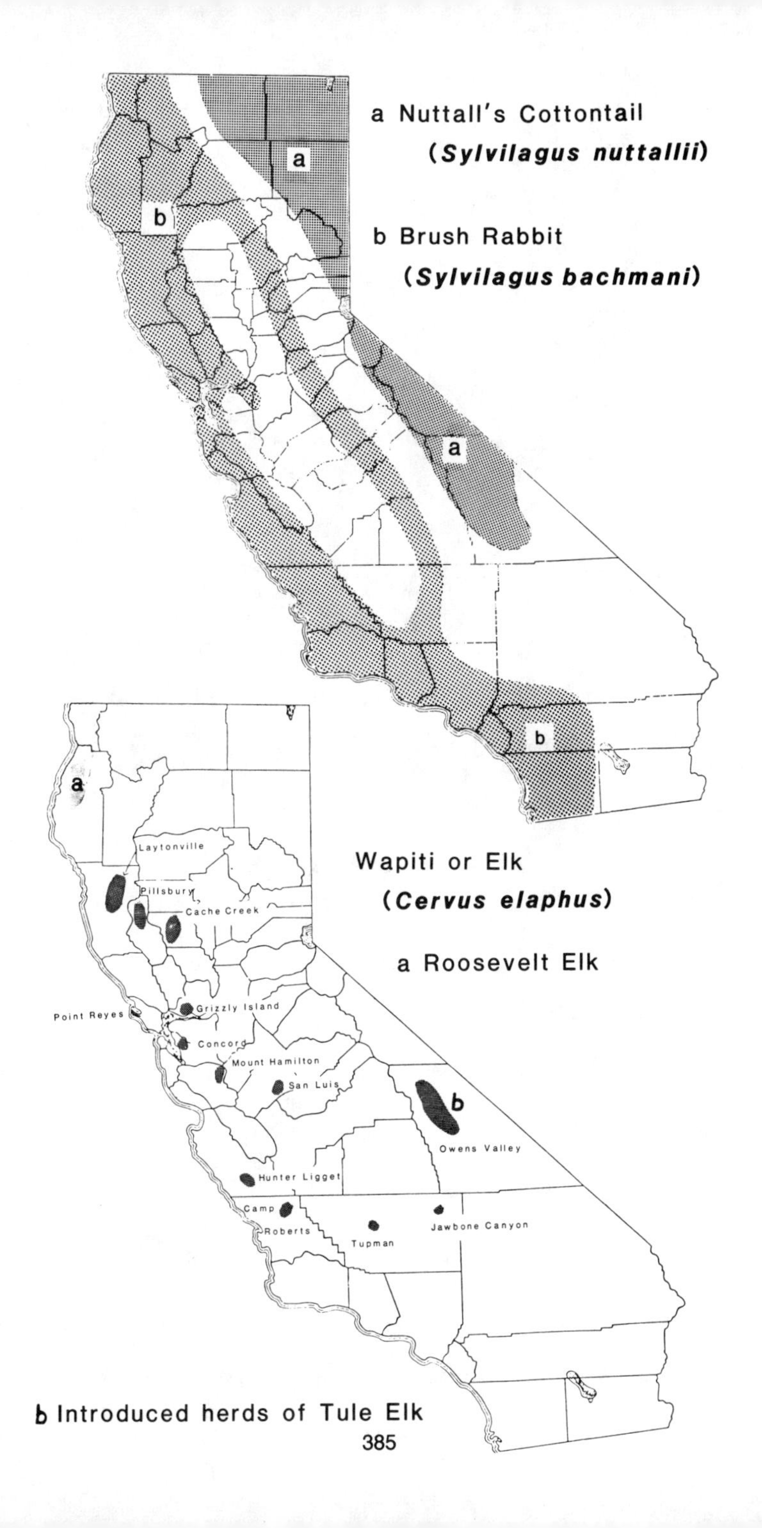
a Nuttall's Cottontail
(*Sylvilagus nuttallii*)
b Brush Rabbit
(*Sylvilagus bachmani*)
a
b
a
b
Wapiti or Elk
(*Cervus elaphus*)
a Roosevelt Elk
a
Laytonville
Pillsbury
Cache Creek
Grizzly Island
Point Reyes
Concord
Mount Hamilton
San Luis
b
Owens Valley
Hunter Ligget
Camp
Roberts
Tupman
Jawbone Canyon
b Introduced herds of Tule Elk

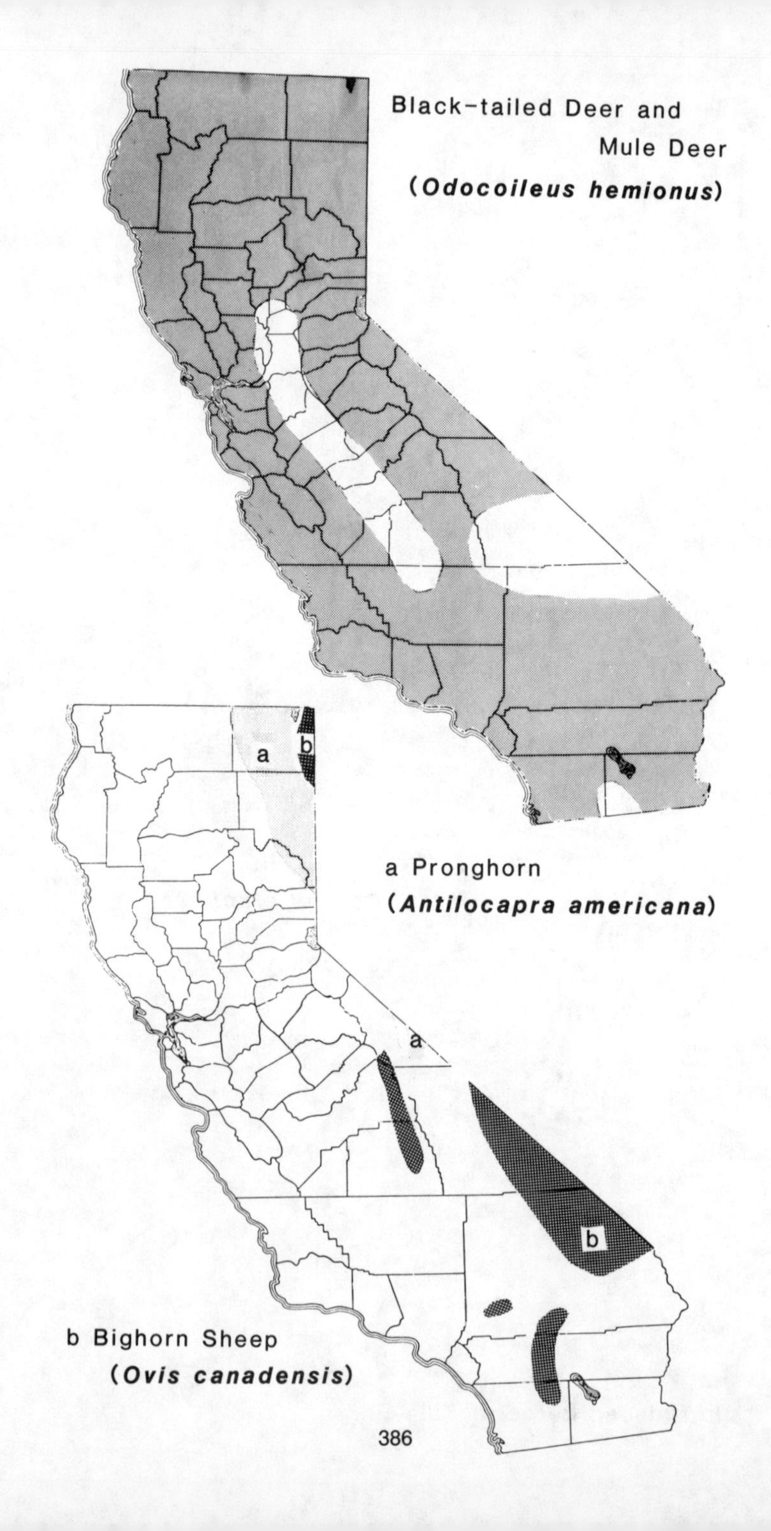

Black-tailed Deer and
Mule Deer
(*Odocoileus hemionus*)
a
b
a Pronghorn
(*Antilocapra americana*)
a
b
b Bighorn Sheep
(*Ovis canadensis*)

Glossary

Altricial: The condition at birth in which the newborn is hairless (or nearly so), blind (with eyes unopened), with earflap closed, and unable to walk. See **precocial.**

Alveolus (pl. **alveoli**): A blind, enlarged end to a duct or tube, such as the alveoli into which milk is secreted (see fig. 6).

Auditory bulla (pl. **bullae**): A bulbular bony covering, enclosing the inner ear, at the bottom of the rear end of the skull (see fig. 4).

Baculum: A bone within the penis of many species of mammals.

Blastocyst: An early embryonic stage consisting of a hollow ball of cells.

Bradycardia: A slowed heartbeat, occurring usually with a drop in body temperature, as in hibernation.

Brown fat: A type of fat with a rich supply of nerves and blood vessels. Capable of providing an infusion of heat into a hibernating mammal or a cold-stressed mammal.

Calcar: A small bone extending from the heel in bats. It provides for an extension of the skin between the leg and the tail (see fig. 25).

Cheek teeth: The molars and sometimes also the premolars; those teeth behind the canines (or behind the incisors in rodents and lagomorphs).

Coprophagy: The reingestion of soft fecal material which is semidigested, as in rabbits and gophers; it makes available for absorption nutrients that have been digested by intestinal microbes, as well as protein from the microbes themselves.

Copulation: Mating.

Coronoid process: A dorsal projection on the lower jaw (see fig. 4).

Crepuscular: Active at dawn and at dusk with a nocturnal rest period.

Deciduous: Shed periodically, as in the horny covering of the horns of the Pronghorn.

Delayed development: Embryonic development that proceeds at a very slow but continuous rate.

Delayed fertilization: Fertilization that occurs weeks or months after mating, as a result of sperm having been stored in the reproductive tract of the female. Characteristic of some species of bats.

Delayed implantation: A condition in which the embryo does not immediately attach to the uterus but remains for some weeks or months in a state of arrested development. See **embryonic diapause.**

Dentition: The number and kind of teeth, expressed as a dental formula indicating the number of upper and lower incisors, canines, premolars, and molars.

Dermal: Referring to the skin.

Dichotomous: Branching into two more or less equal parts, as opposed to smaller branches from a continuous main stem.

Digitigrade: The condition in which a mammal normally walks on its toes, as does a dog or a cat.

Dispersal: Movement away from the place of birth to make a home at another site, usually before sexual maturity.

Dormancy: A resting condition characterized by slowed breathing and heartbeat and a reduced body temperature.

Echolocation: Determination of an object's position by the echo returning from it; a bat can determine the distance, size, and flight speed of an insect by the reflected sound waves which the bat emitted.

Embryonic diapause: The resting state of an embryo, characterized by a delay of weeks or months, before it implants or attaches to the uterus of the mother. Characteristic of many species of the weasel family and some other mammals.

Endocrine glands: Glands which secrete a hormone into the circulatory system.

Entrainment: The process by which a regular physiological process, such as sleep or a seasonal activity, is timed by an environmental factor such as light.

Estivation: A period of dormancy occurring in the summer.

Estrogen: A hormone from the ovaries (and also the adrenal cortex); it stimulates the development of female characteristics and functions in reproduction.

Estrous cycle: The periodic release of one or more eggs and preparation of the uterus to receive an embryo.

Estrus: The point in the estrous cycle when the female is receptive to sexual advances of a male; a time commonly referred to as "heat."

Feral: The condition in which a domestic animal has reverted to a wild existence.

Fertilization: The union of the genetic elements of an egg and a sperm.

Flehmen: The act in which many mammals raise their heads and simultaneously raise their upper lips and expose the mouth to air. This action apparently allows odors to reach the vomeronasal organ at the anterior part of the roof of the mouth. See **vomeronasal organ.**

Follicle-stimulating hormone (FSH): A hormone from the anterior pituitary gland that stimulates the ovary to develop one or more eggs or ova.

Fossorial: Burrowing in habit, as a pocket gopher or mole.

Gestation: The period of embryonic development.

Gonad: The reproductive organ in which eggs and sperm are made; also the site of production of such hormones as estrogen, progesterone, and androgens. Testes in males and ovaries in females.

Gonadotropin: A hormone (such as follicle-stimulating hormone or luteinizing hormone) which stimulates the gonads; produced by the anterior pituitary gland.

Gonadotropin-release hormone (Gn-RH): A hormone which stimulates the anterior pituitary gland to release gonadotropins; produced by the hypothalamus.

Hibernation: A state of dormancy, in the winter, in which most body activities are greatly slowed and body temperature drops.

Home range: The area over which a mammal moves for food on a regular basis.

Homing: The tendency for an animal to return to its home; generally shown by displacement experiments in which an animal is taken from its home, marked by a band or some other identifying mark, and then released.

Hormone: A chemical messenger. Manufactured at one site (an endocrine gland), a hormone circulates throughout the body and stimulates activity (circulation and cellular activity) in a certain area, the "target organ."

Hypothalamus: A structure at the base of the brain. It receives stimuli from both outside and inside the animal and mediates many daily and seasonal activities. It is a major regulator.

Implantation: The attachment of an embryo to the uterus. At this site the placenta subsequently develops.

Incisive foramina: A pair of slits in the roof of the mouth through which two branches of the olfactory nerve enter the mouth. See **flehmen.**

Induced ovulation: Ovulation (the release of an egg from the ovary) as the result of mating.

Infraorbital foramen: An opening or canal between the snout and the eye region through the bone directly in front of the eye cavity.

Interfemoral membrane: A fold of skin connecting the hind legs and frequently enclosing the tail in bats.

Lactation: The process of giving milk; also suckling.

Lactogenesis: The manufacture of milk by the mammary glands.

Luteinizing hormone (LH): A hormone from the anterior pituitary gland. It stimulates the release of the egg from the ovary.

Mammary glands: The structures which, in female mammals, produce milk.

Mammogenesis: The development of mammary glands.

Mantle: The area over the shoulders and behind the head.

Marsupium: The pouch in which the nipples lie and which houses the young of marsupials.

Migration: A movement from one area to another with a subsequent return.

Molar: A tooth which develops behind the premolars; molars are not preceded (in time) by milk teeth.

Mortality: The rate at which individuals die over a specified period of time.

Movement: Wandering from one place to another, usually in search of food or a mate, within the home range.

Myoepithelial strands: Small contractile fibers about the alveoli in mammary glands (see fig. 6).

Natality: Rate of birth.

Navigation: The ability of an animal to determine direction from environmental cues.

Neuroendocrine loop: A physiological communication involving both nervous and hormonal messages.

Olfaction: The detection of or sensitivity to odors.

Orientation: The determination of an animal's geographic position from environmental cues.

Oviduct: The small tube which carries eggs from the ovary to the uterus.

Ovulation: The process in which one or more eggs are released from the ovary or ovaries.

Ovum (pl. **ova**): An egg.

Oxytocin: A hormone from the posterior pituitary (made in the hypothalamus and released through the posterior pituitary); it stimulates the release of milk from the alveoli of the mammary glands and also induces maternal behavior.

Palmate: Flattened and radiating from a central point, as the palm of a hand or a flattened antler of a deer.

Parturition: Birth.

Pelage: Fur.

Photocycle: The complete light/dark period or the annual change in light/dark periods; thus either a daily or an annual photocycle.

Photoperiod: The amount of light and darkness within a 24-hour period.

Pinna (pl. **pinnae**): The external ear, or earflap.

Pituitary gland: An endocrine gland lying at the base of the brain, ventral from the hypothalamus. It consists of three parts, or "lobes," which are quite different in function.

Placenta: The structure which exchanges nutrients and oxygen as well as waste products and carbon dioxide between the embryo and the mother; the "afterbirth."

Placental lactogen: A hormone, produced by the placenta, which stimulates the growth and development of the mammary tissue during pregnancy.

Plague: A bacterial disease of mammals, usually transmitted by fleas.

Plantar: The flat underside of a foot.

Plantigrade: A type of foot, or manner of walking, in which the entire plantar surface is placed on the ground, as in a bear or Raccoon (or a human).

Pleistocene: The last geological period, characterized by alternating "ice ages" and mild interglacial periods.

Polygynous: A mating arrangement involving one male and several or many females, as in fur seals and sea lions.

Postauricular: Behind the ear.

Postmandibular foramen: A hole on the inner surface of the lower jaw (see fig. 10).

Postorbital process: A lateral projection from the top of the skull, just behind the eyes (see fig. 4), characteristic of squirrels.

Precocial: The condition at birth in which the newborn is fully furred, has its eyes open, and is usually capable of walking and feeding on solid food within a few minutes or few hours after birth. See **altricial.**

Prehensile: Capable of holding, such as the somewhat prehensile tail of an opossum.

Premolar: A tooth (or teeth) between the canine and molar; preceded (in time) by a milk tooth.

Preputial gland: A gland which releases a fluid into urine, giving it a distinctive odor.

Primer: A scent which affects sexual development of another individual of the same or the opposite sex.

Progesterone: A hormone produced by the ovary, especially after ovulation, and also produced by the placenta in most species of mammals.

Prolactin: A hormone from the anterior pituitary. It has many functions but is especially critical in the production of milk in many mammals.

Releaser: A scent which elicits an immediate behavioral response in another individual.

Ruminant: A mammal with a four-chambered stomach, each chamber having a different function, holding food in different stages of digestion and physical breakdown.

Sagittal crest: The ridge or ridges at the top of the skull in many mammals, sometimes at the juncture of the parietal bones (see fig. 4).

Sperm: Mature sex cells of the male.

Sphenoidal fissure: An opening in the skull at the rear of the eye cavity (see fig. 96).

Target organ: The structure or organ which is sensitive to a given hormone.

Territory: A part of a home range which the occupant will defend against intruders of the same species.

Thermoregulation: Control of body temperature.

Torpidity: A condition in which breathing rate, heartbeat, and body activity decline with a simultaneous drop in body temperature.

Tuberculate: A condition of premolar and molar teeth in which the surface is provided with rounded projections (tubercles).

Tularemia: A highly infectious bacterial disease of many mammals.

Ultrasonic: Sound waves with a frequency above 20,000 cycles per second (or hertz); above the range of hearing of most humans.

Ungulate: Hooved mammal, such as a pig, deer, or horse.

Unicuspid: A tooth with a single projecting surface.

Uterus: The part of the reproductive tract of the female in which the embryo develops.

Vestigial: A nonfunctional structure which does not fully develop, in contrast to its development in species in which it is functional.

Vomeronasal organ: An odor-sensitive recess in the roof of the mouth of most mammals but not in humans.

Zygomatic arch: A slender bone on the side of the skull, about the outside of the eye (see fig. 4).

Zygomatic plate: The flat, more or less vertical bone at the anterior side of the eye cavity.

INDEX

Page numbers for species accounts are in italics.

Designer: Rick Chafian
Compositor: G & S Typesetters, Inc.
Text: 10/12 Times Roman
Display: Helvetica
Printer: Consolidated Printers, Inc.
Binder: Mountain States Bindery & Consolidated Printers, Inc.